# OXFORD CHEMISTRY MASTERS

## Series Editors

RICHARD G. COMPTON
University of Oxford

STEPHEN G. DAVIES
University of Oxford

JOHN EVANS
University of Southampton

# OXFORD CHEMISTRY MASTERS

# Radical Reactions in Organic Synthesis

SAMIR Z. ZARD
*Département de Synthèse Organique, Ecole Polytechnique, Palaiseau, France*

# OXFORD
UNIVERSITY PRESS

Great Clarendon Street, Oxford OX2 6DP

Oxford University Press is a department of the University of Oxford.
It furthers the University's objective of excellence in research, scholarship,
and education by publishing worldwide in

Oxford New York

Auckland Bangkok Buenos Aires Cape Town Chennai
Dar es Salaam Delhi Hong Kong Istanbul Karachi Kolkata
Kuala Lumpur Madrid Melbourne Mexico City Mumbai Nairobi
São Paulo Shanghai Taipei Tokyo Toronto

Oxford is a registered trade mark of Oxford University Press
in the UK and in certain other countries

Published in the United States
by Oxford University Press Inc., New York

A catalogue record for this title is available from the British Library

Library of Congress Cataloging in Publication Data
(Data available)
ISBN 0 19 850241 9 (Hbk)
0 19 850240 0 (Pbk)

10 9 8 7 6 5 4 3 2 1

Typeset by Newgen Imaging Systems (P) Ltd., Chennai, India
Printed in Great Britain
on acid-free paper by
The Bath Press, Avon

# Preface

The aim of this book is to provide an overview of radical reactions as they pertain to organic synthesis and to help in dispelling any lingering notion that radicals are 'unruly, only fit for producing polymers and tars'. It is intended as an introductory text for synthetic organic chemists wishing to gain a feel for the factors governing the reactivity of radicals and an idea for the scope and potential of radical-based transformations. The contents correspond to an expanded version of a graduate course I first gave at the Université Paris-Sud in Orsay and now give at Ecole Polytechnique in Palaiseau.

Following a brief introduction, the main theoretical concepts are discussed within the context of the chemistry of stannanes. The various other classes of radical-based methods are then introduced, in an order roughly reflecting the change in the way the radicals are generated and the nature of the process, namely whether or not a closed chain is involved. This, somewhat unusual, approach will hopefully allow an easier and perhaps better understanding of the way kinetic and thermodynamic considerations impinge upon the design and performance of a given radical reaction. The crucially important role of radicals in many biological areas (biosynthesis, metabolism, ageing, oncogenesis, etc.), in polymer synthesis, in combustion and explosions, etc. lies beyond the scope of this book, and is not discussed, even if, for the purposes of comparison and illustration, examples are occasionally mentioned.

Radical chemistry in organic synthesis is a vast and rapidly growing field of endeavour and a short introductory text can only hope to cover a tiny fraction. The selection of the examples was guided by personal taste and by a desire to display as much variety as possible within a limited space, yet keeping clarity constantly in mind. It is certainly not complete and I apologize to those, very numerous, whose work is not mentioned. This is especially true in the case of stannane-based chemistry, which has dominated the area for the past four or five decades. Processes employing organotin derivatives are discussed in only two chapters, despite the hundreds of examples that have appeared in recent years. Paradoxically, the tremendous power of stannanes as partners in radical chains has been both a blessing and a curse. The efficiency, versatility, and exceptional scope of stannane-based technology has resulted in such widespread use in academia that 'synthetic radical chemistry' is often equated with that of organotin reagents. This narrow, but unfortunately quite prevalent view has hampered considerably the application of radical processes to the large scale preparation of pharmaceuticals and agrochemicals because of the imagined problems of toxicity and purification. If this is indeed true for stannanes and, at least in terms of toxicity, organomercurials, it is not true for many of the other methods. One of the goals of this book is to attract the attention to the many 'tin-free' radical processes which, even if they do not exhibit the broad applicability of stannanes, are nevertheless exceedingly useful synthetic tools.

I have dedicated this book to the memory of Professor Sir Derek Barton. It is under his guidance that I entered the field of radical chemistry. I hope that I have been able to highlight adequately his numerous original and highly creative contributions. I express my appreciation to Professor Henri Kagan who, 15 years ago, trusted me with a course on radical chemistry. This forced me to collect my thought on the subject and try to see it from the student's perspective. In this respect, I also acknowledge the strong influence of Professor Nguyen Trong Anh with whom I had the good fortune of sharing a memorable teaching experience.

The writing of this book coincided with a rather difficult time. My first thoughts of gratitude go to my wife and children who stood steadfast throughout this tough period and to Professor Pierre Potier for his considerable help over the years. I also record here my heart-felt appreciation to my students and collaborators who, hardships notwithstanding, kept their enthusiasm and high spirits and contributed so much to the progress of our research projects. Some examples from their work are described herein. Last, but not least, I am greatly indebted to Drs Issam Hanna and Béatrice Sire for taking the time to carefully read through the manuscript and for their many helpful comments and corrections. Of course, any errors that remain are entirely mine.

Palaiseau
2003

*To the memory of Sir Derek H. R. Barton*

# Contents

# 1 Introduction and some general concepts

## 1.1 A brief history

In chemistry, as in other branches of science, experiment often precedes theory. Accidental discoveries thus bring fresh light and reveal aspects of the subject hitherto unsuspected. And radical chemistry is no exception. It is therefore perhaps useful to outline, in a very broad manner, the way the field developed over the past century or so, viewed from the perspective of the synthetic organic chemist.

In the early days of organic chemistry, when structural and mechanistic concepts were still hazy, it was noticed that reactions that could in principle lead to a trivalent carbon species produced dimers instead. For instance, the reaction of metallic sodium with methyl bromide or iodide would produce ethane instead of 'methyl', and the Kolbe electrolysis (ca 1850) of solutions of sodium acetate gave rise to carbon dioxide and ethane. It is not surprising, therefore, that the opinion that eventually prevailed amongst the organic chemists of the time was that a trivalent carbon did not exist. That is until Gomberg stumbled upon triphenylmethyl (for a fascinating account, see: McBride 1974). As a post-doctoral associate in the laboratory of Victor Meyer, Gomberg was the first to prepare tetraphenylmethane and, on his return to the University of Michigan, embarked on the synthesis of hexaphenylethane by a similar approach hinging on a Wurtz-type coupling. However, when chlorotriphenylmethane was reduced with various metals (zinc dust proved best), the colourless compound he obtained analysed for $C_{38}H_{30}O_2$ instead of the expected $C_{38}H_{30}$. Somehow, oxygen got incorporated into the product. After several unsuccessful attempts, Gomberg finally repeated his experiment under a carbon dioxide blanket—it must be remembered that in those days argon or nitrogen cylinders were not available and reactions were usually carried under air. The result was strikingly different: another white crystalline material was isolated, which analysed correctly for hexaphenylethane, but which gave yellow solutions that were rapidly discoloured by oxygen, leading ultimately to the compound he first obtained. The inescapable conclusion from Gomberg's pioneering work was that a free radical was involved in these transformations, as summarized in Scheme 1.1.

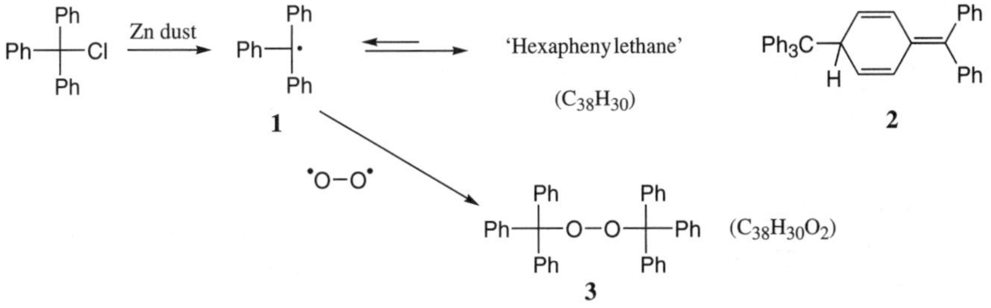

**Scheme 1.1** Attempted synthesis of hexaphenylethane by Gomberg

The white crystalline substance that analysed for hexaphenylethane was shown nearly 70 years later to have in fact structure **2** (a structure that was incidentally proposed by P. Jacobson in 1904!); hexaphenylethane itself has thus never been isolated. Compound **2** is in equilibrium in solution with the yellow-coloured triphenylmethyl radical **1**, which reacts rapidly with triplet oxygen (a biradical itself) to give the colourless peroxide **3**. The paper of Moses Gomberg appeared in 1900 and represents a landmark in radical chemistry; it is also famous for its concluding sentence: 'This work will be continued and I wish to reserve the field to myself.' Although the significance of this work was recognized at the time, its immediate impact on mechanistic and synthetic organic chemistry remained limited. It is amusing to note that a few years earlier, Gomberg made tetraphenylmethane by heating phenylazotriphenylmethane (PAT), a reaction that also involves triphenylmethyl radicals (Scheme 1.2). The thermal decomposition of azo derivatives had previously been studied by Thiele (Walling 1986). Azo compounds, and especially azo-bis-isobutyronitrile (AIBN), are now extensively used to initiate radical chain processes.

The great majority of carbon-centred radicals undergo very fast dimerization (and also disproportionation) reactions but triphenylmethyl and triplet oxygen belong to a class of so-called *persistent* (as opposed to *transient*) radicals. Such radicals, because of steric hindrance (as for triphenylmethyl) or for special electronic reasons (as for molecular oxygen), recombine relatively slowly or not at all *with species of their own kind*. Triphenylmethyl radicals can hence exist as discrete species in solution and oxygen molecules are perfectly tolerant of each other; but these persistent radicals rapidly react with other radicals if given the chance (in Chapter 7, we shall use the persistent radical effect to explain why triphenylmethyl and phenyl radicals, generated in the thermal decomposition of PAT, recombine to give selectively tetraphenylmethane, the cross-coupling product). Thus, oxygen combines exceedingly rapidly, and usually irreversibly, with triphenylmethyl and most other radicals it encounters in a reaction medium. Although this reactivity of triplet oxygen can be used to advantage in some synthetic applications as will be shown later, it is in most cases a nuisance and oxygen has to be removed from the system by deaerating the solvent and working under an inert atmosphere. It may be noted, as a curiosity, that perchlorotriphenylmethyl is so hindered that it is completely dissociated and does not react even with oxygen.

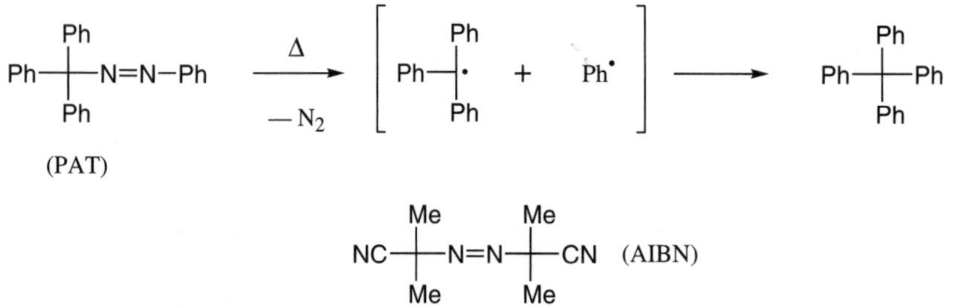

Scheme 1.2 Thermal decomposition of diazo compounds

Perhaps the main reason why radical chemistry did not really catch on in the early decades of the century was that most of the reactions used by organic chemists could be nicely rationalized and classified according to an emerging ionic theory that was both powerful and coherent. 'Homolysis, even between consenting adults, is grounds for instant dismissal from this Department' (Ingold 1996) is a revealing quotation of C. K. Ingold who, along with Sir Robert Robinson, is considered to be the main figurehead in the development and propagation of the ionic mechanistic picture (and father of the 'curly arrow'). Another anecdote illustrating the poor esteem in which radicals were held at the time concerns a manuscript submitted by F. C. Koelsch in 1932 describing a persistent radical of structure **4** (Fig. 1.1), which was rejected by a referee on the grounds that the properties of this compound could not be those of a free radical, presumably because, like perchlorotriphenylmethyl, it was resistant to reaction with oxygen. An electron spin resonance spectrum (esr) on the original sample performed 20 years later confirmed the free radical nature of the substance (now known as a Koelsch radical) and the paper was resubmitted—and accepted (Koelsch 1957).

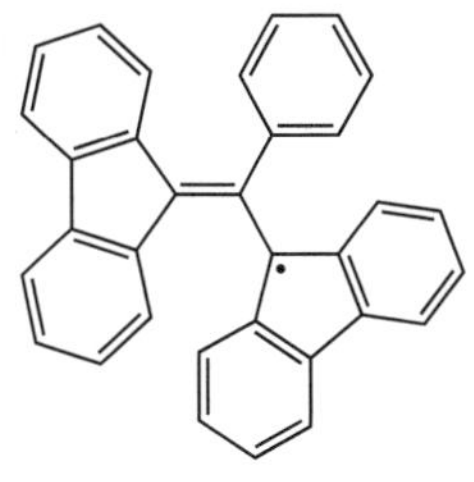

Although the intermediacy of radicals in gas phase reactions received strong support from such famous experiments as Paneth's thermal decomposition of tetramethyllead (Paneth and Hofeditz 1929), their importance as reactive intermediates in solution remained totally unappreciated . Indeed, the influence of the 'ionic theory' was so strong that it was sometimes stretched to extremes to accommodate certain experimental facts. One such instance is the addition of hydrogen bromide to olefins, which did not seem to obey the Markovnikov rule: the regiochemistry of the addition varied from one experiment to another in a puzzling manner. M. S. Kharasch, along with a few other chemists, speculated that olefins could exist as an equilibrium (Scheme 1.3) between two distinct 'electromers' (electronic isomers) and the position of this equilibrium depended on how the olefin was 'prepared': addition of HBr would then lead to the two isomeric bromides in a ratio that reflected this equilibrium (Mayo 1986). This of course is not correct according to modern bonding theory; the so-called 'electromers' represent in fact extreme and hypothetical resonance structures.

Nevertheless, to test this hypothesis, Kharasch asked one of his graduate students, Frank Mayo, to study the addition of HBr on samples of allyl bromide that were 'prepared' in different manners, that is, subjected to various treatments such as heating, irradiation, etc. for various lengths of time and under different conditions in order to modify the ratio of the supposed 'electromers' (Mayo 1986). Allyl bromide was selected because the two isomeric bromides, 1,2- and 1,3-dibromopropane,

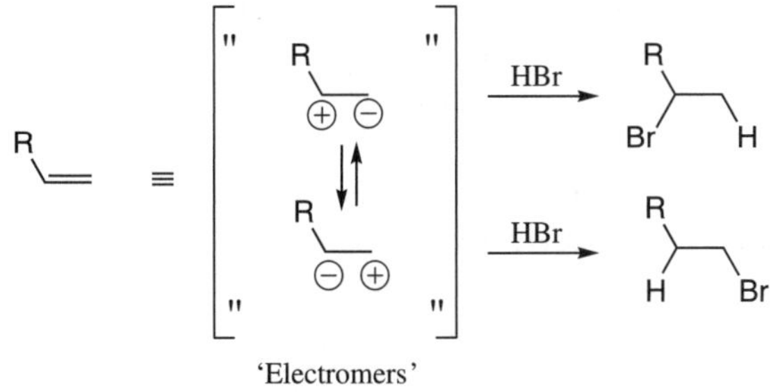

**Fig. 1.1** A Koelsch radical

'Electromers'

**Scheme 1.3** The 'electromer' view

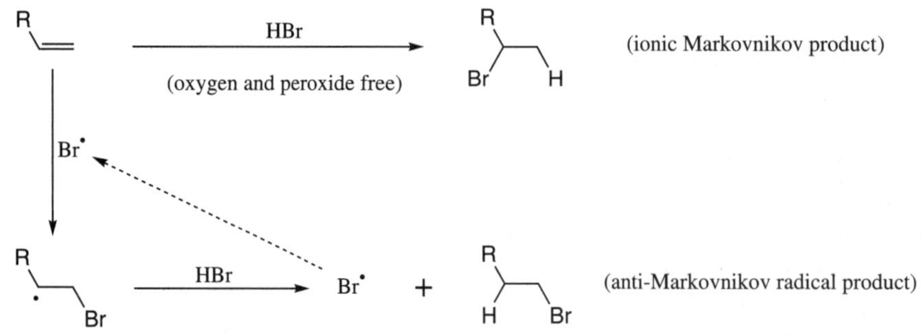

Scheme 1.4 Addition of hydrogen bromide to an olefin

could be separated by careful fractional distillation (NMR and GC–MS did not exist in the 1930s!). However, no logical pattern emerged after several hundred experiments were laboriously completed. To explain the erratic results, Kharasch one day chided his student by saying: 'You know, Mayo, maybe it's the phases of the moon'! As with Gomberg, the reactions were routinely done under air, and the possible influence of oxygen was finally raised and tested, with spectacular effect. In the absence of oxygen (and of any adventitious peroxide), the Markovnikov product (1,2-dibromopropane) prevailed, whereas in its presence or when a peroxide was deliberately added the other isomer (1,3-dibromopropane) was dominant. The proposed mechanism (Scheme 1.4), corresponds to one of the earliest radical chain mechanisms. Adventitious oxygen or peroxide serves to generate a small amount of bromine atoms needed to trigger the radical chain process and, depending on the exact experimental conditions, the competition between the ionic and radical pathways will determine the ratio of isomeric bromides obtained. Whence the discrepancy in the results from one experiment to the other. Because of the chain nature, a small amount of oxygen (initiator) can have a large effect on the outcome.

The influence of this work and later consecutive studies by Kharasch and his students (Mayo, Walling, etc.) on the subsequent development of radical reactions in organic synthesis and perhaps even more in polymer chemistry (with, in parallel, the group of Hermann Mark) is immense. The advent of Second World War imposed on the Allies the urgency to replace the increasingly unavailable natural rubber (needed among others things to make wheels for army vehicles and aeroplanes) by a synthetic substitute and revealed the strategic importance of a variety of other newly discovered plastic materials such as polyethylene for radar and polymethyl methacrylate (Plexiglas) for aircraft manufacture. A vast research programme in polymer chemistry was implemented from which many of the concepts governing radical reactions emerged. Kharasch, too, had difficulties with referees rejecting his papers but he was influential enough to be able to convince the board of the American Chemical Society to start the Journal of Organic Chemistry where he could get his work published with less hassle, being on the board of editors (D. H. R. Barton, private communication).

The Kharasch-type reactions, discussed in detail in Chapter 6, and other, often non-chain, free radical processes that were studied on both sides of the Atlantic, especially by Hey and Waters (1937) in the UK, involved usually simple molecules. This chemistry was therefore largely ignored by most organic chemists engaged in the then budding but rapidly spreading area of partial and total synthesis of complex

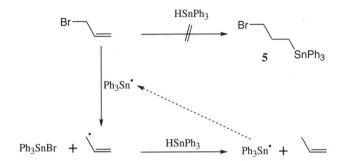

**Scheme 1.5** Reaction of triphenylstannane with allyl bromide

natural products. Allylic bromination with *N*-bromosuccinimide (the Wohl–Ziegler reaction) was for a long time perhaps the only synthetically significant radical reaction (whose mechanism was nevertheless not well understood) that an undergraduate student would encounter. In this context, the reporting in 1957 by the group of van der Kerk (van der Kerk *et al.* 1957) of the reduction of halides with organotin hydrides (stannanes), the prototype of the tin-based radical chemistry that was to prove of such extraordinary power many years later, went practically unnoticed. This reaction also owes its discovery to accident. Many organotin compounds had been found to possess a range of useful insecticide and anti fungal activity. Some were very potent anti-fouling agents that could be usefully applied to protect ship hulls against overgrowth of barnacles and other marine molluscs. Hydrostannylation of alkenes and alkynes (which, incidentally, often proceeds by a Kharasch-type radical addition) was an established way for elaborating organotin derivatives, and van der Kerk and his group attempted the addition of triphenyltin hydride to the double bond of allyl bromide (Scheme 1.5). The expected adduct **5** was not obtained; instead, the reaction produced triphenyltin bromide and propene in excellent yield. The mechanism turns out to be a highly efficient radical chain, again initiated by oxygen (Kuivila 1968). Oxygen and, by coincidence, allyl bromide have thus played a key role in revealing two synthetically very important radical processes.

The ability of trialkyl- and triaryltin hydrides to reduce halides under especially mild conditions was found to extend to various other groups. But it is perhaps the deoxygenation of alcohols through their xanthates and similar thiocarbonyl derivatives, invented by Barton and McCombie (1975) and applied with success to structures of incredible complexity, that struck the imagination of organic chemists. Also of tremendous importance is the realization at about the same time (even though known by polymer chemists much earlier) that carbon–carbon bond formation, and especially 5-membered ring construction, was particularly easy using radicals (Walling *et al.* 1966; Julia 1971).

The renaissance in radical-based methods, that was to hit the synthetic chemistry scene with such great force less than a decade later, was also being quietly but no less importantly prepared by several groups of physical organic chemists who, in the meantime, were painstakingly and, in many ways, ingeniously, measuring the absolute rates for the most useful elementary steps (Griller and Ingold 1980). As will become shortly apparent, a knowledge of rate constants, or at least a feel for their order of magnitude, is essential for gaining mastery of a given sequence of radical transformations and for understanding the influence of the various experimental parameters on the outcome. It is almost certainly the lack of reliable rate constants that has most hampered the development of radical reactions in organic synthesis and entertained the ill-deserved reputation that such processes were unruly, fit only for producing tars.

## 1.2   Timescales: the importance of reaction rates

The main difficulty in handling radical reactions is associated with the fact that two free radicals, unless they belong to the special but very limited class of persistent radicals, interact with each other at essentially diffusion-controlled rates, that is with a rate constant around $10^9$–$10^{10}$ $M^{-1}$ $s^{-1}$ in a non-viscous solvent. The energy barrier for such a process is quite small, of the order of 1–2 kcal mole$^{-1}$. Primary and secondary radicals give mostly dimers, whereas tertiary radicals tend to disproportionate (two *t*-butyl radicals for example would produce one molecule each of isobutane and isobutylene). The existence of this 'self-reaction' of radicals imposes severe constraints on the design of synthetically useful processes. In ionic chemistry, one can generate, say, the enolate of a ketone before lunch and return to add the electrophilic partner with a good deal of confidence that the enolate has not reacted with itself during one's absence. This is not the case with free radicals: once created, their lifetime in the medium is only a tiny fraction of a second. The synthetic chemist wishing to capture his radical R$^\bullet$ with a reagent X–Y for example is thus faced with the difficult situation summarized in Scheme 1.6.

The desired step leading to R–X must be faster than the dimerization of the radical (or disproportionation; for simplicity, only the dimerization is shown); it must also compete with the reaction of the radical with the solvent. Practically all reactions between a radical and a non-radical entity have a rate constant many orders of magnitude smaller than that of diffusion, so how is it possible to win against a reaction that is diffusion controlled? The solution to this problem is found by noticing that the rate of the unwanted dimerization (eqn 1) depends on the *square* of the concentration of the radical species, in contrast to that of the desired reaction (eqn 2). If the steady-state concentration of the radical species is kept very low, for example at $10^{-8}$ M (which is roughly the concentration detectable by esr) then, taking a value of $10^9$ $M^{-1}$ $s^{-1}$ for $k_1$, one obtains a rate of $10^9$ $[10^{-8}]^2$ M $s^{-1}$ or $10^{-7}$ M $s^{-1}$. If the concentration of the trap X–Y is 1 M, then the rate of capture will be equal to: $k_2$ $[10^{-8}][1]$ M $s^{-1}$, and this term has to be greater than $10^{-7}$ M $s^{-1}$, which means that $k_2$ must be greater than 10 $M^{-1}$ $s^{-1}$ (Curran 1991). This is a minimum value and, generally, if $k_2$ is smaller than $10^2$ $M^{-1}$ $s^{-1}$, it will be difficult to control the system since the competing reaction with the solvent, which we have neglected in this rudimentary calculation, can become serious. For example, the rate constant for the addition of

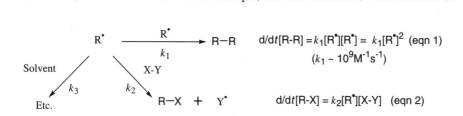

$$d/dt[R\text{-}R] = k_1[R^\bullet][R^\bullet] = k_1[R^\bullet]^2 \text{ (eqn 1)}$$
$$(k_1 \sim 10^9 M^{-1}s^{-1})$$

$$d/dt[R\text{-}X] = k_2[R^\bullet][X\text{-}Y] \text{ (eqn 2)}$$

**Scheme 1.6** Competing radical reactions

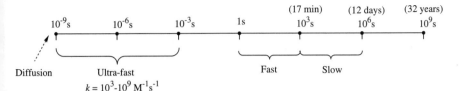

**Fig. 1.2**  Timescale

aliphatic radicals to benzene, a very common solvent for radical reactions, has been estimated to be $10^2$ $M^{-1}$ $s^{-1}$ (Citterio *et al.* 1977; Griller *et al.* 1981). Fortunately, this addition is often reversible. In other words, the concentration of the radical species must be kept as low as is feasible and the rate constant $k_2$ for the desired process must be as large as possible. The situation is a little bit like riding a bicycle, the faster one moves, the easier it is to keep one's balance.

The need to know, or at least to be able to crudely estimate the rate constants for the various steps involved in a radical sequence is obvious from the above analysis. The scale shown in Fig. 1.2 is perhaps useful in allowing a comparison of and giving a feel for the time intervals corresponding to the rates of various chemical processes. For a chemist, a reaction that is complete in around 20 min when all the ingredients have been mixed is considered to be fast; if it takes a few hours to a few days, it is deemed slow but still practical. Such a duration corresponds to rate constants ranging from 1 to $10^{-6}$ $M^{-1}$ $s^{-1}$, which is where rate constants for most ionic reactions lie (extremely fast proton transfers are an obvious exception). The lifetime of ionic interactions is thus generally of a length that is on a human timescale, that is, seconds, minutes, or hours. A chemist can therefore, from personal experience, from reading the literature etc. acquire a direct feel for the reactivity of the various ionic species towards a given functional group under given experimental conditions. He can hence, to a certain extent, predict their selectivity without the need for precise kinetic data. In fact, reaction rate constants are very seldom mentioned in publications involving ionic or organometallic reactions. Rate constants for radical processes in contrast are several orders of magnitude greater, spreading for most reactions from $10^2$ to $10^8$ $M^{-1}$ $s^{-1}$ (considering only bimolecular processes) and corresponding to a half-life of the intermediates of milli-, micro- or even nanoseconds. Such short time intervals are beyond human experience and can only be measured with sophisticated equipment or by comparison with other fast processes whose absolute rate constants are known. It is only relatively recently that reliable absolute rate constants have become available. One can therefore understand the bewilderment and eventual discouragement of earlier chemists trying to apply radical reactions and obtaining erratic results that were difficult to rationalize. The contribution of kineticists in clarifying the relative reactivity of radical intermediates and allowing their domestication for organic synthesis can hardly be overestimated.

The rapidity of elementary radical steps should not be construed as implying a lack of selectivity since the rate constants for these ultra-fast reactions still nevertheless span nearly 9 orders of magnitude! Moreover, even if the *individual elementary radical steps* are much faster than most ionic processes, it must be realized that the *macroscopic* duration of a radical reaction will depend on how fast the radicals are being generated in the first place, and also on the concentration and whether a chain mechanism is operating or not. In an explosion, involving highly complex branched radical chains, the reaction may be complete in less than a millisecond— with dramatic results; synthetically more useful reactions can be designed so as to take from a few minutes to a few hours or even longer.

## 1.3   Further comparison with ionic reactions

The need for a synthetic organic chemist to have some knowledge of reaction rates in order to be able to exploit radical processes intelligently is compensated by the fact that a given free radical *is the same species whatever the method used to generate it* and, being neutral and little solvated means that the rates will depend little on the nature of the solvent (Kanabus-Kaminska *et al.* 1989). Thus, variations in reaction rates with the solvent rarely exceed a factor of 10. This is illustrated by the rate of decomposition of the three diazo derivatives in Table 1.1, which varies by a factor of less than three on going from cyclohexane to cyclohexanol, even for the last derivative where a slightly polarized transition-state might be expected (Luedtke *et al.* 1987).

   Thus, rate constants measured in one solvent system can be transposed with reasonable confidence to another medium. The rare exceptions include for example, hydrogen abstraction by chlorine atoms where the selectivity depends significantly on the nature of the solvent. This is apparently due to the formation of some kind of complex between the highly reactive chlorine atoms and the solvent (especially with aromatic solvents) causing a variation in reactivity of the chlorine atoms from one solvent to another (Dneprovskii *et al.* 1998; Ingold *et al.* 1990). In another instance, it was found that a 'dramatic' 200-fold slowing down of hydrogen abstraction from phenol by cumyloxy radicals (PhCMe$_2$O$^\bullet$) occurred on going from carbon tetra-chloride to *t*-butanol, and this was ascribed to hydrogen bonding which makes the hydroxy hydrogen of the phenol less available (Avila *et al.* 1995). This exceedingly rare two orders of magnitude effect of the solvent on a radical reaction pales next to the effect of a solvent change in a typical ionic reaction. An S$_N$2 displacement for example, can be accelerated by a factor of $10^6$ by replacing a protic solvent like methanol with a polar aprotic medium such as DMF or DMSO. Rate modifications

**Table 1.1**  Relative rates of decomposition of azo compounds in different solvents

| X | Temperature °C | Relative rates | | | |
|---|---|---|---|---|---|
| | | Cumene | Cyclohexane | DMF | Cyclohexanol |
| Ph | 65 | 1.0 | 0.76 | 0.87 | 0.90 |
| CN | 80 | 1.0 | 0.97 | 1.14 | 1.16 |
| CN; MeO | 70 | 1.0 | 1.64 | 2.30 | 2.70 |

of the same order of magnitude are found in $S_N1$ type solvolytic processes, on going from a non-polar solvent to a protic medium of high dielectric constant.

Another advantage that accrues from the neutral nature of free radicals is that, very often, reactions can be run under very mild, non-acidic or basic conditions. Moreover, the absence of a counter-ion and of a sticky layer of solvating molecules around the reactive species translate in practice into a lesser sensitivity to steric factors as compared with ionic intermediates. This is further aided by the fact that radical reactions often proceed through an early ('loose') transition state where the newly formed bonds are still relatively long. Troublesome quaternary centres are therefore generally more easily constructed using radical methods.

Free radicals are also on the whole less prone to Wagner–Meerwein type rearrangements, which sometimes plague cationic processes. On the other hand, hydrogen atom transfers are more frequent than hydride shifts and can be a source of complications but, like Wagner–Meerwein rearrangements, can also be employed to spectacular effect. $\beta$-Eliminations of certain groups such as hydroxy groups, acetates, etc. involving cleavage of a carbon–oxygen or (but less generally) carbon–nitrogen bond, which occur easily with anionic intermediates, are usually too slow in the case of radicals (of course, this 'slowness' is relative to other competing reactions open to the radical species). This feature makes radical-based methods especially useful in the carbohydrate field where hydroxy and ester groups are found in great abundance. It is also perhaps worth underlining the fact that whereas O–H bonds in carboxylic acids and alcohols (and to a lesser extent an N–H bond of an amine or an amide) are labile under ionic conditions (heterolytic dissociation) and must frequently be protected, they seldom interfere with radical processes since these, like C–O bonds, are strong links that do not break easily by homolysis (phenols represent a special case, however). In many situations, cumbersome and yield-lowering protection–deprotection steps become unnecessary. The utility of these features for organic synthesis will emerge in the remaining chapters, where actual examples are presented and discussed.

## 1.4 A few words on the structure, stability, and reactivity of radicals

Before proceeding with a description of the applications of radical processes in synthesis, a few comments need to be made concerning the structure, stability, and reactivity of radicals, even if some of these aspects will be discussed in greater detail later.

As shown in Fig. 1.3, the odd electron in a radical can occupy an essentially pure p-orbital (a $\pi$-radical) or an orbital containing some s-character such as $sp^3$, $sp^2$, or

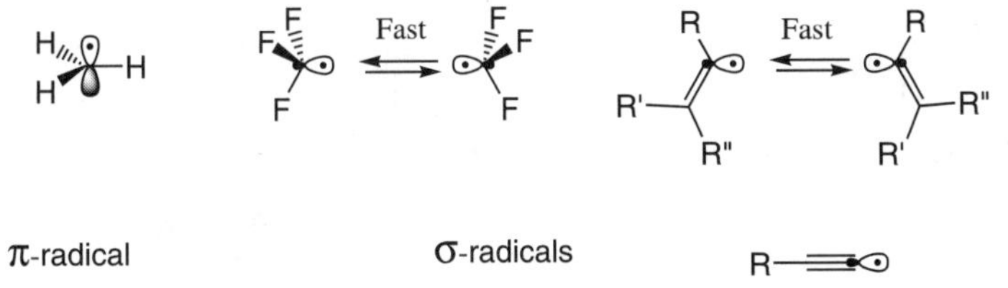

$\pi$-radical          $\sigma$-radicals

**Fig. 1.3** Structure of radicals

sp (a σ-radical). Methyl and most simple aliphatic and alicyclic radicals belong to the π-type (cyclopropyl and bridgehead radicals are notable exceptions; Walborsky 1981), whereas those substituted with electronegative elements tend to acquire partial s-character (Kochi 1975). Thus, the trifluoromethyl radical appears to have an almost pure $sp^3$ hybridization. The extent of s-character can be estimated by measuring hyperfine coupling constants in an esr spectrum, especially the $^{13}$C coupling in isotopically labelled radicals: the greater the coupling the greater the s-character. Vinylic and aromatic radicals are usually σ-radicals (Galli *et al.* 1997). Because of their partial s-character, σ-radicals tend to be more electrophilic in general than π-radicals, other factors being equal. In an s-orbital, the electron has a non-zero probability of being found near the positively charged nucleus, hence the greater coupling constant with a $^{13}$C nucleus and a greater electrophilic character (this incidentally also explains the dramatic increase in acidity—more than 20 p$K$ units—of an acetylenic hydrogen in comparison with a simple alkane hydrogen). Esr spectroscopy can provide a wealth of other information on radical intermediates such as energy barriers to rotation and preferred conformations in open chain and cyclic structures (Ingold and Walton 1989).

π-Radicals are flat; σ-radicals invert very rapidly so that any stereochemical information contained in the precursor is lost at the radical stage. The stereochemical information can be restored, however, if other fixed centres in the molecule can force the radical to react preferentially from one side of the molecule. Such a situation will obtain in many of the examples discussed in the following chapters. In some special cases a 'memory' effect may be observed. These usually involve cyclic radicals which are captured before a change of conformation (through ring flipping) has had time to take place, as illustrated by the hypothetical transformation in Scheme 1.7 (Buckmelter *et al.* 1998). Thus, rupture of the C–X bond gives an anomeric-type radical which, because of the chair shape of the molecule, reacts with a radical trap preferentially from the β-face. Ring-flip produces the other chair form which reacts preferentially from the α-face leading to the enantiomeric product. If ring flipping is faster than the capture of the radical by Y–Z, then conformational equilibrium is attained and a racemic mixture of products is observed. In contrast, if interception of the radical occurs *before* complete conformational equilibration, then a non-racemic

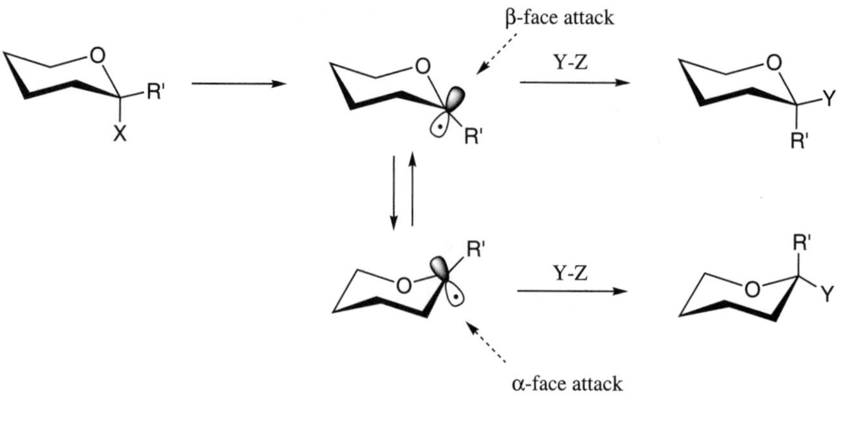

**Scheme 1.7**  'Memory' effect

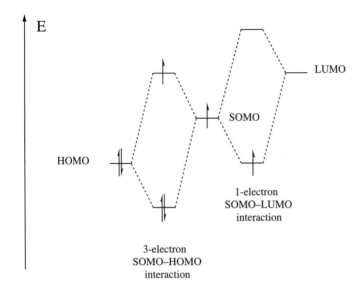

**Fig. 1.4** Orbital interactions of unpaired electron with empty and filled orbitals

mixture is produced, the ratios being controlled by the relative rate constants, the concentration of the X–Y reagent, and exact experimental conditions.

The orbital that holds the single electron is referred to as the SOMO (*Singly Occupied Molecular Orbital*). Its interaction with both an empty or filled orbital is stabilizing: in the former case it is a one-electron interaction, and the electron is stabilized; in the latter, it is a three-electron interaction, so two electrons are stabilized and one is destabilized, resulting generally in an overall stabilization of the system. These orbital interactions are shown in the highly simplified diagram in Fig. 1.4. Radicals are therefore ambiphilic species that are stabilized by both electron-withdrawing *and* electron-releasing substituents. Those that are flanked by one substituent of each type are called capto-dative radicals (Viehe *et al.* 1989) and seem to enjoy a special stabilization but its extent is still the subject of some controversy (Welle *et al.* 1997).

This ambiphilic character means that a radical displays a much wider reactivity pattern than the corresponding anion or cation since it can in principle react with both electron-rich or electron-poor substrates, even if the relative reaction rates may vary greatly. In the absence of electrostatic effects, which play a very significant role in ionic reactions, orbital interactions, also referred to as polar effects, acquire a special importance in understanding the reactivity of radicals (more on this in the next chapter).

When dealing with radicals, it is also important to distinguish between *thermodynamic stability* and *kinetic lability*. A methyl radical is thermodynamically less stable than a benzyl radical, the estimated difference being of the order of 20 kcal mole$^{-1}$ (Tsang 1985; Clark and Wayner 1991; Lehd and Jensen 1991; Brocks *et al.* 1998). However, two methyl radicals and two benzyl radicals will recombine at essentially the same, diffusion-controlled rate. In this respect, they are both *transient* species, even though the formation of ethane will be much more exothermic than that of diphenylethane. The difference in reactivity between a methyl and a benzyl radical becomes more apparent in reactions with non-radical entities, where the rates are much lower than diffusion. *Transient* radicals can, in a sense, be made *persistent* if they can be separated from each other as in an argon matrix or in empty space. Alternatively, the steric hindrance around the radical centre can be made so large that two such radicals are unable to react with each other as with Koelsch's radicals above. Such radicals are truly *persistent*. In rare cases, the persistence can be due to a special electronic configuration as for triplet oxygen, or nitric oxide (•NO), or Frémy's salt (potassium nitrosodisulfonate). The last compound was reported in 1845 by Edmond Frémy, a professor at Ecole polytechnique. It is very probably the first manmade persistent radical.

## 1.5 Chain reactions

One further aspect that needs to be briefly addressed before moving to the synthetic applications of radical processes concerns the fact that once a radical is generated, then its reaction with a non-radical entity will always lead to a new radical as shown in the equations in Scheme 1.8. The first radical, represented as In• (from *Initiator*), produced by thermal or photochemical dissociation of some precursor, or by a redox process involving an electron transfer, serves as a trigger or initiator (whence the term initiation step or steps) and generates in turn a radical from the main reactant. This new radical can then undergo a number of typical elementary radical reactions such as substitution (called an $S_H2$ for second order *Homolytic Substitution* by analogy with an $S_N2$ process) or addition to a multiple bond (the reverse is a fragmentation or $\beta$-cleavage, or again $\beta$-scission) depending on the reagent or radical trap placed in the medium. The only way to 'get rid' of the radical is by a reaction that is essentially the reverse of the initiation, that is, by a recombination of radicals or by a redox process that converts the radical into an anionic or cationic species. Such transformations represent termination steps. From a synthetic standpoint, it is necessary to direct the overall process into providing the desired product cleanly, and this is best achieved if the cascade of reactions is designed as a closed chain. The intermediate steps then constitute the propagation part of the system, the last step in the propagation sequence producing the same radical species implicated in the first one (Y• in the typical sequence in Scheme 1.8).

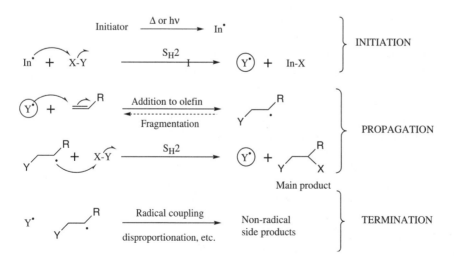

**Scheme 1.8**
Elementary radical reactions

**Scheme 1.9** Alternative arrow formalism

Chain processes, as illustrated by the peroxide initiated hydrogen bromide addition to olefins described earlier, represent perhaps the most elegant, even if by no means the only way, to control a sequence of radical reactions. If the individual propagation steps are fast, then the steady-state concentration of the various radical species will remain low and termination steps will become negligible. And of course, only a small amount of initiation will be needed to convert all the substrate cleanly into product (in non-chain systems, the 'initiation' and 'termination' have to be stoichiometric). The most important chain reactions will be discussed in Chapters 2–6; synthetically useful non-chain and redox processes will form the subject matter of Chapters 7 and 8.

By analogy with ionic reactions, where curly arrows are used to depict the movement of an electron pair, we shall use, as in Scheme 1.8, half-headed curly arrows for radicals since only one electron is involved. However, it must be remembered that the arrow is drawn starting from the radical to the non-radical partner without implying, as would be the case in ionic chemistry, that the radical is acting as a 'nucleophile'. Another, more realistic but also more cumbersome depiction that can be found in the literature makes use of two one-headed arrows, as illustrated by the radical addition to an olefin in Scheme 1.9.

With a sketchy historical background and a few basic elements in hand, we shall next turn to the study of the radical chemistry of organotin derivatives (organostannanes) as it applies to organic synthesis and use the concepts that will emerge as the foundation upon which the remainder of the material will be structured.

## References

Avila, D. V., Ingold, K. U., Lusztyk, J., Green, W. H., and Procopio, D. R. (1995). Dramatic solvent effects on the absolute rate constants for abstraction of the hydroxylic hydrogen from *tert*-butyl hydroperoxide and phenol by the cumyloxyl radical. The role of hydrogen bonding. *Journal of the American Chemical Society*, **117**, 2929–2930.

Barton, D. H. R. and McCombie (1975). A new method for the deoxygenation of secondary alcohols. *Journal of the Chemical Society, Perkin Transactions 1*, 1574–1585.

Brocks, J. J., Beckhaus, H.-D., Beckwith, A. L. J., and Rüchardt, C. (1998). Estimation of bond dissociation energies and radical stabilization energies by ESR spectroscopy. *Journal of Organic Chemistry*, **63**, 1935–1943.

Buckmelter, A. J., Powers, J. P., and Rychnovsky, S. D. (1998). Nonequilibrium radical reductions. *Journal of the American Chemical Society*, **120**, 5589–5590.

Citterio, A., Minisci, F., Porta, O., and Sesana, G. (1977). Nucleophilic character of the alkyl radicals. 16. Absolute rate constants and the reactivity selectivity relationship in the homolytic aromatic alkylation. *Journal of the American Chemical Society*, **99**, 7960–7968.

Clark, K. B. and Wayner, D. D. M. (1991). Are relative bond energies a measure of radical stabilization energies? *Journal of the American Chemical Society*, **113**, 9363–9365.

Curran, D. P. (1991). Radical addition reactions. In *Comprehensive organic synthesis*, Vol. 4 (ed. B. M. Trost and I. Fleming) pp. 715–777. Pergamon Press, Oxford.

Dneprovskii, A. S., Kutzestov, D. V., Eliseenkov, E. V., Fletcher, B., and Tanko, J. M. (1998). Free radical chlorinations in halogenated solvents: are there *any* solvents which are truely non complexing? *Journal of Organic Chemistry*, **63**, 8860–8864.

Galli, C., Guarnieri, A., Koch, H., Mencarelli, P., and Rappoport, Z. (1997). Effect of substituents on the structure of the vinyl radical: calculations and experiments. *Journal of Organic Chemistry*, **62**, 4072–4077.

Griller, D. and Ingold, K. U. (1980). Free radical clocks. *Accounts of Chemical Research*, **13**, 317–323.

Griller, D., Marriott, P. R., Nonhebel, D. C., Perkins, M. J., and Wong, P. C. (1981). Homolytic addition to benzene. Rate constants for the formation and decay of some substituted cyclo-hexadienyl radicals. *Journal of the American Chemical Society*, **103**, 7761–7763.

Hey, D. H. and Waters, W. A. (1937). Some organic reactions involving the occurrence of free radicals in solution. *Chemical Reviews*, **21**, 169–208.

Ingold, K. U. and Walton, J. C. (1989). Probing ring conformations with EPR spectroscopy. *Accounts of Chemical Research*, **22**, 8–14.

Ingold, K. U., Lusztyk, J., and Raner, K. D. (1990). The unusual and the unexpected in an old reaction. The photochlorination of alkanes with molecular chlorine in solution. *Accounts of Chemical Research*, **23**, 219–225.

Ingold, K. U. (1996). Introduction to 'C. K. Ingold: master and mandarin of physical organic chemistry'. *Bulletin for the History of Chemistry*, **19**, 1.

Julia, M. (1971). Free radical cyclization. *Accounts of Chemical Research*, **4**, 386–392.

Kanabus-Kaminska, J.-M., Gilbert, B. C., and Griller, D. (1989). Solvent effects on the thermochemistry of free radical reactions. *Journal of the American Chemical Society*, **111**, 3311–3314.

Kochi, J. K. (1975). Configurations and conformations of transient alkyl radicals in solution by electron spin resonance spectroscopy. In *Advances in Free Radical Chemistry*, Vol. 5, (ed. G. H. Williams), pp. 189–317. Elek Science, London.

Koelsch, F. C. (1957). Syntheses with triarylvinylmagnesium bromides. $\alpha,\gamma$-Bisdiphenylene-$\beta$-phenylallyl, a stable free radical. *Journal of the American Chemical Society*, **79**, 4439–4441.

Kuivila, H. G. (1968). Organotin hydrides and organic free radicals. *Accounts of Chemical Research*, **1**, 299–305.

Lehd, M. and Jensen (1991). Improved radical stabilization energies. *Journal of Organic Chemistry*, **56**, 884–885.

Luedtke, A., Meng, K., and Timberlake, J. W. (1987). Carbonyl group stabilisation of radicals: solvent effects on azoalkane decomposition. *Tetrahedron Letters*, **28**, 4255–4258.

Mayo, F. R. (1986). The evolution of free radical chemistry at Chicago. *Journal of Chemical Education*, **63**, 97–99.

McBride, J. M. (1974). The hexaphenylethane riddle. *Tetrahedron*, **30**, 2009–2023.

Paneth, F. and Hofeditz, W. (1929). Über die Darstellung von freiem Methyl. *Berichte der Deutschen Chemischen Gesellschaft*, **62**, 1335–1347.

Tsang, W. (1985). The stability of alkyl radicals. *Journal of the American Chemical Society*, **107**, 2872–2880.

van der Kerk, G. J. M., Noltes, J. G., and Luijten, J. G. A. (1957). Investigations on organotin compounds. VII. The addition of organotin hydrides to olefinic double bonds. *Journal of Applied Chemistry*, **7**, 356–365.

Viehe, H. G., Janousek, Z., and Merényi, R. (1989). Captodative substituent effect in synthesis. In *Free radicals in synthesis and biology* (ed. F. Minisci). NATO ASI Series Vol. 260, pp. 1–26. Kluwer Academic Publishers, Dordrecht.

Walborsky, H. M. (1981). The cyclopropyl radical. *Tetrahedron*, **37**, 1625–1651.

Walling, C. (1986). The development of free radical chemistry. *Journal of Chemical Education*, **63**, 99–102.

Walling, C., Cooley, J. H., Ponaras, A. A., and Racah, E. J. (1966). Radical cyclizations in the reaction of trialkyltin hydrides with alkenyl halides. *Journal of the American Chemical Society*, **88**, 5361–5363.

Welle, F. M., Beckhaus, H.-D., and Rüchardt, C. (1997). Thermochemical stability of $\alpha$-amino-$\alpha$-carbonylmethyl radicals and their resonance as measured by ESR. *Journal of Organic Chemistry*, **62**, 552–558.

# 2 General principles: chain reactions based on stannane chemistry

## 2.1 A simple chain reaction based on organotin hydrides

Chain reactions involving organotin hydrides are by far the most commonly employed in organic synthesis. A wide variety of carbon- and heteroatom-centred radicals can thus be generated and captured under conditions of sufficient mildness to be compatible with many of the highly complex intermediates encountered in modern synthesis. The propagation steps are generally fast enough that the steady-state concentration of the intermediate radicals remains automatically low, and unwanted radical–radical interactions are greatly diminished.

We have seen, in Chapter 1, how van der Kerk and his group stumbled upon the simplest of organotin-hydride-mediated reactions: the replacement of a halogen with a hydrogen. The process may be written in a more general form (Scheme 2.1), since a number of functional groups, represented by **X**, can in fact be reduced by a similar mechanism. For the sake of clarity, tributyltin hydride (or tributylstannane) is used in the scheme as it is the most common reagent, but the same will apply for other organotin hydrides, with perhaps some relatively minor variations in the reaction rates. But before illustrating the synthetic potential of this process, it is necessary to discuss in more detail the individual steps as well as some practical aspects that will often pertain to some of the other radical methods described in this book.

In the original experiment, adventitious oxygen acted as the initiator. This is however not a convenient or an easily reproducible way to trigger the chain reaction. A small amount of a peroxide or, even better, a diazo derivative such as AIBN is far preferable. AIBN is a nicely crystalline substance which decomposes upon heating into isobutyronitryl radicals and nitrogen (see Table 1.1) with a half-life of about an hour at 80°C (this happens to be nearly—and conveniently—the boiling point of

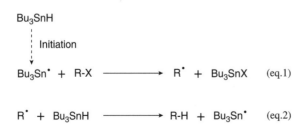

**Scheme 2.1** Radical chain reduction using *n*-tributyltin hydride

**Table 2.1** Approximate rates for the reaction of various R–X with tri-*n*-butylstannyl radicals

| R–X | Approximate rate constant, $K$ $(M^{-1} s^{-1})$ |
| --- | --- |
| Alkyl iodides | $\geq 10^9$ |
| Aryl and vinyl iodides | $10^7$–$10^8$ |
| Alkyl bromides | $10^7$–$10^8$ |
| Aryl and vinyl bromides | $10^5$–$10^6$ |
| Alkyl chlorides | $10^2$–$10^4$ |
| $\alpha$-Chloroesters | $10^5$–$10^6$ |
| Alkylphenyl selenides | $10^5$–$10^6$ |
| Alkylphenyl sulfides | $10^2$–$10^4$ |

benzene or cyclohexane). As with most other unimolecular fragmentations, the half-life of AIBN diminishes rapidly with increasing temperature: it drops to only a few minutes at 110 °C, the temperature of refluxing toluene (for graphs giving the variation of the half-lives of various initiators with temperature, see: Walling and Huyser 1963). In some cases, where a lower reaction temperature is desired, it is possible to use light (UV or sun lamp) or a combination of triethylborane and oxygen (Nozaki *et al.* 1987). Triethylborane readily autoxidizes, liberating reactive ethyl radicals which initiate the chain process. One practical way of conducting such reactions is to equip the reaction flask with a toy balloon filled with argon and let oxygen from the atmosphere slowly diffuse into it and hence into the reaction medium (Yorimitsu *et al.* 1998). It must be remembered that the rate of decomposition of the initiator determines the number of chains that are started and that the steady-state concentration of the intermediate radicals will depend on the efficiency of the propagation steps so that the experimental conditions (especially temperature, concentration, and mode of addition of the reagents) and the type and amount of initiator must therefore be selected and eventually modified with these considerations in mind. It is also advisable to avoid contact between the refluxing solvent and any rubber septa since additives in the rubber can leach into the medium and act as inhibitors of the chain reaction.

The first propagation step (eqn. 1) concerns the reaction of the substrate **R–X** with the tributylstannyl radical. **R–X** can be a halide (but not fluoride), a chalcogenide, an isocyanide, a nitro group or some more complicated moiety such as a xanthate. The rate of this step depends on **X**, the weakness of the R–X bond (Galli and Pau 1998), and the stability of the ensuing radical R• (roughly speaking, the more stable the radical R•, the weaker the R–X bond); some typical rate constants are collected in Table 2.1 (Curran *et al.* 1991). As one would expect from the respective bond strengths, the order for the halides is R–I> R–Br >>R–Cl for the same R group, and selenides react faster than sulfides for the chalcogenides. In general, when the bond strengths are comparable, monovalent halogenides tend to be more reactive than divalent chalcogenides (cf. bromides and selenides), presumably for steric reasons (the cone of approach is wider for the former). For the same **X**, aliphatic derivatives are much more reactive than vinylic or aromatic substrates, and substituents such as a carbonyl group enhance the reactivity (cf. alkyl chlorides and $\alpha$-chloroesters), reflecting the stability of the corresponding radicals. It can be seen on inspection of the table that the absolute rates are high enough to belong to the ultra-fast domain but still sufficiently different to allow for a considerable amount of selectivity. Rates for such ultra-fast elementary radical processes are usually established by competition experiments, ultimately against other, often unimolecular reactions, whose absolute rate has been determined directly by a sophisticated kinetic technique such as laser flash photolysis. These reference reactions provide a convenient scale and are termed radical clocks or radical 'horlogerie' (Griller and Ingold 1980; Newcomb 1993).

Absolute rates for hydrogen abstraction from the stannane, the second propagation step (eqn 2), have also been measured. This is indeed an exceedingly efficient process, as indicated by the rates compiled in Table 2.2 (Johnston *et al.* 1985; Newcomb 1993; Garden *et al.* 1996; Chatgilialoglu and Newcomb 1999). These rates are on the higher end of the scale with relatively little spread. Aromatic, vinylic, and cyclopropyl radicals react fastest, at close to the diffusion limit, whereas aliphatic radicals are some 100 times slower, with not much difference between primary, secondary or tertiary species. Benzylic radicals on the other hand are the least reactive. It can also be seen that, for a given radical, triphenyltin hydride is a little better as a hydrogen atom donor than the more commonly used tributyl reagent.

Tributyl- and triphenyltin hydrides are now commercially available, the former being much cheaper as its preparation involves merely distilling a mixture hexabutylditin oxide and poly(methylhydrosiloxane) [$Me_3Si(SiHMeO)_nSiMe_3$] (Hayashi *et al.* 1967). These hydrides may also be obtained by reduction of the corresponding triorganotin halide with lithium aluminium hydride or sodium borohydride (the deuteride and tritide analogues are prepared in this way). These hydrides are slowly

**Table 2.2** Rates for the reaction of some carbon-centred radicals with organostannanes

| Radical | Stannane | Rate constant ($M^{-1}s^{-1}$; °C as subscript) |
|---------|----------|---------------------------------------------------|
| $CH_3{}^{\bullet}$ | $Bu_3SnH$ | $k_{30} = 1.2 \times 10^7$ |
| $RCH_2{}^{\bullet}$ | $Bu_3SnH$ | $k_{30} = 2.7 \times 10^6$ |
| $RCH_2{}^{\bullet}$ | $Ph_3SnH$ | $k_{25} = 5 \times 10^6$ |
| $Me_2CH^{\bullet}$ | $Bu_3SnH$ | $k_{30} = 1.5 \times 10^6$ |
| $Me_3C^{\bullet}$ | $Bu_3SnH$ | $k_{30} = 1.7 \times 10^6$ |
| cyclo-$C_3H_5{}^{\bullet}$ | $Bu_3SnH$ | $k_{30} = 8.5 \times 10^7$ |
| $Me_2C{=}CH^{\bullet}$ | $Bu_3SnH$ | $k_{30} = 3.5 \times 10^8$ |
| $C_6H_5{}^{\bullet}$ | $Bu_3SnH$ | $k_{30} = 5.9 \times 10^8$ |
| $PhCH_2{}^{\bullet}$ | $Bu_3SnH$ | $k_{25} = 3.6 \times 10^4$ |

decomposed by exposure to oxygen and light, and the quality of the reagent may be variable, especially in the case of the tributyltin hydride which is a liquid and therefore more sensitive to alteration on storage. Chlorinated solvents must be avoided (including CDCl$_3$!) as they can react violently with tin hydrides. The use of hot protic solvent such as methanol or ethanol must also be shunned since they can decompose the hydride by an ionic mechanism leading to the liberation of hydrogen; *t*-butanol, however, appears to be suitable. The main practical difficulty when working with stannanes, apart from the somewhat nauseating stench which seems to cling forever to the glassware, is the purification of the product and the complete removal of organotin residues. Various tricks have been contrived to this end such as precipitation as organotin fluorides (Milstein and Stille 1978), partitioning between methanol or acetonitrile and petroleum ether (Leibner and Jacobus 1979), treatment with DBU (Curran *et al.* 1991), sodium hydroxide (Renaud *et al.* 1998), etc. In any case, it is advisable to use freshly (flash-) distilled material; better yields are invariably obtained and the purification is often easier. In some instances, the stannane can be used in catalytic amounts in conjunction with another reducing agent such as the poly(methylhydrosiloxane) mentioned above (Grady and Kuivila 1969; Lopez *et al.* 1997), sodium borohydride (Corey and Suggs 1975), or cyanoborohydride (Stork and Sher 1986). Polymer supported reagents (Weinschenken *et al.* 1975; Ueno *et al.* 1982; Harendza *et al.* 1993; Dumartin *et al.* 1994) and stannanes with nitrogen (Clive and Yang 1995; Suga *et al.* 1999) or fluorous or polyaromatic substituents (Curran and Hadida 1996; Curran 1998; Gastaldi and Stien 2002) have also been proposed to simplify purification. Finally, organotin derivatives must be considered to be potentially toxic and appropriate care exercised while handling them.

## 2.2   Some examples

Reductions with organotin hydrides can be performed on substances that are a far cry from the original allyl bromide in terms of complexity or fragility (Neumann 1987). The highly efficient reduction of the bromo penicillinate in Scheme 2.2 is one such instance (Aimetti *et al.* 1979).

The stereoselectivity is determined by the cup-shape of the intermediate radical in the hydrogen abstraction step, which takes place from the least hindered convex or *exo* face, and not by the stereochemistry of the starting bromide *since both epimers give the same intermediate radical*. Steric effects at the radical level are almost always the dominating factors in controlling the stereoselectivity. In terms of chemoselectivity, there is no need to protect the free hydroxyl group and no harm is

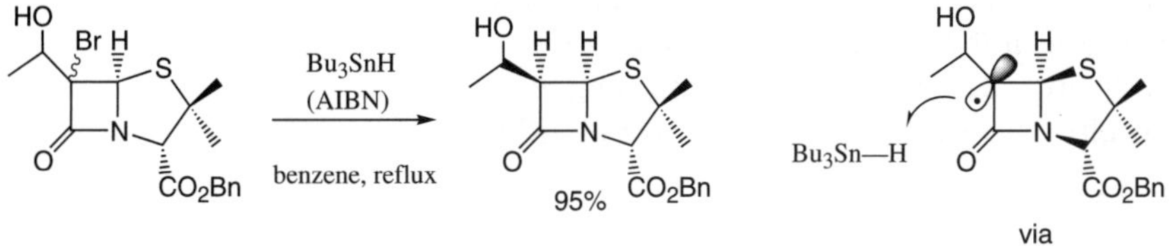

**Scheme 2.2**
Debromination of a penicillin derivative

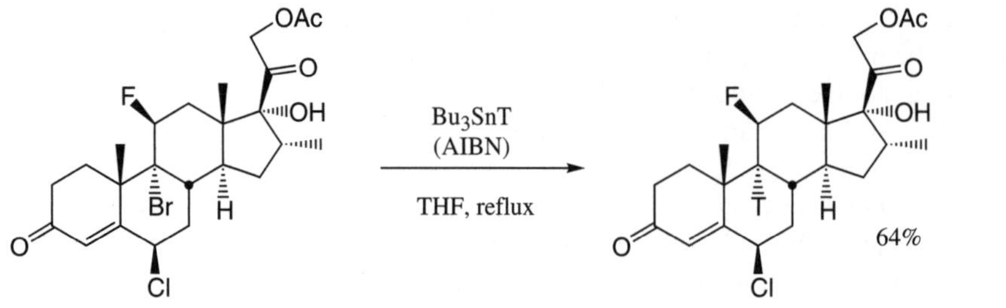

**Scheme 2.3** Selective tritiation of a corticosteroid

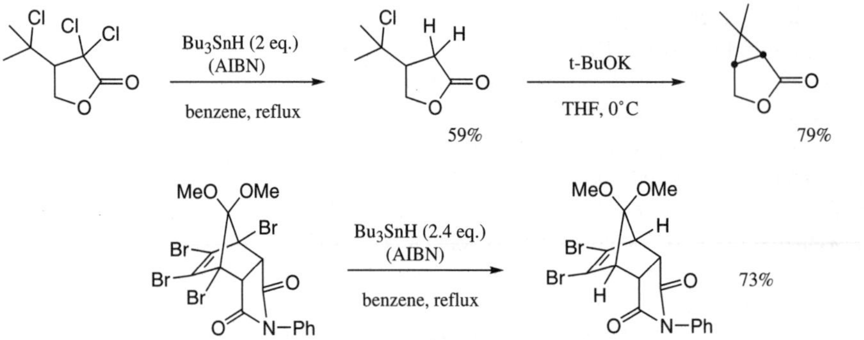

**Scheme 2.4** Selective dechlorination and debromination

done to the cyclic sulfide because of the important difference in the rate of reaction with stannyl radicals between a bromide and a sulfide (Table 2.1). $\beta$-Scission of the sulfide could have occurred had the ensuing olefin not been too strained (more on $\beta$-scissions in the next chapter).

Another example of selective reduction and compatibility with various functional groups is demonstrated by the debromination of the corticosteroid in Scheme 2.3, where neither the chloride nor the fluoride are affected (Parnes and Pease 1979). In this case, tributyltin tritide was used to prepare the tritiated analogue. Organotin deuterides or tritides turn out to be exceedingly useful reagents for the site selective labelling of complex organic molecules.

High chemoselectivity can sometimes be achieved with the same halogen, if the ensuing radicals differ sufficiently in stability, since this is directly related to the strength of the R–X bond. The reduction of the trichloride in Scheme 2.4, itself prepared by a Kharasch-type radical reaction (see Chapter 6), with two equivalents of tributyltin hydride gives the tertiary chloride, a precursor for *cis*-chrysanthemic acid (Takano *et al.* 1984). This selective reduction reflects the greater stability of a radical that is conjugated to a carbonyl group as compared with a simple tertiary radical. The last chlorine could of course have been removed by adding an extra equivalent of hydride. In the second example in Scheme 2.4, the bridgehead bromines can be selectively reduced away without affecting the vinylic ones, reflecting the greater difficulty in generating a vinylic radical (Khan and Prabhudas 1999).

## 2.3   The Barton–McCombie and related deoxygenation of alcohols

Alcohols cannot normally be deoxygenated by direct reaction with an organotin hydride. A more generally applicable method, devised by Barton and McCombie (1975), involves the use of a thiocarbonyl derivative of the alcohol such as a xanthate (Scheme 2.5).

The addition of tributylstannyl radicals to the highly radicophilic thiocarbonyl group is fast, and was later shown to be also reversible (Barton *et al.* 1986). The slow, rate-determining step is the $\beta$-scission of the relatively strong C–O bond. Indeed, the intermediate adduct radical has been captured by an internal olefin (Bachi and Bosch 1988; Rhee *et al.* 2003). Xanthates are readily made from alcohols by treating the corresponding alcoholate with carbon disulfide followed by methyl iodide. The sequence is usually carried out in one pot, in some cases in a two-phase system in the presence of a phase-transfer catalyst. The Barton–McCombie reaction has found numerous applications because the hydroxy function is ubiquitous in natural products and in synthetic intermediates (for reviews, see: Hartwig 1983; Crich and Quintero 1989). The two examples in Scheme 2.6 are illustrative of

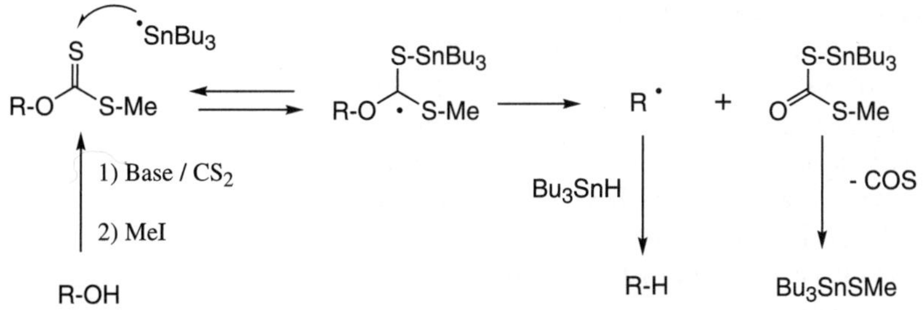

**Scheme 2.5**  Barton–McCombie deoxygenation via xanthates

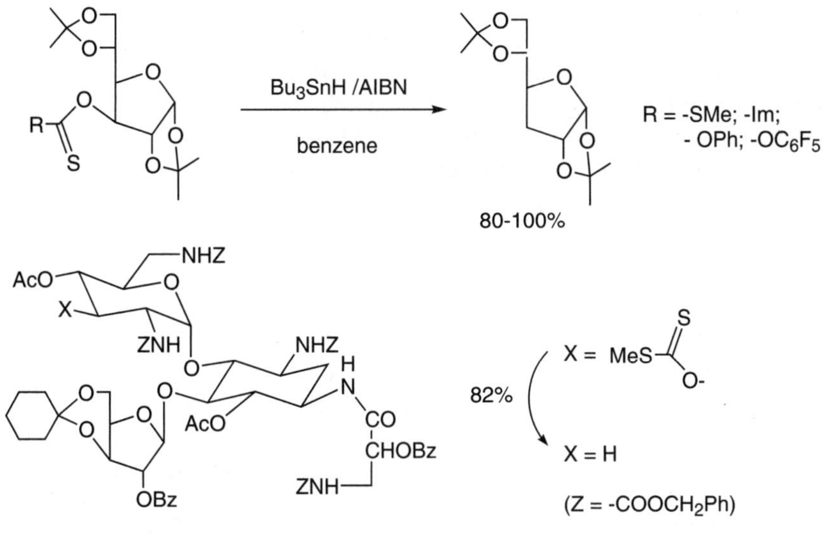

**Scheme 2.6** Deoxygenation of carbohydrates

its tremendous utility: clean, expedient deoxygenations on such complex structures are beyond the reach of classical ionic methods.

Thiocarbonyl imidazolides and thiocarbonates can also be used for deoxygenation (variation in R in the first example in Scheme 2.6). They are often more easily introduced than xanthates and have a similar reactivity towards organotin hydrides but can be more expensive for large scale work. With xanthates as well as with the more recent precursors, secondary alcohols are the best substrates. Thiocarbonyl derivatives of tertiary alcohols are difficult to handle because of their tendency to undergo Chugaev elimination leading to olefins; however, in cases where for various reasons the elimination is retarded, then deoxygenation can be readily performed (Barton *et al.* 1993). Another solution is to initiate the process using $Et_3B/O_2$ at low temperature.

For primary alcohols, the problem is not in making the precursor but inherent in the fragmentation step. Unlike the reduction of halides and chalcogenides, where a simple $S_H2$ is involved, the reaction of thiocarbonyl compounds is a two-step process: a fast bimolecular and reversible addition to the thiocarbonyl group followed by a unimolecular fragmentation. In the case of a xanthate of a primary alcohol the fragmentation is comparatively slow because it leads to a less stable primary radical. Thus, the fragmentation step enters into competition with other pathways open to the intermediate adduct radical, especially capture by the stannane (bimolecular process), followed by collapse to a thioformate (Scheme 2.7) and further reduction (Barton *et al.* 1986; Nicolaou *et al.* 1995; Bensasson *et al.* 1997).

One way around this difficulty is to exploit the large difference in the dependence on temperature and concentration between a bimolecular and a unimolecular fragmentation process. In a bimolecular reaction, both partners have to meet, and an increase in temperature has a limited positive effect on the probability of a favourable collision. In contrast, in a unimolecular transformation, the molecule can acquire sufficient energy to overcome the activation barrier by colliding with any other molecule (solvent, vessel walls, etc.), and an increase in the thermal agitation can considerably improve the chances of the desired transformation. This is especially important when a fragmentation occurs because of a large increase in the entropy term, which is multiplied by the temperature (remember $\Delta G^{\neq} = \Delta H^{\neq} - T\Delta S^{\neq}$). In the Barton–McCombie reaction, the fragmentation is the rate-determining step and an increase in temperature should therefore preferentially favour this step as compared with bimolecular side reactions. Dilution also disfavours bimolecular processes but not unimolecular fragmentations for similar reasons: solvent

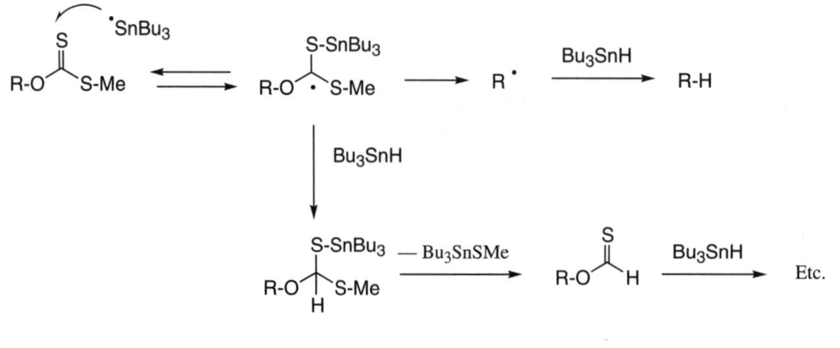

**Scheme 2.7** Capture of the intermediate adduct in the Barton–McCombie reaction

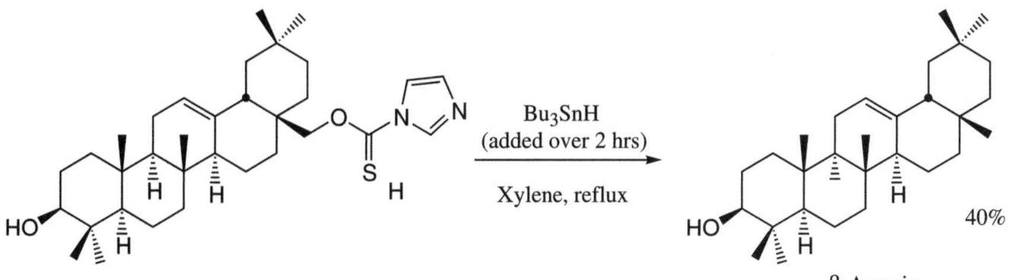

**Scheme 2.8**
Deoxygenation of a primary alcohol via a thiocarbonyl imidazolide

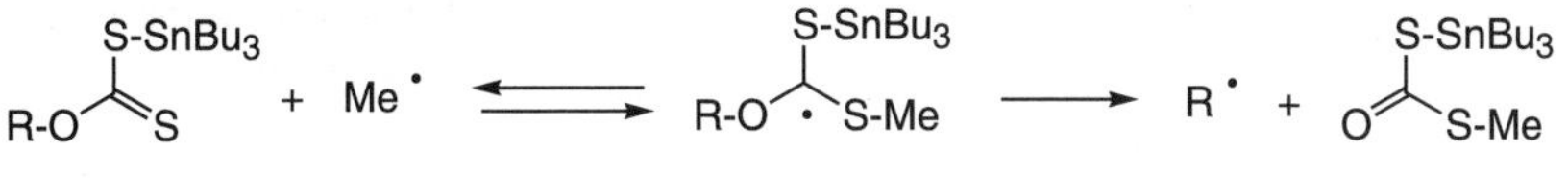

**Scheme 2.9** Dual pathways in the fragmentation of the intermediate adduct in the case of primary alcohols

molecules can participate in productive collisions in the latter case but not in the former. Thus, for primary alcohol derivatives, an increase in temperature and in dilution will enhance the desired pathway. This is indeed the case: when tributyltin hydride was added slowly (thus ensuring that the concentration of the stannane remains low) to a refluxing solution of erythrodiol monothiocarbonyl imidazolide in xylene, a moderate yield of $\beta$-amyrin was obtained (Scheme 2.8; Barton *et al.* 1981).

Xanthates are generally less suitable for primary alcohol deoxygenation because of a competing (but reversible) fragmentation to give a methyl radical by cleavage of a C–S bond. A methyl radical is less stable than a primary radical but this is compensated by the fact that a C–S bond is weaker than a C–O bond. The two processes being unimolecular fragmentations, a change in the reaction conditions will have a similar effect on both (Scheme 2.9).

The reasoning we have used to analyse the Barton–McCombie reaction for primary alcohol derivatives might seem perhaps pedantic, but the same considerations apply in many other related situations where several pathways involving bimolecular and unimolecular steps are in competition. It is therefore crucial to be able to predict, at least qualitatively, the effect of a modification in the experimental variables on the product distribution in order to bias the system in the desired direction. This is demonstrated in a dramatic way by yet another, lesser known, deoxygenation of alcohols via the corresponding phenyl selenocarbonates (Pfenninger *et al.* 1980). When this method was applied to the pregnane derivative **1** (Scheme 2.10), and the reaction performed at room temperature by irradiation, mostly formate **4** was obtained. At room temperature, the expulsion of carbon dioxide from alkoxycarbonyl radical **2** to give **3** is too slow to compete with hydrogen abstraction from the stannane. At 80 °C, in refluxing benzene, an almost equal mixture of the formate **4** and the desired alkane **5** was produced. Finally, when the reaction was conducted at 144 °C in refluxing xylene, the alkane became the dominant product. The effect of temperature in speeding up preferentially a unimolecular fragmentation process to the point where it overtakes the bimolecular hydrogen abstraction is quite clear, the concentration of the reactants being the same in all three cases.

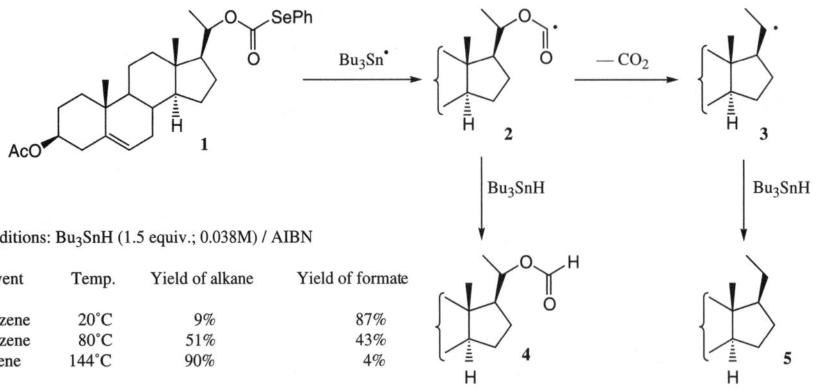

Conditions: Bu₃SnH (1.5 equiv.; 0.038M) / AIBN

| Solvent | Temp. | Yield of alkane | Yield of formate |
|---------|-------|-----------------|------------------|
| Benzene | 20°C | 9% | 87% |
| Benzene | 80°C | 51% | 43% |
| Xylene | 144°C | 90% | 4% |

**Scheme 2.10**
Decarboxylation of a selenocarbonate

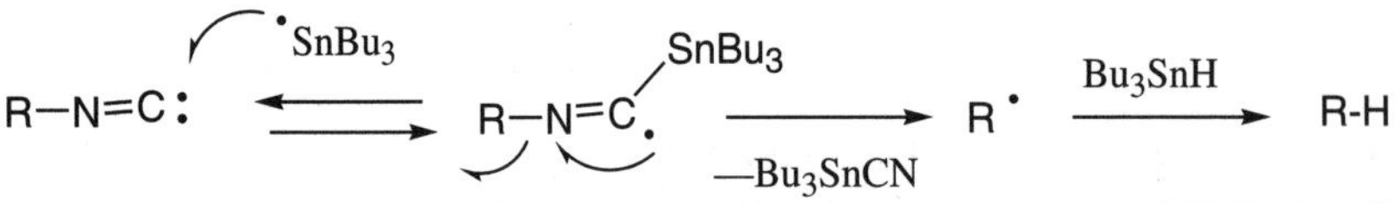

**Scheme 2.11**
Mechanism of deamination via isocycyanides

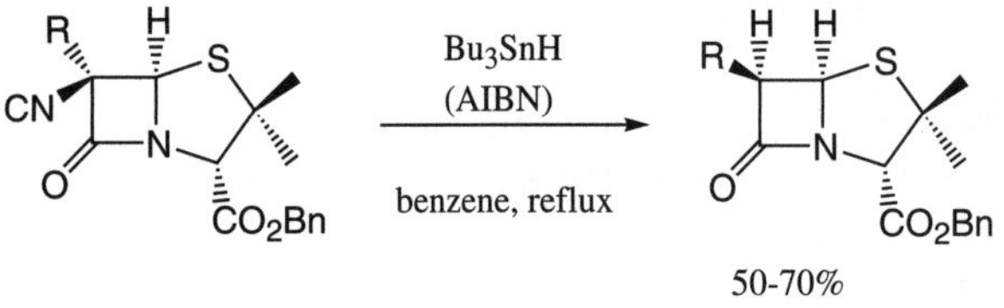

50-70%

R = PhCH₂-
= Me₂C(OH)-
= MeO₂CCH₂-
= MeO₂CCH₂CH₂-

**Scheme 2.12**
Deamination of β-lactam isocyanides

## 2.4  Reduction of C–N bonds

Amines, too, cannot be reduced directly with organotin hydrides and must first be converted into an isonitrile. The method is therefore limited to primary amines (Saegusa *et al.* 1968; Barton *et al.* 1980*a*). The mechanism is outlined in Scheme 2.11 and involves a reversible addition of stannyl radicals to the isonitrile group followed by a slower β-scission.

Primary amines attached to all types of aliphatic carbon centres (primary, secondary, or tertiary) can be deaminated. The reduction of 6-isocyanopenicillin derivatives depicted in Scheme 2.12 is representative (Ivor John *et al.* 1979; see also Aratani *et al.* 1985). Quenching of the intermediate carbon radicals by the stannane takes place from the least hindered *exo*-face to give the stereochemistry shown, as in the related reduction in Scheme 2.3 above.

In the case of polyamines, the regioselectivity of the deamination can be controlled because the rate-determining step is rupture of the C–N bond. The stability of the ensuing carbon radical will therefore decide which isonitrile group is reduced first. For example, the tetraisocyano- derivative of neamine (Scheme 2.13) can be

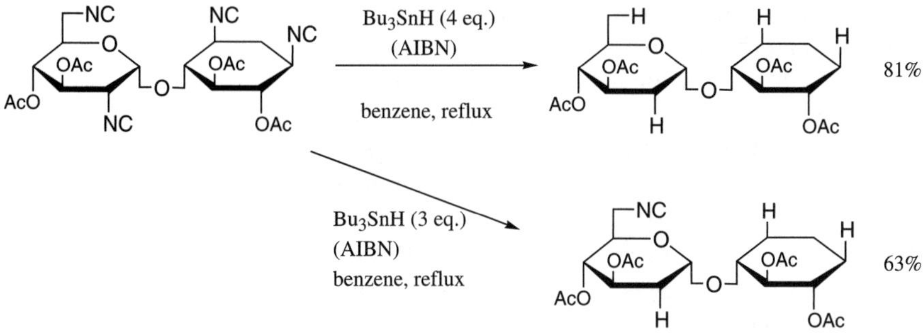

**Scheme 2.13** Selective deamination of neamine

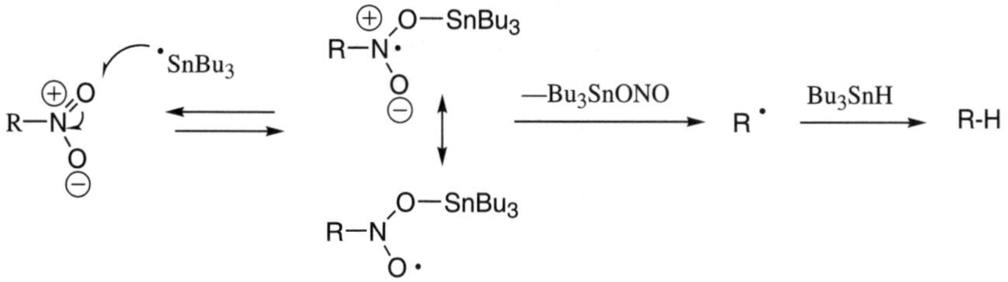

**Scheme 2.14** Mechanism of radical denitration

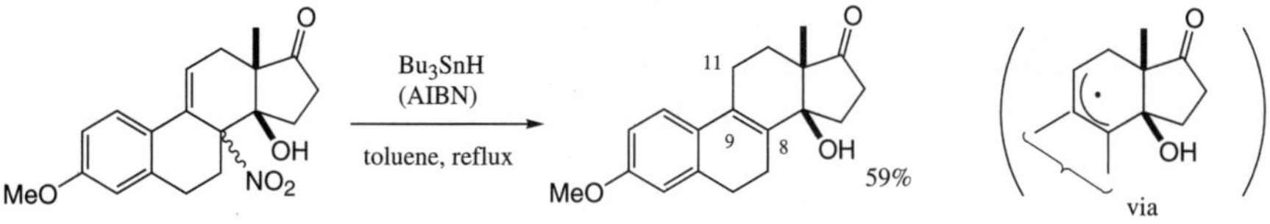

**Scheme 2.15** Reductive denitration of a nitrosteroid

deaminated to give a totally or partially deaminated product depending on the amount of tributyltin hydride used. The least reactive primary isonitrile is the one left behind in the second example, from which the amine group can be regenerated (Barton *et al.* 1980*b*; Barton and Motherwell 1981).

Another group that reacts with tributyltin hydride with ultimate cleavage of a C–N bond is the aliphatic nitro group (Tanner *et al.* 1981; Ono *et al.* 1985). The accepted mechanism, displayed in Scheme 2.14, involves a slow fragmentation step following a reversible addition of the stannyl radical to one of the oxygens of the nitro group.

Nitro compounds are less reactive than isocyanides, and only those leading to a tertiary, benzylic, or allylic radicals can be effectively reduced. The reductive denitration of the nitrosteroid in Scheme 2.15 is one such example (Barlaam *et al.* 1995). The intermediate allylic radical reacts with the stannane at the least hindered secondary C-11 terminus, leaving the olefin in the 8,9-position.

The fact that stannyl radicals add to an oxygen centre in this reaction raises the question of why other oxygen containing groups (ketones, esters, amides, etc.), present in some of the previous examples we have seen, are not affected. The answer is that even though stannyl radicals can (and do) react with various centres in a given molecule, these interactions occur with widely different rates and, perhaps more importantly, are in many cases reversible. Unless the addition to a carbonyl group is followed by a fast irreversible step such as a fragmentation then, macroscopically, no change is observed. It is possible to take advantage of the moderately efficient and reversible addition of stannyl radicals to a carbonyl group. It is known for example that benzoates (Khoo and Lee 1968) and oxalates (Dolan and MacMillan 1985) that can fragment to resonance-stabilized radicals will undergo reduction in useful yields when subjected to tributyltin hydride. These variations are less general but complement nevertheless the Barton–McCombie reaction, especially for the deoxygenation of tertiary and allylic alcohols, where access to the thiocarbonyl derivative is problematic. Another application is the generation of nitrogen-centred radicals starting with benzoates of oximes and hydroxamic acids discussed below. It is worth underlining once again that one of the major strengths of stannane-based methodology is the variety of functional groups that can be implicated in a given radical process, allowing the conception and implementation of numerous transformations.

## 2.5   Additions to olefins: general considerations

The reactions we have so far considered are simple reductions, that is the replacement of various **X** groups with hydrogen without alteration of the carbon skeleton. If the intermediate carbon radical could be intercepted by a carbon–carbon bond forming process, *before hydrogen atom transfer from the stannane occurs*, then the architecture of the substrate could be profoundly modified and its complexity rapidly increased.

But attractive as this may seem for organic synthesis, tampering with the basic chain reaction has some serious pitfalls that must be overcome. The main difficulty is the rapid multiplication of competing pathways as shown by the manifold in Scheme 2.16, and not all the possibilities have been included!

First, the stannyl radicals generated following the initiation step now have the possibility of adding to the olefin instead of abstracting the X group. Fortunately, this

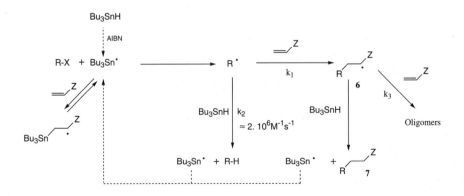

**Scheme 2.16** Reaction manifold for capture of intermediate radical by an olefin

addition is strongly reversible (we shall exploit this property later) but it can be problematic in the case of alkynes where the reverse process is much slower, and the vinylic radical from the addition especially reactive. For the present, we shall neglect this potential side reaction. A much more serious difficulty arises at the level of radical R$^{\bullet}$: the desired pathway is addition to the olefin but this step is in competition with the irreversible and, as we have seen (Table 2.2), fast hydrogen abstraction from the stannane. Moreover, when we reach adduct **6**, the problem is reversed: now it is a second addition to the olefin that competes with the desired hydrogen abstraction leading to **7**.

We are thus faced by two contradictory requirements: if we lower the concentration of the stannane (e.g. by adding it slowly to the reaction mixture) and increase that of the olefinic trap to favour the formation of adduct **6**, then the latter will also 'see' a lot more olefin than stannane and formation of telomers will be the most likely outcome. If, on the other hand, we increase the concentration of the stannane, then radical R$^{\bullet}$ will be reduced before it has a chance to add to the olefin, and the alkane R–H will dominate under these conditions.

This seemingly hopeless task can in practice be approached in two ways. The first is to differentiate as much as possible between the rate of addition of R$^{\bullet}$ to the olefin as compared to the rate of addition of the adduct radical **6** to the same olefin in order to have $k_1 \gg k_3$. As we shall see, this may be accomplished in some cases by an appropriate choice of the **Z** group. The second approach is to have the olefin tethered to the starting material so that the addition becomes *intramolecular*. We return therefore to a situation where the desired step (now a cyclization) is unimolecular and the competing reaction bimolecular and, as we have seen, higher dilution will tilt the balance in favour of the former. The latter route is by far the easiest and certainly the most popular way of creating new C–C bonds using organostannanes and radical chemistry in general. We shall first however examine the case of intermolecular additions.

## 2.6   Intermolecular additions to olefins: polar and steric effects

If R$^{\bullet}$ is a radical possessing nucleophilic character (i.e. high-lying SOMO as is the case with simple aliphatic or alicyclic radicals) and **Z** is a strongly electron withdrawing group, then the addition would be expected to be especially favourable because of the complementary polar characteristics of the two partners. The adduct radical, in contrast, will have an electrophilic character due to the presence of the **Z** group geminal to the radical centre; its reaction with an electrophilic olefin should be slower because the two reactants are now both electrophilic in nature. It should therefore be possible to speed up the first addition and to slow down the path to telomerization. But are these expectations borne out by facts?

The interaction between two molecules can be expressed as a combination of attractive and repulsive terms reflecting polar (electrostatic and orbital) interactions, and steric effects (usually repulsive). In the case of radicals, which are not charged species, the relative difference in the electrostatic (or coulombic) term is small or negligible, and polar effects become dominated by orbital interactions. Frontier orbitals thus play a prominent role, and the reactivity can indeed be often understood,

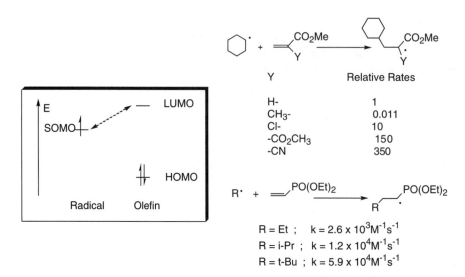

| Y | Relative Rates |
|---|---|
| H- | 1 |
| CH$_3$- | 0.011 |
| Cl- | 10 |
| -CO$_2$CH$_3$ | 150 |
| -CN | 350 |

R = Et ;  k = 2.6 x 10$^3$M$^{-1}$s$^{-1}$
R = i-Pr ;  k = 1.2 x 10$^4$M$^{-1}$s$^{-1}$
R = t-Bu ;  k = 5.9 x 10$^4$M$^{-1}$s$^{-1}$

**Scheme 2.17**  Radical addition to electron-poor olefins

*at a qualitative level*, by simply looking at the energy gap between the SOMO of the radical and the LUMO or the HOMO of the non-radical partner (Scheme 2.17). If the radical has a high-lying SOMO, then it is considered to have a *nucleophilic character* and the SOMO–LUMO interaction will be more important in comparison with the SOMO–HOMO (singly occupied orbitals normally lie in the energy domain between filled and empty orbitals). For the addition of a given 'nucleophilic' radical to an olefin, the rate should therefore increase as the energy level of the LUMO is lowered (i.e. as the olefin is made more electrophilic). The SOMO–LUMO energy gap becomes smaller making this particular orbital interaction more favourable.

This may be seen by examining the variation in the relative rates of addition of the cyclohexyl radical to various substituted acrylates (Scheme 2.17): there is a quite considerable 350-fold increase in going from methyl acrylate (absolute rate: approximately $10^5$ M$^{-1}$s$^{-1}$) to methyl cyanoacrylate (Giese 1983). Conversely, for a given electrophilic olefin, if the energy level of the SOMO of the radical is increased, the rate will increase, as shown by the rates of addition of ethyl, isopropyl, and *t*-butyl radicals to diethyl vinylphosphonate (Giese 1983). The variation is only 20-fold, but in this case, the increase in nucleophilicity of the radicals (reflected by their ionization potentials; Wayner *et al.* 1988) is somewhat offset by an increase in steric bulk on going from ethyl to *t*-butyl. The LUMO of an alkyne is generally higher than that of the corresponding alkene; addition of a nucleophilic radical to an alkyne is about 5 times slower (Giese and Lachhein 1982).

If the SOMO of the radical is low-lying, then the radical is considered to have *electrophilic character* and it is the SOMO–HOMO interaction that becomes more important (Scheme 2.18). This is illustrated by the 100-fold increase in the rate of addition to ethylene as the SOMO is lowered by going from an ethyl to a trifluoromethyl radical (Tedder and Walton 1976). Alternatively, the SOMO can be kept constant and the HOMO raised as in the S$_H$2 reaction of the 5-hexenyl radical with diphenyl disulfide, diselenide, and ditelluride (Russell and Tastoush 1983).

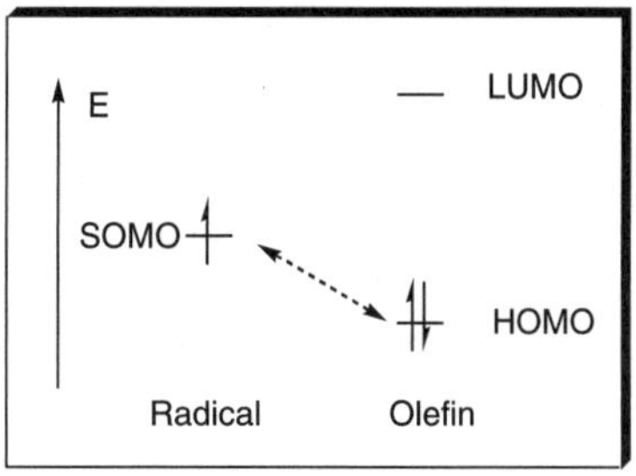

$$R^{\bullet} \; + \; = \quad \longrightarrow \quad R \diagup^{\bullet}$$

R = Et ;      $k = 3.5 \times 10^4 M^{-1}s^{-1}$
R = Me ;      $k = 4.6 \times 10^4 M^{-1}s^{-1}$
R = $CH_2F$ ;  $k = 2.8 \times 10^5 M^{-1}s^{-1}$
R = $CF_3$ ;   $k = 3.5 \times 10^6 M^{-1}s^{-1}$
(rate constants at 164°C)

$$\diagup\!\!\!= \diagdown^{\bullet} \; + \; Ph\text{-}X\text{-}X\text{-}Ph \quad \longrightarrow \quad \diagup\!\!\!= \diagdown^{X\text{-}Ph} \; + \; Ph\text{-}X^{\bullet}$$

X = S ;      $k = 7.6 \times 10^4 M^{-1}s^{-1}$
X = Se ;     $k = 1.2 \times 10^7 M^{-1}s^{-1}$
X = Te ;     $k = 4.8 \times 10^7 M^{-1}s^{-1}$

**Scheme 2.18**  Radical addition to electron-rich olefins

These sets of kinetic data are intended, on one hand, to give an idea of the absolute rate constants for some representative radical processes so that they can be situated on the timescale of molecular interactions and, on the other, to provide a feel for the order of magnitude of the effect of various substituents on a given radical reaction so as to allow for better synthetic planning. However, even if this element-ary theoretical picture often does lead to a correct *qualitative* prediction of the effects of substituents, it must be kept in mind that radicals are *ambiphilic* species. Looking at only one interaction is a gross oversimplification of reality (Héberger and Lopata 1998; Fischer and Radom 2001). Furthermore, we have implicitly assumed that radical additions have early transition states, but enthalpic factors (e.g. stabilization of adduct radical by the substituents on the olefin), which have been completely neglected in the foregoing discussion, do have some influence on the rates (Fischer and Paul 1987; Walbiner *et al.* 1995; Zytowski and Fischer 1996; Beckwith and Pool 2002).

Polar effects manifest themselves in other radical processes besides additions to olefins. Hydrogen atom transfers for example are also accelerated when the radical and the substrate have matching polarities (Russell 1973). Chlorine atoms, alkoxy, and tri-fluoromethyl radicals abstract hydrogen from methane faster than do methyl radicals, even though all these reactions are essentially thermoneutral. This can only be under-stood by invoking polar effects similar to the ones relating to additions to olefins.

Now how about steric factors? Free radicals are neutral species that are little solvated. They are not shackled by counter-ions (whence the term *free*; this is obvi-ously not the case for radical anions or radical cations). In addition, most radical reactions are fast processes with an early and 'loose' transition state; steric hindrance *around the radical centre* will therefore have a less pronounced effect as compared with typical ionic reactions. Tertiary radicals, for example, still interact with each other at nearly diffusion-controlled rates (albeit to give mostly disproportionation), and will often usefully add to olefins, allowing the generation of quaternary centres. Of course, if the steric bulk is taken to an extreme, then the radical becomes so shielded as to loose much of its kinetic lability. This is the case with many persistent radicals (triarylmethyls; Koelsch radicals, etc.) which are sterically prevented from interacting with each other.

| Y | Relative Rates |
|---|---|
| H | 1000 |
| CH₃- | 11 |
| t-Bu- | 0.05 |

**Scheme 2.19** Steric effect of the β-substituent on the olefin

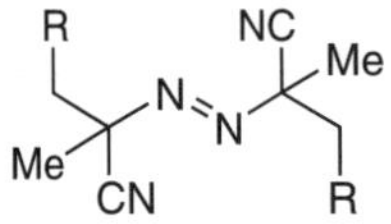

AIBN : R = H

**8**    : R = *t*-Bu

**Fig. 2.1** Two diazo compounds with very different half-lifes

In contrast, substituents on the β-terminus of the olefinic trap exert a strong retarding effect, as shown by the relative rates of addition of cyclohexyl radicals to β-substituted acrylates displayed in Scheme 2.19 (Giese 1983). The more sterically congested the olefinic trap, the more difficult it will be to avoid competing side reactions (mainly premature hydrogen atom abstraction from the stannane). Moreover, other factors being equal, a radical will add from the least hindered side of an olefin. Sometimes steric factors can even override the polar effect of the substituents.

Steric encumbrance does not always have a deleterious effect on the rate of radical processes. When the rate determining step is a *unimolecular fragmentation*, release of steric strain can considerably speed up the collapse. For example, sterically congested diazo derivative **8** decomposes approximately a 100 times faster than AIBN at 80°C (Fig 2.1), yet the thermodynamic stability of the ensuing radicals is quite similar (Overberger *et al.* 1954; Engel 1980). Rate accelerations in bimolecular processes caused by steric hindrance have also been observed, albeit much more rarely. Thus, more hindered aryl bromides were found to react faster with tributylstannyl radicals because steric repulsion causes a lengthening (and hence weakening) of the C–Br bond (Crich and Recupero 1998).

## 2.7  Intermolecular additions to olefins: some synthetic applications

Most of the examples of stannane-mediated intermolecular additions (often called Giese reactions) involve a radical with 'nucleophilic character' and an electrophilic olefin, generally used in some excess. Thus, slow addition of tributyltin hydride (and AIBN) to a mixture of the xanthate derived from diacetone glucose and a 5-fold excess of acrylonitrile in refluxing toluene gives a reasonable yield 40% (75% by GC) of adduct **9** (Giese *et al.* 1984). A small amount (10%) of the prematurely reduced derivative **10** can be isolated, with the balance presumably made up of dinitrile **11**, arising from double addition to acrylonitrile (Scheme 2.20). If a greater excess of acrylonitrile is used, a decrease in premature reduction and an increase in telomerization would be expected, and vice versa. In this case, a 5-fold excess of acrylonitrile appears to be a reasonable compromise.

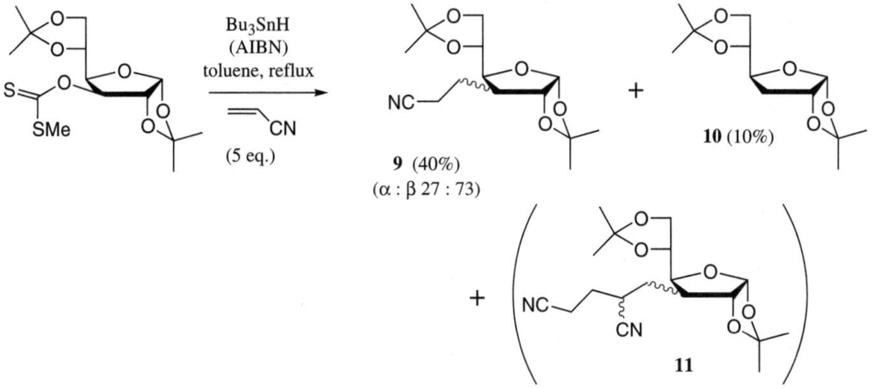

**Scheme 2.20**
Deoxygenative radical
addition to acrylonitrile

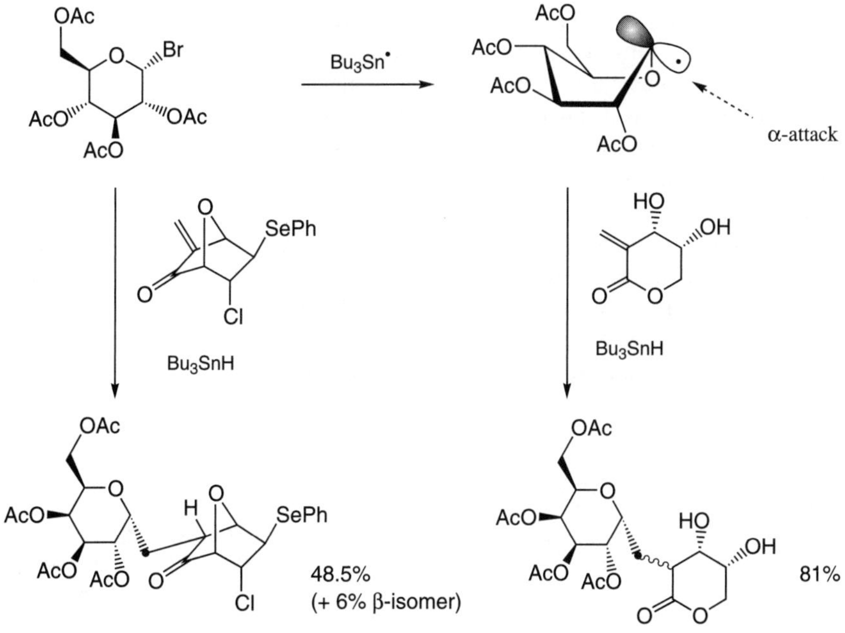

**Scheme 2.21** Addition
of an anomeric radical

The electrophilic trap need not be a simple olefin. The two examples of C–C bond creation at the anomeric position of tetraacetyl glucopyranose pictured in Scheme 2.21 are quite spectacular because such transformations are exceedingly difficult to accomplish using ionic intermediates. Esr shows the intermediate carbon radical on the sugar adopting a boat conformation resulting in the predominant formation of the $\alpha$-adduct (to the extent of about 10:1). In the first example involving an unprotected, carbohydrate-like lactone trap, the hydrogen abstraction step is less selective (Giese and Dupuis 1983; see also Giese and Witzel 1986). In the second example, which underscores once again the selectivity in generating a radical when a bromide, a chloride, and a selenide are simultaneously present in the reactants, the stereoselectivity in the reduction step is high because delivery of the hydrogen has to occur from the least hindered *exo*-face of the [2.2.1]oxabicycloheptane structure (Mampuya Bimwala and Vogel 1992).

In Scheme 2.22, an example of addition of an 'electrophilic', cyanomethyl radical to the electron-rich enol ether moiety on a sugar is shown (Giese 1986). The intermediate carbon radical is quenched from the least hindered face to give the observed isomer. Such combinations are less common—but synthetically as useful—than ones involving a 'nucleophilic' radical and an electrophilic olefinic trap, at least in the intermolecular case. The situation will be reversed when we examine Kharasch-type reactions in Chapter 6.

## 2.8   Intramolecular additions to olefins—rates and regiochemistry

Attaching the olefinic entity to the radical precursor simplifies considerably the kinetic picture. The linearity and ensuing loss of convergence in the synthetic plan is more than offset by the tremendous efficiency and versatility of the cyclization process. Since oligomerization of the olefin is eliminated as a complicating factor, the main competing pathway becomes the premature reduction with stannane (Scheme 2.23). This is a bimolecular process that is disfavoured by increasing dilution in contrast to the unimolecular cyclization step, which is not affected. In practice, this is accomplished by slow addition of the stannane to the reaction medium, using for example a syringe pump.

But even if the situation is much easier to control than in the intermolecular case, there are still some limitations. The very fast hydrogen abstraction from the stannane makes it a formidable opponent and not all ring sizes can be constructed with the same ease, even using high dilution techniques. The stannane is needed in the last propagation step after cyclization, and if its concentration is decreased beyond

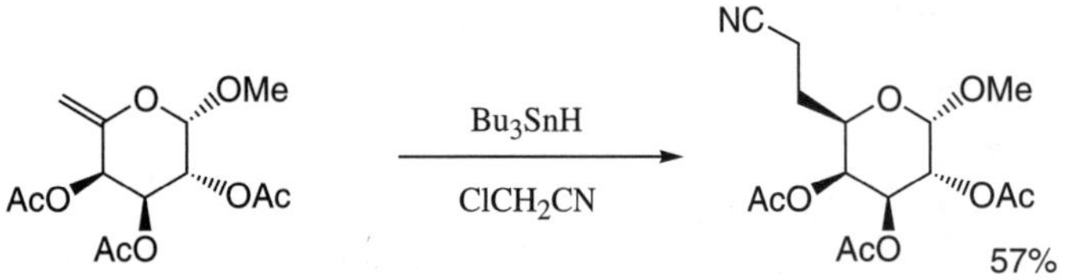

**Scheme 2.22**
Intermolecular addition of an electrophilic radical

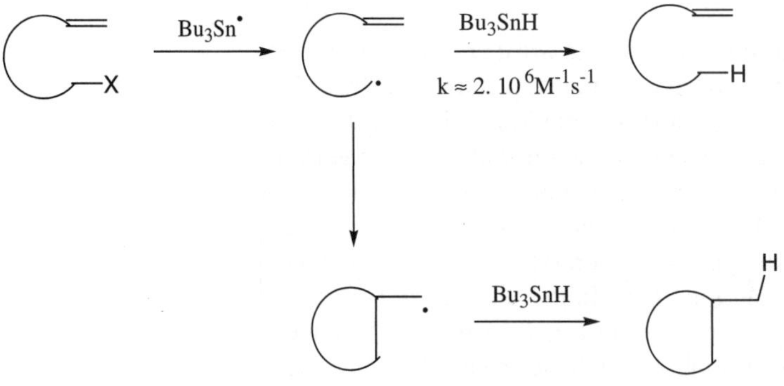

**Scheme 2.23**
Intramolecular capture of the intermediate radical

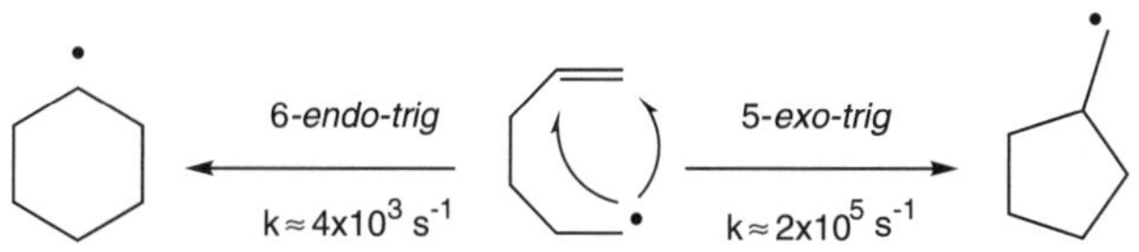

6-*endo-trig*

$k \approx 4 \times 10^3 \ s^{-1}$

5-*exo-trig*

$k \approx 2 \times 10^5 \ s^{-1}$

**Scheme 2.24** 5-*Exo* versus 6-*endo* ring closure

a certain limit, other side reactions start interfering: unwanted rearrangements, interactions with the solvent, radical–radical terminations etc. It is therefore essential to have at least an approximate idea of the rates of cyclization, which of course will depend on the ring size, the substituents, and on the nature of the reacting radical centre. Absolute rate constants have been measured for many of the prototype processes that are of interest to the synthetic organic chemist (Beckwith and Schiesser 1985).

Cyclizations to give 5-membered rings are among the fastest, and certainly by far the most widely used ring forming process in radical chemistry. Closure of the 5-hexenyl radical, the simplest example, has been thoroughly studied and often serves as a standard clock against which the rates of other radical reactions are measured (c.f. the use of the 5-hexenyl radical in Scheme 2.18). This rearrangement can in principle lead either to a cyclopentylmethyl radical by a *5-exo-trig* ring closure or to the more stable cyclohexyl radical through the alternative *6-endo-trig* mode (Scheme 2.24). It is the former, with a rate of $2 \times 10^5 \ s^{-1}$ at 25 °C, that predominates by a factor of about 50, indicating that the process is under kinetic control.

It turns out that for small rings (3–7), the *exo* mode is preferred in general (the Beckwith–Baldwin rules). This is due to the need for a good overlap between the SOMO of the radical and the $\pi$ system of the olefin. The angle of approach of the radical to the double bond which has been estimated to be of the order of 107° (Beckwith and Schiesser 1985; Spellmeyer and Houk 1987) is very similar to the Bürgi–Dunitz angle for the addition of nucleophiles to a carbonyl group. The strain ensuing from such a stereoelectronic requirement disfavours more the 6-*endo* transition structure as compared with the 5-*exo* mode. This vector analysis approach is analogous to that for intramolecular reactions of nucleophiles and the same nomenclature has been adopted for radical cyclizations: the number indicates the size of the ring being formed; *exo* or *endo* indicates that the unsaturation being attacked is within or outside the ring being formed; *trig* or *dig* (from trigonal and digonal) indicates the centre being attacked is sp$^2$ or sp, respectively (Baldwin 1976).

A nearly 50-fold preference for the *5-exo* as compared with a *6-endo* mode of ring closure of the 5-hexenyl radical, impressive as it may seem, represents only a difference of 1.7 kcal mol$^{-1}$ in the activation energy. A minor modification in the size, nature, or disposition of the substituents in the substrate can easily lead to further enhancement or even a reversal of this preference. For example, the simple introduction of a methyl group in the 5-position hinders strongly the 5-*exo* approach but barely affects the 6-*endo* mode as shown by the rates given in Scheme 2.25. In contrast, placing the methyl group on the radical site has no significant effect on either ring closures (cf. discussion on steric hindrance above). Decreasing the natural tetrahedral angle by introducing for example two geminal methyl group which repeal each other or by replacing a methylene with an oxygen (lone pair repulsion) greatly favours the smaller 5-membered ring—the so called Thorpe–Ingold effect—by compressing

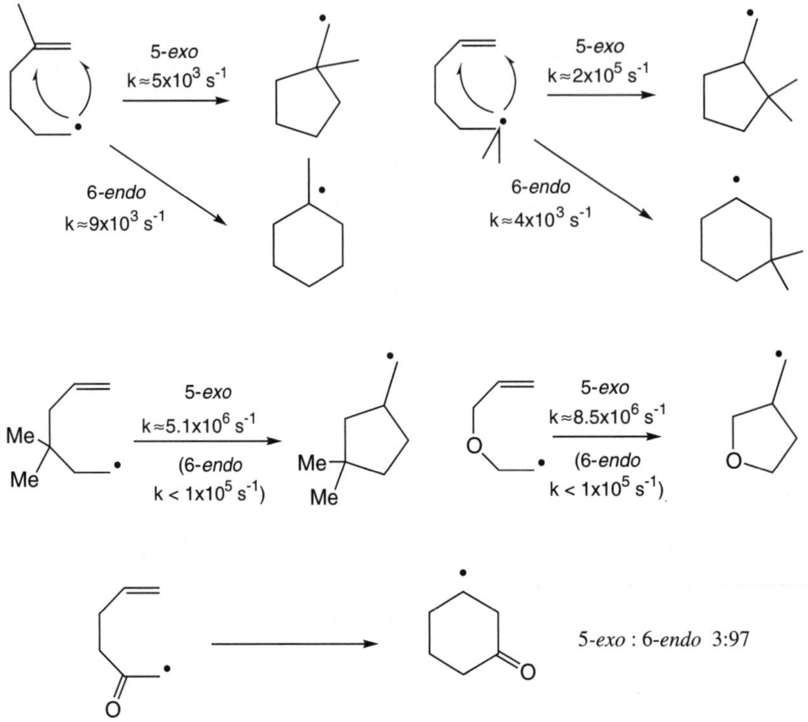

**Scheme 2.25** Various modes of ring closure

the two reacting centres close together and perhaps also by increasing the population of the right rotamer (Scheme 2.25). Alternatively, a ketone next to the radical centre will flatten the molecule and widen the angle to 120°, thus favouring the 6-*endo* mode of ring closure (Curran and Chang 1989; Broeker and Houk 1991).

The formation of smaller rings such as cyclopropanes and cyclobutanes is hampered by the high reversibility of the cyclization step (and its sluggishness in the case of the 4-membered ring) due to ring strain. The ring-opening process, however, can be turned into synthetic advantage as will be shown in the next chapter. Closures to cyclohexanes (by a 6-*exo* mode) and to cycloheptanes (7-*exo* or 7-*endo*), are in general relatively slow in comparison to 5-membered ring formation owing to increasingly unfavourable entropic factors. Further complications can arise because of the possibility of radical translocation by abstraction of an allylic hydrogen (Scheme 2.26). Nevertheless, efficient cyclizations of this type can sometimes be achieved by a judicious choice of the substrate.

## 2.9   Intramolecular additions to olefins: synthetic applications

Notwithstanding the simplicity of the cyclizations shown above, whose sole utility is as prototypes for kinetic studies, the intramolecular addition of radicals to olefins translates in practice into a synthetic tool of incredible power and versatility. The examples that follow have been selected to cover as much variety as possible but

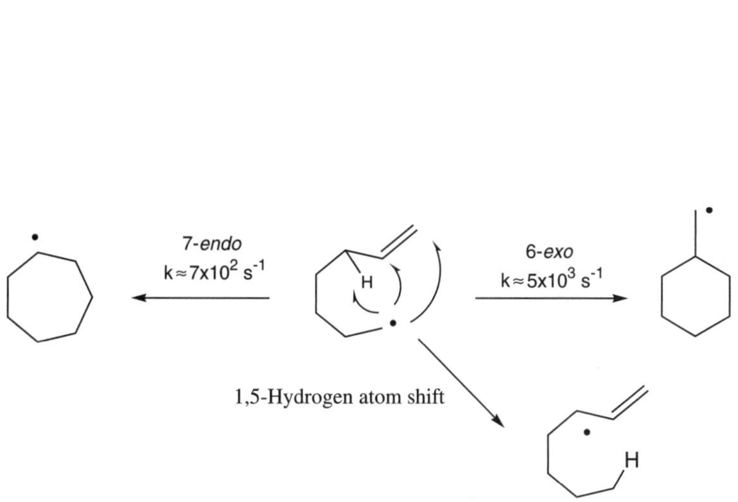

Scheme 2.26 *6-Exo* versus *7-endo* cyclization and allylic hydrogen atom abstraction

represent only a tiny fraction of the hundreds of cyclizations that can be found in the literature (a fairly exhaustive compilation has recently been published by Giese *et al.* 1996; see also Bowman *et al.* 2000).

In a pioneering and now classical study, Stork (Stork *et al.* 1983; Stork 1990), simultaneously with Ueno (Ueno *et al.* 1982), have exploited the ease of 5-membered ring formation to transfer stereochemical information from an allylic alcohol to control the configuration of the adjacent carbon. The allylic alcohol is first converted into the mixed bromoacetal, followed by reductive cyclization with tributyltin hydride. In the example drawn in Scheme 2.27 (Ladlow and Pattenden 1984), a bicyclic structure containing angular methyl and isobutoxy groups is obtained. The configuration of the newly created chiral centre is unambiguously defined since the junction of the two fused rings is necessarily *cis*.

The concept of using a tether for controlling the stereochemistry was later extended to include silicon atoms (Stork and Kahn 1985; Stork and Sofia 1986). The elaboration of the CD ring system of steroids displayed in Scheme 2.28 represents one application. 'Protection' of the allylic hydroxy group as its bromomethyl dimethylsilyl ether provides a precursor for an efficient reductive cyclization. Now, not only two new chiral centres have been established with total control but, because the final hydrogen atom transfer from the stannane occurs from the least hindered *exo*-face, the thermodynamically less stable trans hydrindane is generated in the process. This approach represents an elegant and general solution to the construction of *trans* hydrindane structures, a difficult and long-standing problem in steroid total synthesis. The silicon containing ring can be usefully cleaved with potassium *t*-butoxide in DMSO or by the Tamao reaction leaving behind either a methyl or a hydroxymethyl group, in addition to the original hydroxy function.

In another variation, the temporary silicon chain is made to contain the radical trapping element whereas the radical is generated on the main structure. This

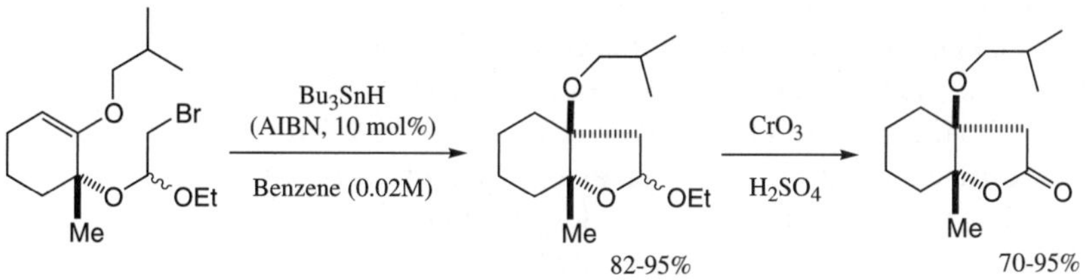

Scheme 2.27 Cyclization of a bromoacetal

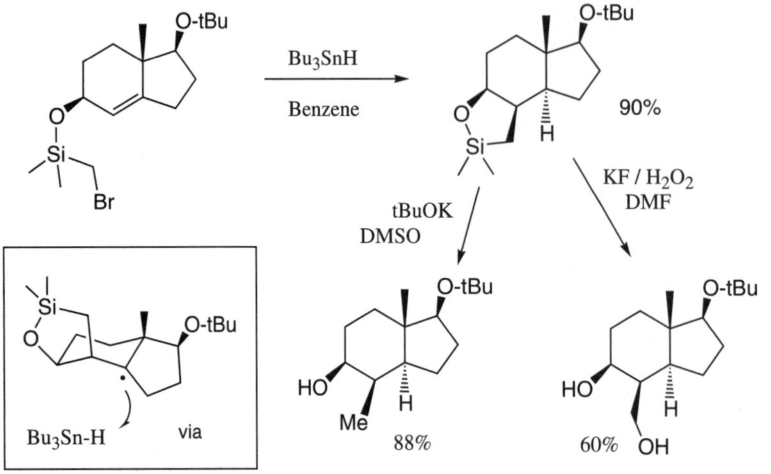

**Scheme 2.28**   The use of a silicon tether for controlling relative stereochemistry

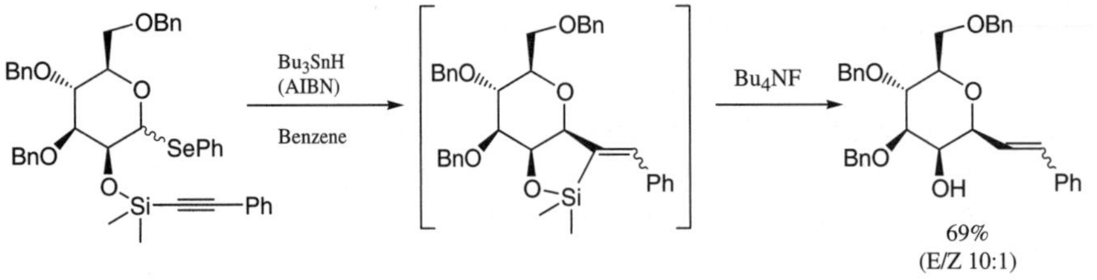

**Scheme 2.29**   Transfer of a vinyl group via a silicon tether

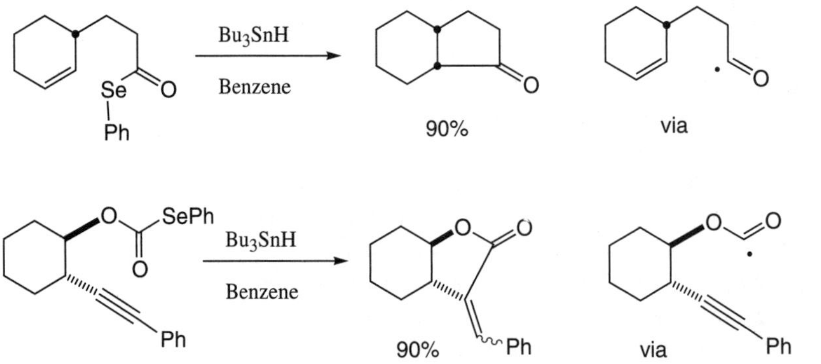

**Scheme 2.30**   Ring closure of acyl radicals

approach may be used to attach a styrene unit in a stereospecific manner on a carbohydrate for example, as illustrated by the transformation in Scheme 2.29 (Stork *et al.* 1991). The initial radical adduct is not isolated but treated with fluoride to remove the silicon. The styryl group is interesting because it is easily cleaved to an aldehyde which can act as a spring-board to a host of other derivatives.

The radical centre can itself bear a functionality. The generation and capture of acyl and alkoxycarbonyl radicals using the corresponding selenide precursors is illustrated by the reactions in Scheme 2.30. In the first (Boger and Mathvink 1988),

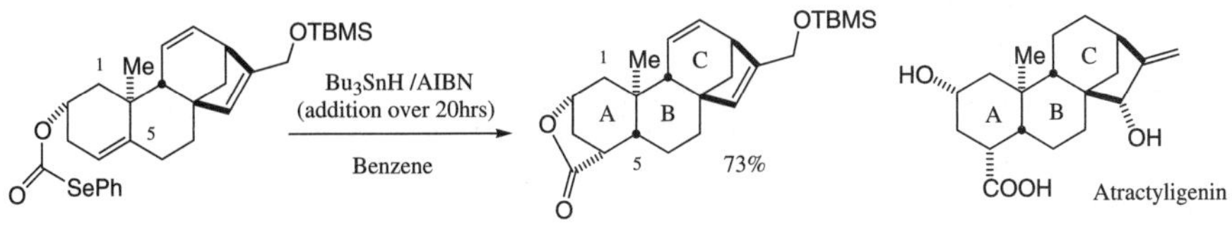

**Scheme 2.31** Acyl radical cyclization in the synthesis of atractyligenin

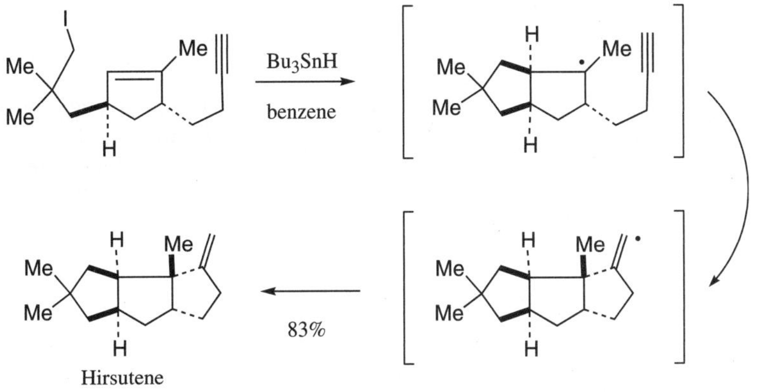

**Scheme 2.32** Tandem cyclizations in the synthesis of hirsutene

which leads to a cyclopentanone, the ring junction is created in the cyclization step and must therefore be *cis*. In the second transformation (Bachi and Bosch 1986), an alkoxycarbonyl radical adds to an alkyne; the ring junction is *trans*, reflecting the stereochemistry of the two vicinal substituents in the cyclohexane precursor (which was prepared by opening of cyclohexane epoxide with lithium phenylacetylide, whence the *trans* stereochemistry). This general route to *trans*-fused lactones gives two geometrical isomers in this case because quenching of the intermediate vinylic radical can take place from either side.

A bridging lactone can also be constructed, as in one of the key steps in the synthesis of atractyligenin depicted in Scheme 2.31 (Singh *et al.* 1987). Quenching of the radical at C-5 by Bu₃SnH following the cyclization step has to occur from the least hindered side, resulting in a *trans* junction between rings A and B.

One is not limited to the construction of only one ring. A judicious positioning of the unsaturations in the substrate can allow the simultaneous elaboration of several rings, and this possibility has been exploited in a superbly conceived synthesis of hirsutene, outlined in Scheme 2.32, and featuring a tandem radical cyclization (Curran and Rackiewicz 1985). It is by far the shortest route to this terpene. The stereochemistry of the ring junctions is dictated by the initial disposition of the chains adorning the central cyclopentane template.

An elegant, short synthesis of (±)-α-cedrene (Scheme 2.33) was accomplished by combining the efficiency and mildness of the aldol addition of a nitronate to an aldehyde (the Henry reaction) with the power of a tandem radical cyclization (Chen *et al.* 1993). Radical generation from the tertiary nitro group proceeds smoothly, in preference to attack on the sulfide group. The two consecutive 5-*exo* cyclizations are followed by expulsion of a phenylthiyl radical to give nor-cedrenone, after oxidation

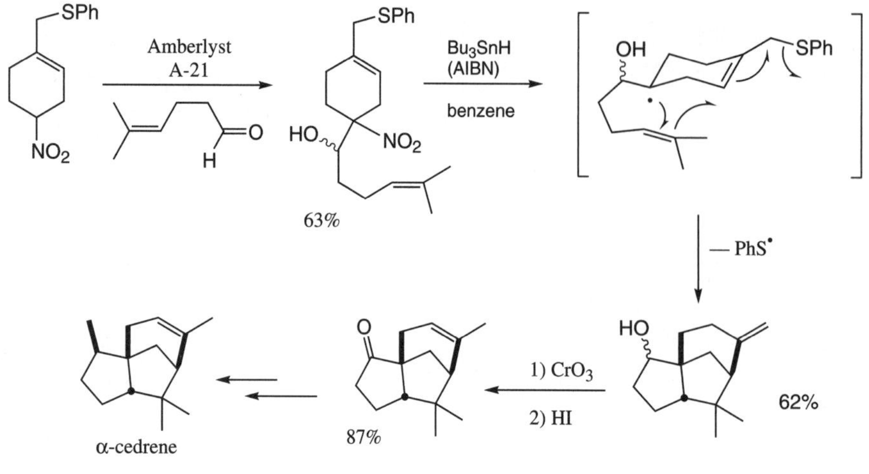

**Scheme 2.33** Ring closures and fragmentation starting from a tertiary nitro precursor

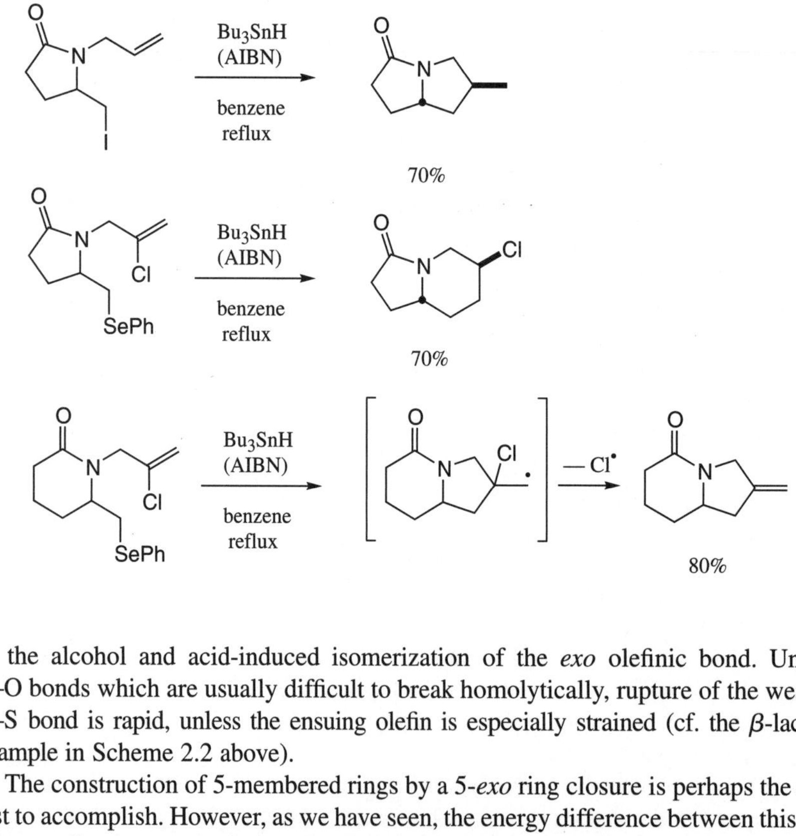

**Scheme 2.34** Effect of substituents and adjacent rings on the mode of cyclization

of the alcohol and acid-induced isomerization of the *exo* olefinic bond. Unlike C–O bonds which are usually difficult to break homolytically, rupture of the weaker C–S bond is rapid, unless the ensuing olefin is especially strained (cf. the β-lactam example in Scheme 2.2 above).

The construction of 5-membered rings by a 5-*exo* ring closure is perhaps the easiest to accomplish. However, as we have seen, the energy difference between this and other cyclization modes is not very large and sometimes small modifications in the structure of the substrate can completely alter the type of ring size that is obtained. This is nicely illustrated by the cyclization of the iodo- and seleno-lactams in Scheme 2.34. 5-*Exo*-cyclization onto the naked allyl group in the first example leads to the pyrrolizidine structure (Keusenkothen and Smith 1992) whereas an indolizidine skeleton is obtained via a 6-*endo* closure if the allyl group contains a 2-chloro substituent (Knapp *et al.* 1990). This transformation provides, incidentally, another instance of selectivity in the generation of a radical—a selenide is more reactive than a chloride. Steric hindrance by the chlorine group, combined with the (slight) strain arising from stitching together two 5-membered rings succeed in overcoming the preference for the 5-*exo* mode. If the strain factor is alleviated by starting for example with a δ-lactam, then a chlorine substituent is not large enough to prevent the formation of the 5-membered ring. In this last example, elimination of a chlorine atom follows the cyclization to give the olefinic product.

Shrinking to a smaller β-lactam template will tend to strongly favour the formation of the larger ring, as shown by the example in Scheme 2.35, where a 7-*endo* prevails over a 6-*exo* closure, even on an unsubstituted allyl side chain (Bachi *et al.* 1981).

The subtle interplay of strain and steric hindrance on the outcome of radical cyclizations have been exploited in ingenious ways. In one of the earliest uses of radical cyclizations for the synthesis of a natural product, Büchi and Wüest (1979) succeeded in constructing one of the 6-membered rings of β-agarofuran (Scheme 2.36) by a radical cyclization starting with the bridgehead chloride **12** under comparatively concentrated conditions. A small amount of prematurely reduced material **13** was also isolated which, in retrospect, could perhaps have been avoided by extra dilution.

The key step in the more recent synthesis of (+)-cladantholide, outlined in Scheme 2.37, is a cascade in which a 5- and a 7-membered ring are consecutively produced with amazing efficiency and stereo control (Lee *et al.* 1997). The approach of the stannane in the last hydrogen atom transfer step is commanded by the shape of the polycyclic intermediate.

Discouraging the 5-*exo* in favour of a 6-*endo* cyclization by strategically placing methyl substituents can be used to mimic the biogenetic cationic polyene cyclization for the elaboration of steroid and triterpene skeletons (Breslow *et al.* 1962). The same folding of the polyene chain assumed in cationic cyclization dictates the configuration of all but the last of the several asymmetric centres generated in the remarkable transformation portrayed in Scheme 2.38 (Chen *et al.* 1994).

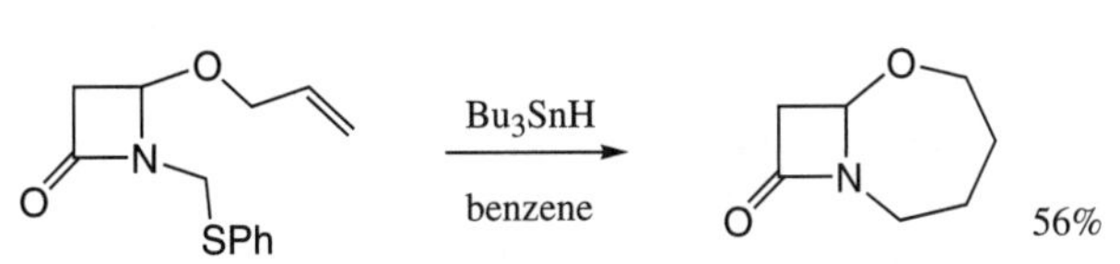

Scheme 2.35 Effect of strain on mode of cyclization

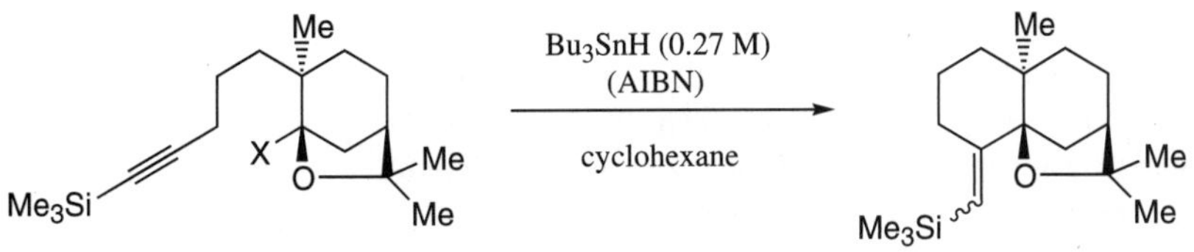

**12** : X = Cl
**13** : X = H (13%)

72% (4:1)

Scheme 2.36 An application of a 6-*exo*-dig cyclization

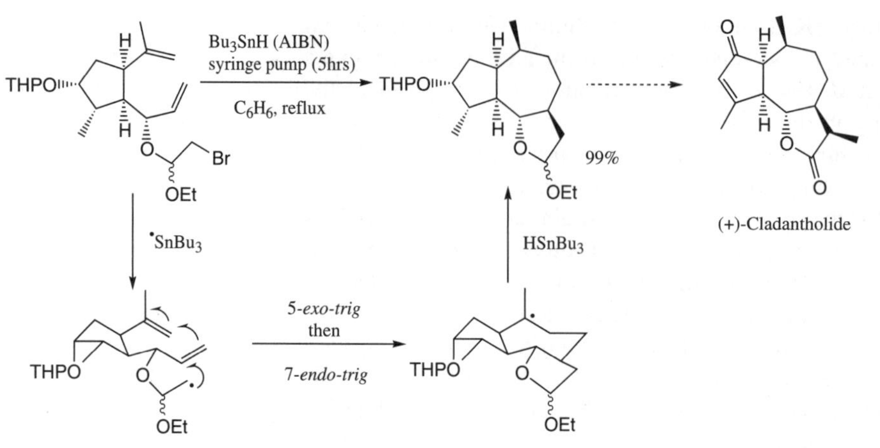

**Scheme 2.37** A 7-*endo* ring closure in the synthesis of (+)-cladantholide

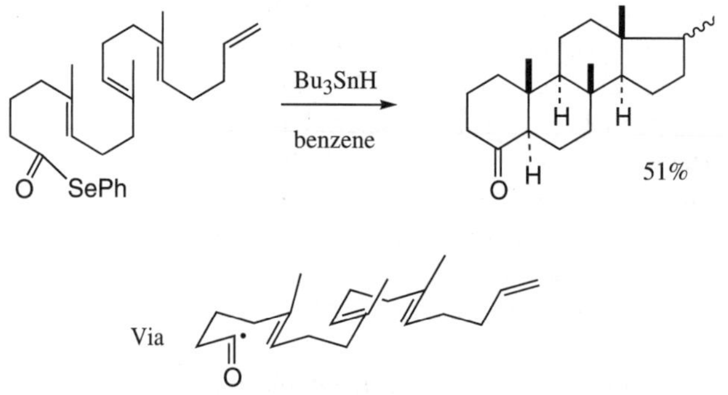

**Scheme 2.38** Radical cyclization of polyenes

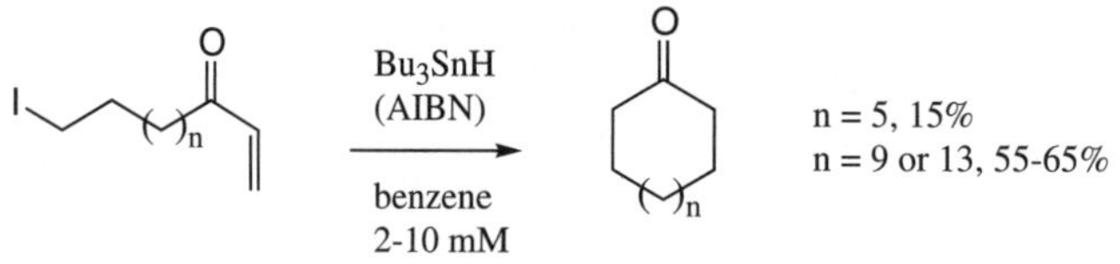

**Scheme 2.39** Large ring formation

Medium and large rings can also be directly produced, but the process is not very general (Porter *et al.* 1988, 1989; Yet 1999). Nevertheless, some interesting examples can be found. High dilution and a reactive iodide precursor are normally required with simple flexible chains (Scheme 2.39), and the internal olefinic trap must be activated. In other words, all the propagation steps must be speeded up as much as possible.

The presence of one or more rigidifying groups in the structure will more or less counteract the negative role of entropy, resulting in a faster cyclization process. This is illustrated in Scheme 2.40 by a rare 14-*endo*-cyclization, a key step in the synthesis of (−)-zearalenone, a substance with mycotoxic properties (Hitchcock and Pattenden 1992). The second example, which concerns a remarkably efficient 10-*endo-dig* ring closure, also owes its success to the considerable rigidity imparted by the two aromatic rings and the alkyne and amide groups (Rodriguez *et al.* 1998). Only one isomer was obtained in this case but the stereochemistry was not determined. Notwithstanding

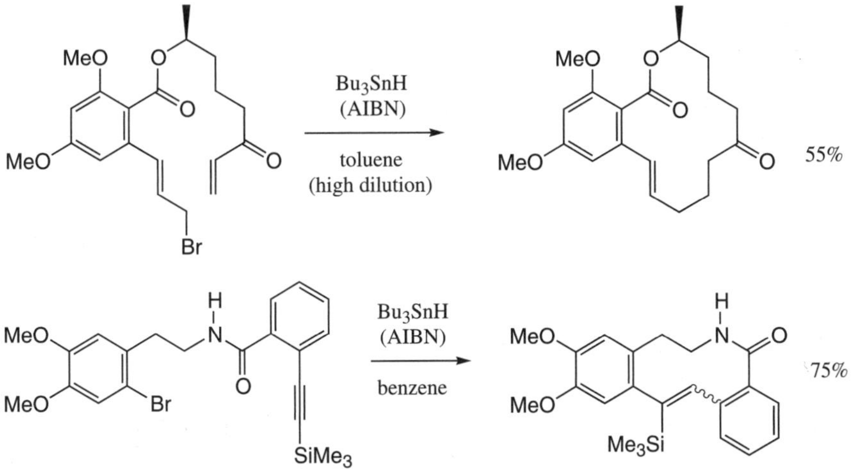

**Scheme 2.40**
Examples of
macrocyclizations

these two examples, access to large rings by direct ring closure remains a difficult problem in radical chemistry; a more satisfying solution, albeit an indirect one, will be discussed in the next chapter.

## 2.10 Some stereochemical aspects of radical cyclizations

Generally speaking, controlling the stereochemistry in radical processes is usually less easy than with ionic or organometallic reactions. The early, comparatively non-polar transition states, and the absence of a counter-ion (and solvation) character-istic of radical transformations limit somewhat the effects that are traditionally exploited to master the stereochemical outcome in an 'ionic' process. In many of the preceding examples, the required rigidity was gained by having a 5- or 6-membered ring acting as a template around which the cyclization reaction was performed. The template is usually part of the target structure as in several of the syntheses we have seen, but it can also be a temporary construction destined to be destroyed at a later stage. In some especially favourable, but less common situations, good control of the stereochemistry can be achieved because the linear chain adopts one particular folding leading to the desired relative or absolute configuration. For simpler struc-tures (e.g. **14** and **16**, Scheme 2.41), the stereoselectivity is not so spectacular. The preferred transitions states are assumed on the basis of molecular mechanics calcu-lations to be chair-like (**14a** and **16a**, respectively), with the bulkiest substituents adopting a pseudo-equatorial disposition (Beckwith 1993; Beckwith and Schiesser 1985). The length of the incipient $\sigma$-bond between C1 and C5 is estimated to be about 2.3 Å. Thus, in the case of 3- and 2-methyl substituted hexenyl radicals drawn in Scheme 2.41 (**14** and **16**, R = Me), it is the *cis-* and *trans-* cyclopentylmethyl isomers **15-cis** and **17-trans** that are the major products, respectively, but the selectivity is modest, (2:1 and 3:2, respectively), because the boat-like alternative structures **14b** and **16b** for the transition state are quite close in energy. Even for the 3-*t*-butyl substituted derivative (**14**, R = *t*-Bu), the ratio is 4.5:1, corresponding

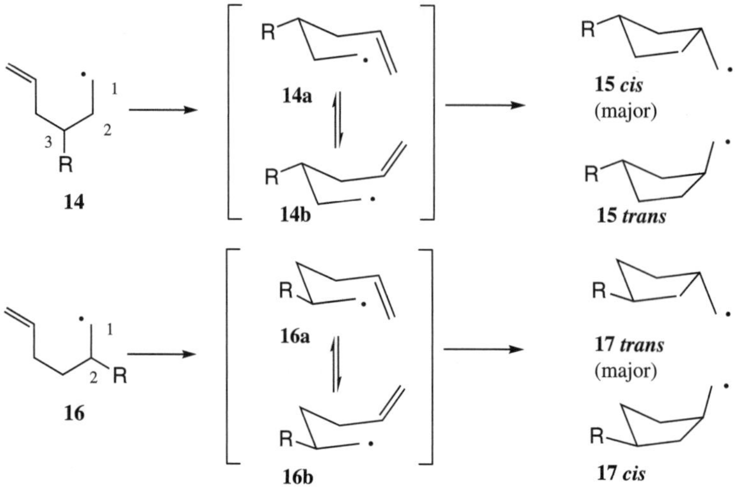

**Scheme 2.41**
Stereochemistry of 5-*exo* ring closure

in this case to a difference ($\Delta\Delta G^{\neq}$) of only 1 kcal mol$^{-1}$ between the chair-like and boat-like transition structures, **14a** and **14b**, respectively (Beckwith and Zimmerman 1991). This small energy difference means that the preferred transition-state can be easily perturbed, by anomeric effects for example (Beckwith and Page 1998; Villar and Renaud 1998) or by the presence of Lewis acids which can complex with basic sites on the molecule (Renaud and Gerster 1998). For the sake of comparison, it is interesting to note that the organolithium derivative corresponding to radical **16** (R=Me) ring-closes with a much stronger preference for the *trans*-isomer (*trans*:*cis* 10:1), and that the selectivity can be significantly modified (to >20:1) by the presence of additives such as diamines (Bailey *et al.* 1991). Poor stereoselectivities (<3:2) have also been observed in 6-*exo* radical cyclizations of simple chains structures (Bailey and Longstaff 2001). As was pointed out above, such cyclizations also suffer from competing 7-*endo* closure and allylic hydrogen abstraction.

The effect of rigidity and deliberate biasing of one or the other of the privileged transition-state conformation on the stereoselectivity can be appreciated by looking at the cyclization of the three carbohydrate derivatives in Scheme 2.42 (RajanBabu 1991). The first, a glucose derivative with a flexible structure, leads to three isomeric products with a modest predominance of the 1,5-*cis* isomer over the two 1,5-*trans* isomers. In contrast, the benzylidene protecting moiety in the second compound, also glucose-derived, brings us back to the case where a ring acts as a template and imparts a greater rigidity on the transition-state. This forces in this instance a boat-like transition structure because otherwise the benzyloxy group in position 4 will find itself cramped underneath the dioxane ring. The reaction hence leads exclusively to the 1,5-*trans* derivative. If the carbon in position 4 is inverted as in the mannose-derived analogue, the chair-like transition becomes more favourable and the 1,5-*cis* cyclopentane is obtained essentially exclusively (99:1). Playing with the substituents and protecting groups is a simple yet a quite promising approach to controlling the stereochemistry of radical cyclizations.

Rigidity can also be attained by exploiting intramolecular hydrogen bonding which can 'freeze' a structure in the non-polar media frequently used for conducting

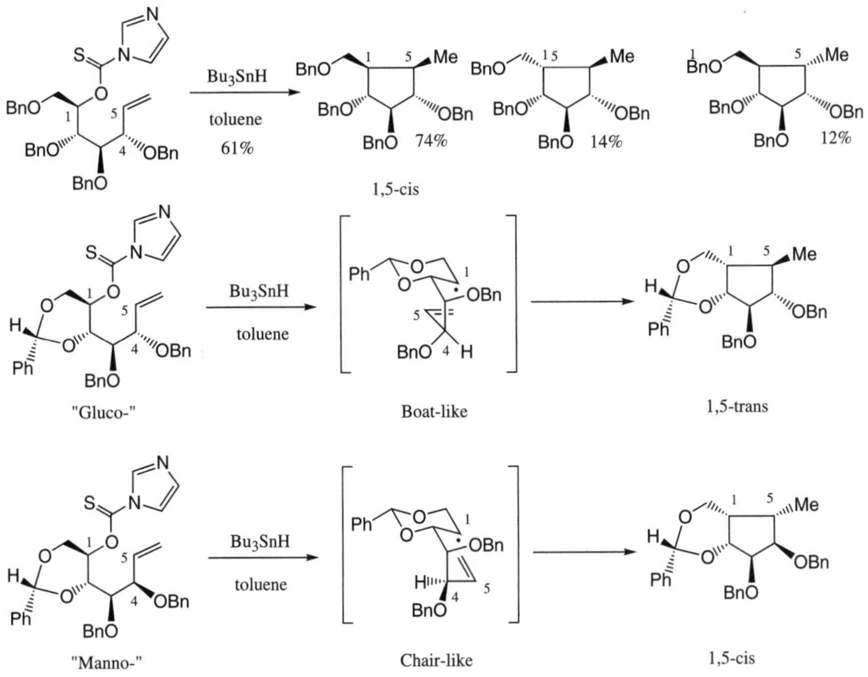

**Scheme 2.42** Effect of adjacent ring on stereochemistry of cyclization

radical reactions. Further stereochemical aspects pertaining to diastereo- and enantio-selectivity will be discussed in the next chapter.

## 2.11 Some further synthetic variations

Essentially limitless variations in the nature of the internal or external trap that can be used and in the way and order in which they can be combined. Indeed, many ingenious schemes have been conceived and implemented (Rhéault and Sibi 2003). Thus, combining one or more cyclization steps with an intermolecular radical addition allows for more convergent synthetic design. Such a strategy is however more difficult to implement because of the complications arising from competition with premature hydrogen abstraction from the stannane discussed above (Section 2.5). Since the external, usually electrophilic olefin has to be used in excess (typically 5-fold), such approaches have often been limited to the use of simple, readily available activated olefins (methyl acrylate, acrylonitrile, etc.). The highly convergent approach to prostaglandin $F_{2\alpha}$ outlined in Scheme 2.43 is an exception: the external olefin is quite elaborate but still needs to be used in some excess (Stork *et al.* 1986). This reaction also illustrates the possibility of using the stannane in catalytic amounts, the stoichiometric reductant being sodium cyanoborohydride. In this transformation, the radical adduct was not purified but thermolysed at $140\,^\circ$C to convert the $\alpha$-silyl ketone to the corresponding enol silyl ether which was then oxidized by palladium acetate to the desired enone. As we shall see in the next chapter, it is possible to obtain the same enone more directly, using the fragmentation method to generate stannyl radicals.

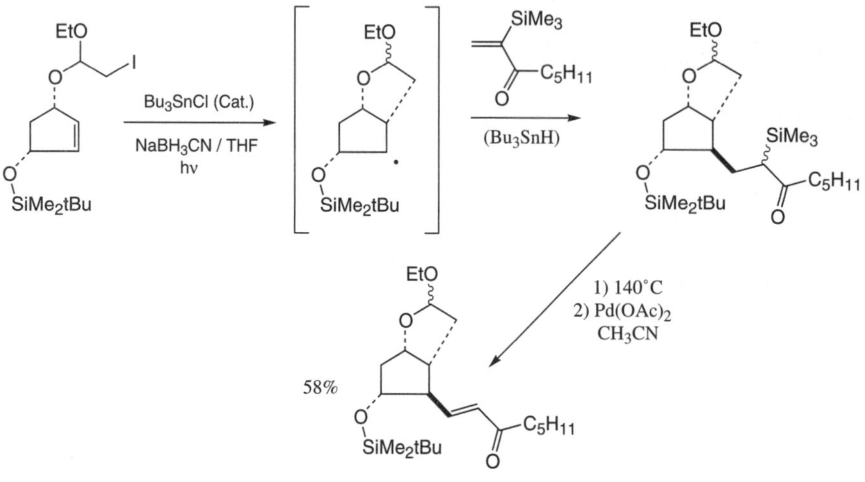

**Scheme 2.43**
Sequential intramolecular and intermolecular radical additions in the synthesis of prostaglandins

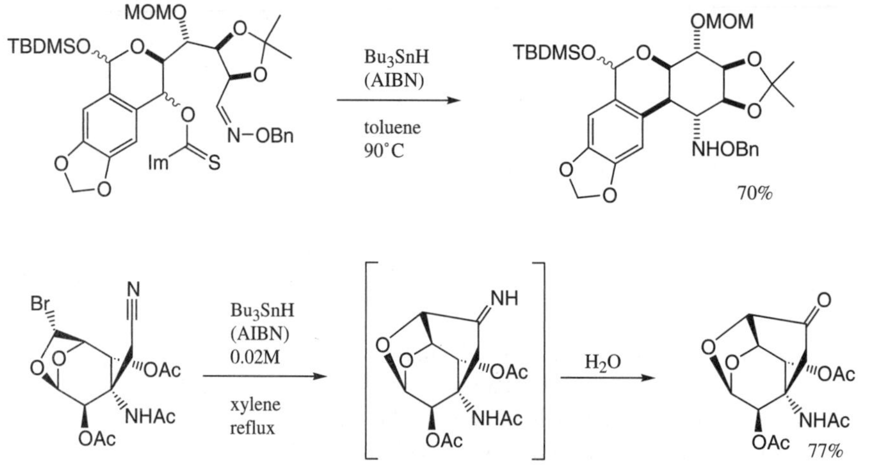

**Scheme 2.44**   Radical addition to carbon–nitrogen multiple bonds

Olefins are not the only traps for radicals. Intramolecular additions to carbonyl groups, oximes, hydrazones, nitriles, etc. are well documented (Fallis and Brinza 1997; Tauh and Fallis 1999) and such additions can be incorporated in interesting radical cascades or tandem processes. Examples of efficient addition to an oxime ether (Keck *et al.* 1995) and to a nitrile (Alonso *et al.* 1993) are portrayed in Scheme 2.44. The former is a key step to 7-deoxy-pancratistatin, whereas the latter was used in an approach to tetrodotoxin starting with a carbohydrate template.

Carbon-centred radicals also add to carbon monoxide (eqn 3, Scheme 2.44) and, under appropriate conditions, this reversible reaction can be harnessed into a synthetically useful process. The rate for the addition ($k_1$) of a primary alkyl radical has been estimated to be $2 \times 10^5$ $M^{-1}s^{-1}$ at 50 °C (Nagahara *et al.* 1995), whereas the reverse fragmentation ($k_{-1}$) is of the order of $1.3 \times 10^4$ $s^{-1}$ and $3.9 \times 10^5$ $s^{-1}$ at 80 °C for a primary and secondary R, respectively (Chatgilialoglu and Lucarini 1995). The equilibrium thus depends on the nature and stability of the radical, and on the pressure of carbon monoxide. With aromatic, vinylic, and primary radicals, the equilibrium will tend to lie to the right whereas with tertiary, benzylic, and other more stabilized radicals, the equilibrium will shift to the left making the carbonylation process difficult. This difference can be exploited, for example by starting with a reactive radical which will readily add carbon monoxide and terminating the sequence with a stabilized radical. This is illustrated by the transformation depicted in Scheme 2.45, wherein the initial radical is primary and the final radical is either tertiary or stabilized by a phenyl or an ester group (Ryu *et al.* 1996).

One interesting variation is to generate a carbon-centred radical using tin hydride technology under an atmosphere of carbon monoxide and in the presence of sodium cyanborohydride (Scheme 2.46, Ryu *et al.* 1997). A catalytic amount of a fluorous stannane reagent is used and the reaction conducted in a two-solvent mixture. The intermediate aldehyde produced is immediately reduced into a primary alcohol and

$$ R^\bullet + CO \underset{k_{-1}}{\overset{k_1}{\rightleftharpoons}} RCO^\bullet \qquad \text{(eq. 3)} $$

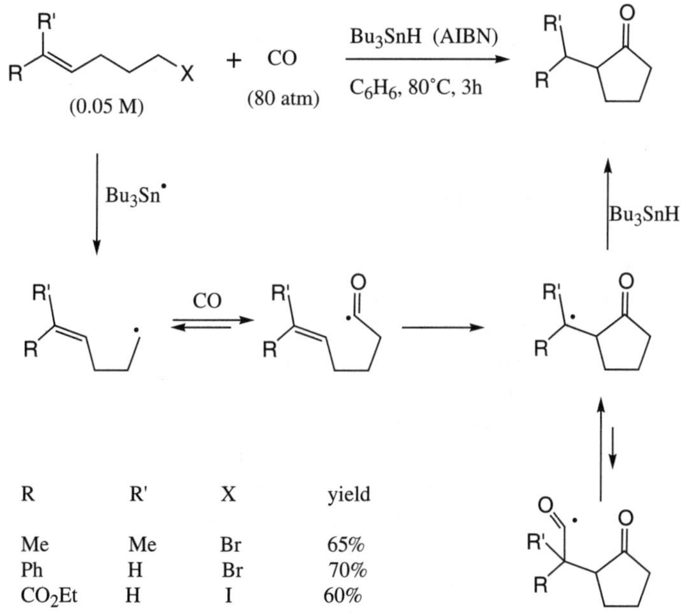

| R | R' | X | yield |
|------|-----|-----|------|
| Me | Me | Br | 65% |
| Ph | H | Br | 70% |
| CO$_2$Et | H | I | 60% |

**Scheme 2.45** Radical additions to carbon monoxide

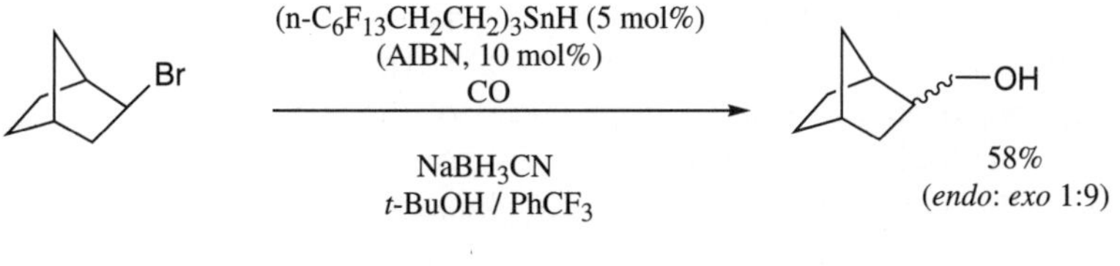

**Scheme 2.46** Reductive radical addition to carbon monoxide

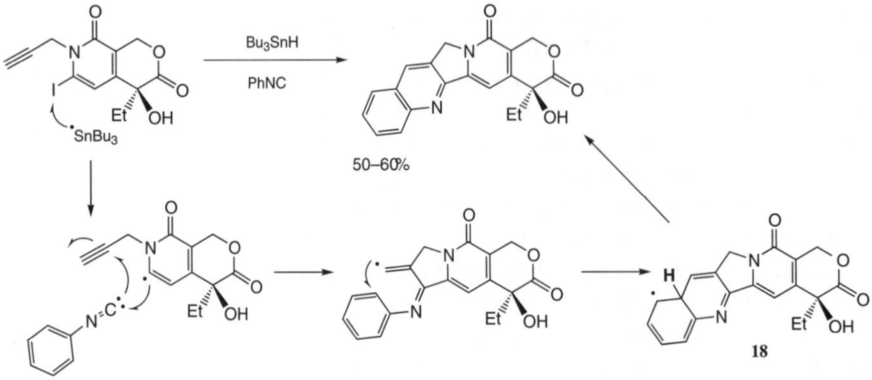

**Scheme 2.47** Radical cascade involving isocyanide in the synthesis of camptothecin

the stannane is continuously regenerated. At the end of the reaction, a three-solvent extraction procedure separates the tin residues into a fluorous phase, the salts into an aqueous phase, and the product into a normal organic phase. This neat contrivance allows the conversion of a bromide for example into the homologous carbinol and makes the separation very easy.

Isonitriles, which are isoelectronic with carbon monoxide, can also be used to capture radicals (Stork and Sher 1986). One interesting application is a recent, elegant synthesis of camptothecin (Curran *et al*. 1995; Ryu *et al*. 1996) where, as shown in Scheme 2.47, two rings are created at the same time. It is not clear how the final aromatization takes place and more than one pathway may be implicated. Such rearomatizations often follow a radical addition to an aromatic ring and there is some debate as to the exact way they occur (Beckwith and Storey 1995). The hydrogen in bold type on the cyclohexadienyl portion in intermediate **18** must be exceedingly easy to abstract. It has been calculated that a C–H bond vicinal to a radical centre is weakened by about 30–40 kcal mol$^{-1}$ (Zhang 1998). This is a considerable amount, especially that in this case, the C–H bond in adduct **18** is also doubly allylic.

Radical reactions in general, whether involving tin hydrides or not, are usually conducted in the careful absence of oxygen which, being a biradical, can interfere with the desired process. Under appropriate experimental conditions, however, it is possible to capture the intermediate carbon radical with molecular oxygen in a synthetically useful manner. The ultimate product is an alcohol and this reaction has been used for example to prepare directly $^{17}O$ and $^{18}O$ labelled alcohols from the appropriate halides, as shown in Scheme 2.48 (Sawamura *et al*. 1997). In this experiment, the cyanoborohydride is used both to regenerate the tributyltin hydride and to reduce the intermediate hydroperoxide.

A more reliable way to create C–O bonds is to use tetramethypiperidine-*N*-oxyl (TEMPO), a persistent oxygen-centred radical which is easier to handle than molecular oxygen. TEMPO and other nitroxyl radicals capture carbon-centred radicals extremely rapidly, with a measured rate of the order of $10^7$–$10^9$ M$^{-1}$s$^{-1}$, which is close to the diffusion limit (Beckwith *et al*. 1988; Chateauneuf *et al*. 1988). The transformation in Scheme 2.49 was used to prepare simplified analogues of the duocarmycin alkylating subunit (Boger and McKie 1995). It is not a chain process and the desired product is obtained through a radical–radical recombination.

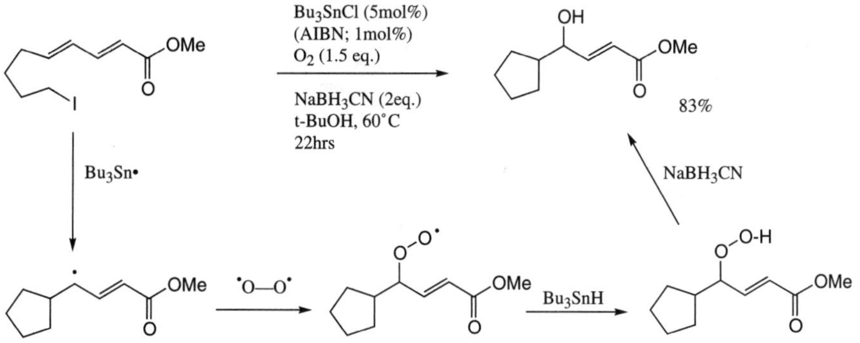

**Scheme 2.48** Radical cascade involving triplet oxygen

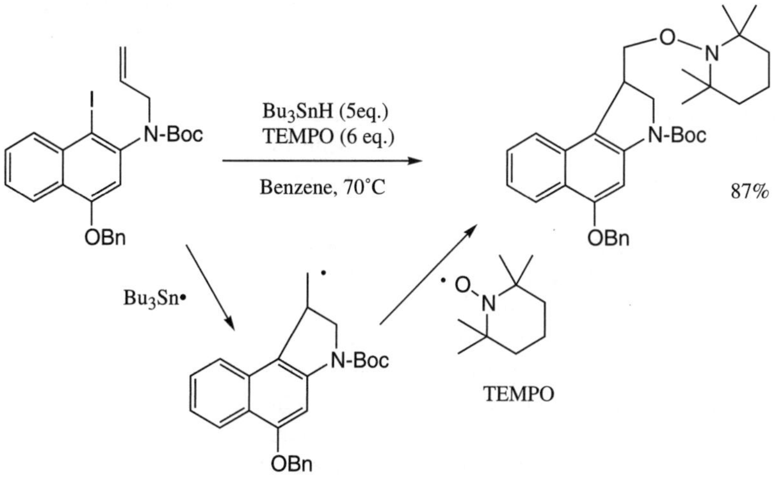

**Scheme 2.49** Carbon–oxygen bond formation by capture of a carbon radical by TEMPO

Normally, radical–radical interactions represent termination steps and are not normally selective, except when a persistent radical, such as TEMPO or molecular oxygen, is involved. The reason for this peculiar but exceedingly important behaviour will be discussed in detail in Chapter 7.

## 2.12   Nitrogen- and oxygen-centred radicals

In addition to carbon radicals, various types of radicals centred on nitrogen and oxygen can be created and captured using stannane technology. The nature of the precursors is largely inspired by the ones used for carbon radical generation. Thus, aminyl radicals can be produced by cleaving the weak nitrogen–sulfur bond in sulfenamides. Aminyl radicals are less reactive than alkyl radicals, and their addition to olefins is reversible. It therefore usually best to couple the aminyl radical cyclization with a fast irreversible process. This can be clearly seen in the elegant experiment summarized in Scheme 2.50 (Bowman *et al.* 1994): under the same experimental conditions, compound **19a** gives a 1 : 4 mixture of uncyclized and cyclized products **20** and **21**, whereas the *N*-allyl analogue **19b** leads to the tetracyclic derivative **23**

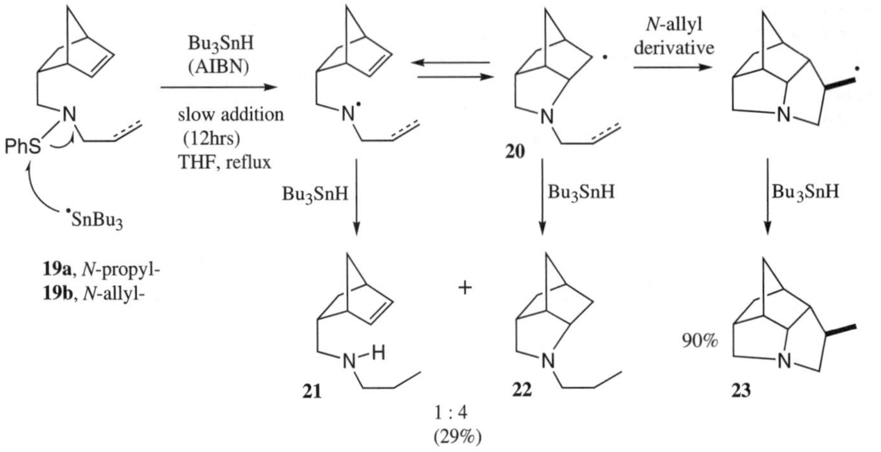

**Scheme 2.50**
Reversible cyclization of a
neutral aminyl radical

exclusively. The reversibility of the first cyclization step was confirmed by generating the intermediate carbon radical **20** directly from another precursor using the organomercury hydride method (Chapter 4) and finding that ring opening occurred partially to give again a mixture of **21** and **22**.

The reactivity of the aminyl radical, as we shall see later, can be greatly enhanced by protonation or complexation with a Lewis acid or a transition metal. Alternatively, one can use the intrinsically more reactive iminyl or amidyl radicals. These and related nitrogen centered radicals can be produced from sulfenamide derivatives (Boivin *et al.* 1990; Esker and Newcomb 1993*a*) or, perhaps more generally, from the readily available thiosemicarbazone precursors, as illustrated by the two examples in Scheme 2.51 (Callier-Dublanchet *et al.* 1995; Zard 1996). The rates of 5-*exo* cyclization of amidyl radicals depend on whether the olefin is on the acyl or *N*-alkyl side chain and have been estimated to be $>10^7$ and $5 \times 10^4$ s$^{-1}$, respectively (Esker and Newcomb 1993*b*); as for 5-*exo* ring closure of iminyl radicals, it appears to be about one order of magnitude slower than the saturated carbon case, that is, of the order of $5 \times 10^4$ s$^{-1}$ (Le Tadic-Biadatti *et al.* 1997). Clean C–N bond formation can therefore be accomplished under mild, neutral conditions.

We have seen that ketones and ordinary esters are not normally affected by triorganotin radicals because the addition to the carbonyl group is relatively slow and reversible. Nevertheless, by coupling this unfavourable step with a subsequent rapid cleavage of a weak bond such as an N–O bond, then another useful route to nitrogen-centred radicals can be devised. This is illustrated in Scheme 2.52 by the easy generation and capture of an iminyl radical from an oxime benzoate to give an azabicyclo-phosphonate derivative (Boivin *et al.* 1995). The rate of $\beta$-scission of the N–O bond has been estimated to be greater than $10^8$ s$^{-1}$ (Wu and Begley 2000).

Oxygen-centred radicals are less accessible than either carbon or nitrogen radicals using organotin chemistry because of a dearth of suitable precursors (Hartung *et al.* 2002). Their use in synthesis is further hampered by their high reactivity and a somewhat unruly behaviour. The construction of tetrahydrofurans by intramolecular addition of an alkoxy radical to an olefin is shown in Scheme 2.53 (Beckwith *et al.* 1989). The rate of cyclization is quite high (estimated to be $5.2 \times 10^8$ s$^{-1}$) but hydrogen

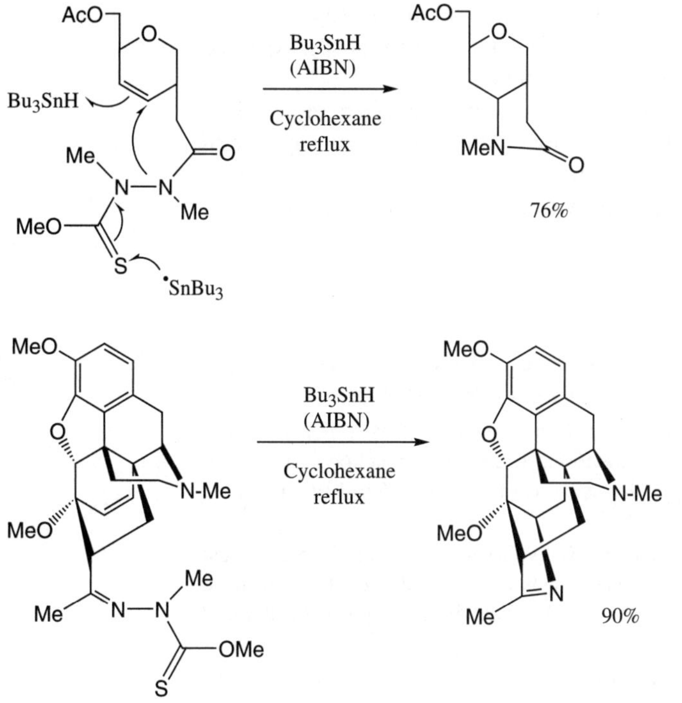

**Scheme 2.51**
Cyclization of amidyl and iminyl radicals

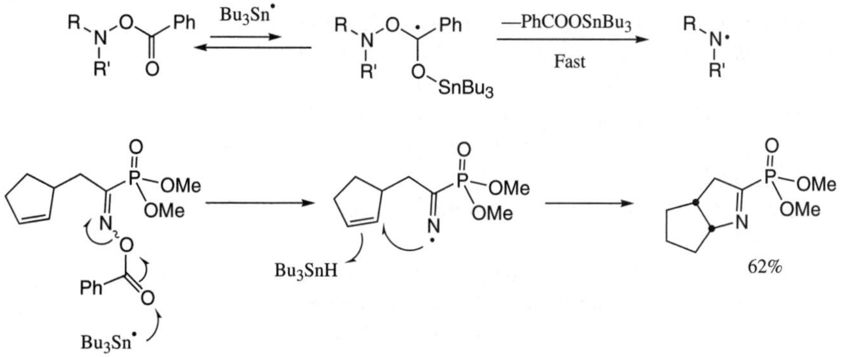

**Scheme 2.52**
*O*-Benzoates of hydroxylamine derivatives as precursors of nitrogen radicals

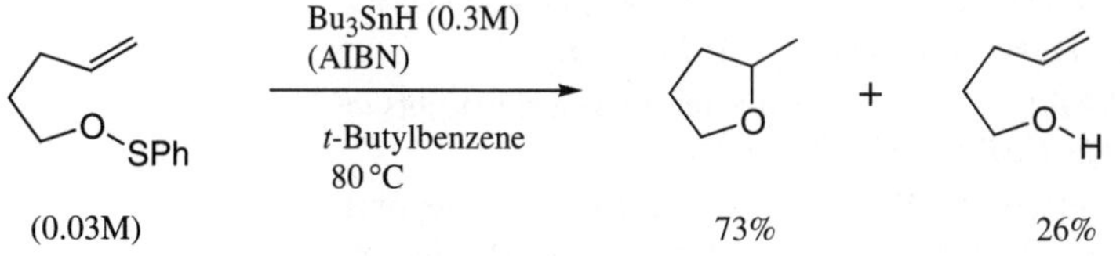

**Scheme 2.53**
Generation and capture of alkoxy radicals

transfer from the stannane is also exceedingly rapid. It is the propensity of alkoxy radicals for fragmentation and hydrogen abstraction that has been mostly exploited in synthesis, as we shall see in later sections of the book.

In this chapter, we have presented some reactions involving organotin hydrides but considered only reduction and addition reactions. Many other pathways are in fact open to the intermediate radicals, and the organotin reagents themselves can be modified. These aspects will constitute the subject matter of the next chapter.

# References

Aimetti, J. A., Hamanaka, E. S., Johnson, D. A., and Kellogg, M. S. (1979). Stereoselective synthesis of 6$\beta$-substituted penicillinates. *Tetrahedron Letters*, **20** 4631–4634.

Aratani, M., Hirai, H., Sawada, K., Yamada, A., and Hashimoto, M. (1985). Synthetic studies on carbapenem antibiotics from penicillins.III. Stereoselective radical reduction of a chiral 3-isocyanoazetidinone. Total synthesis of optically active carpetimycins. *Tetrahedron Letters*, **26**, 223–226.

Alonso, R. A., Burgey, C. S., Rao, B. V., Vite, G. D., Vollerthun, R., Zottola, M. A., and Fraser-Reid, B. (1993). Carbohydrates to carbocycles: synthesis of the densely functionalized carbocyclic core of tetrodotoxin by radical cyclisation of an anhydro sugar precursor. *Journal of the American Chemical Society*, **115**, 666–6672.

Baldwin, J. E. (1976). Rules for ring closure. *Journal of the Chemical Society, Chemical Communications*, 734–736.

Bachi, M. D. and Bosch, E. (1986). Synthesis of $\alpha$-alkylidene-$\gamma$-lactams by intramolecular addition of alkoxycarbonyl free radicals to acetylenes. *Tetrahedron Letters*, **27**, 641–644.

Bachi, M. D. and Bosch, E. (1988). On the mechanism of reductive degradation of dithiocarbonates by tributylstannane. *Journal of the Chemical Society, Perkin Transactions,* **1**, 1517–1519.

Bachi, M. D. and Hoornaert, C. (1981). Free radical annelation in the synthesis of bicyclic $\beta$-lactams. 2. Alternative use of chloro, phenylseleno, and phenylthio-functionalities as radical precursors. *Tetrahedron Letters*, **22**, 2693–2694.

Bailey, W. F. and Longstaff, S. C. (2001). Cyclization of methyl-substituted 6-heptenyl radicals. *Organic Letters*, **3**, 2217–2219.

Bailey, W. F., Khanolkar, A. D., Gavaskar, K., Ovaska, T. V., Rossi, K., Thiel, Y., and Wiberg, K. B. (1991). Stereoselectivity of cyclization of substituted 5-hexen-1-yl lithiums: regiospecific and highly stereoselective insertion of an unactivated alkene into a C–Li bond. *Journal of the American Chemical Society*, **113**, 5720–5727.

Barlaam, B., Boivin, J., El Kaim, L., Elton-Farr, S., and Zard, S. Z. (1995). A short total synthesis of estrone derivatives: a novel 1,3-rearrangement of an allylic nitro group. *Tetrahedron*, **51**, 1675–1683.

Barton, D. H. R. and McCombie (1975). A new method for the deoxygenation of secondary alcohols. *Journal of the Chemical Society, Perkin Transactions,* **1**, 1574–1585.

Barton, D. H. R. and Motherwell, W. B. (1981). New and selective reactions and reagents in natural product chemistry. *Pure & Applied Chemistry*, **53**, 1081–1099.

Barton, D. H. R., Bringmann, G., and Motherwell, W. B. (1980*b*). Reactions of relevance to the chemistry of aminoglycosides. Part 15. The selective modification of neamine by radical-induced deamination. *Journal of the Chemical Society, Perkin Transactions,* **1**, 2665–2669.

Barton, D. H. R., Motherwell, W. B., and Stange, A. (1981). Radical induced deoxygenation of primary alcohols. *Synthesis*, 743–745.

Barton, D. H. R., Parekh, S. I., and Tse, C.-L. (1993). On the stability and radical deoxygenation of tertiary xanthates. *Tetrahedron Letters*, **34**, 2733–2736.

Barton, D. H. R., Crich, D., Lobberding, A., and Zard, S. Z. (1986). On the mechanism of the deoxygenation of secondary alcohols by the reduction of their methyl xanthates by tin hydrides. *Tetrahedron*, **42**, 2329–2338.

Barton, D. H. R., Bringmann, G., Lamotte, G., Motherwell, W. B., Hay-Motherwell, R. S., and Porter, A. E. A. (1980*a*). Reactions of relevance to the chemistry of aminoglycosides. Part 14. A useful radical deamination reaction. *Journal of the Chemical Society, Perkin Transactions, 1*, 2657–2664.

Beckwith, A. L. J. (1981). Regio-selectivity and stereo-selectivity in radical reactions. *Tetrahedron*, **37**, 3073–3100.

Beckwith, A. L. J. and Page, J. M. (1998). Formation of some oxygen-containing heterocycles by radical cyclisation: the stereochemical influence of anomeric effects. *Journal of Organic Chemistry*, **63**, 5144–5153.

Beckwith, A. L. J. and Pool, J. S. (2002). Factors affecting the rates of addition of free radicals to alkenes—determination of absolute rate coefficients using the persistent aminoxyl method. *Journal of the American Chemical Society*, **124**, 9489–9497.

Beckwith, A. L. J. and Schiesser, C. H. (1985). Regio and stereo-selectivity of alkenyl radical ring closure. *Tetrahedron*, **41**, 3925–3942.

Beckwith, A. L. J. and Storey, J. M. D. (1995). Tandem radical translocation and homolytic aromatic substitution: a convenient and efficient route to oxindoles. *Journal of the Chemical Society, Chemical Communications*, 977–978.

Beckwith, A. L. J. and Zimmerman, J. (1991). Cyclisation of the the 3-tert-butylhex-5-enyl radical: a test of transition-state structure. *Journal of Organic Chemistry*, **56**, 5791–5796.

Beckwith, A. L. J., Bowry, V. W., and Moad, G. (1988). Kinetics of the coupling reactions of of the nitroxyl radical 1,1,3,3-tetramethylisoindoline-2-oxyl with carbon centered radicals. *Journal of Organic Chemistry*, **53**, 1632–1641.

Beckwith, A. L. J., Hay, B. P., and Williams, G. M. (1989). Generation of alkoxy radicals from *O*-alkyl benzenesulfenates. *Journal of the Chemical Society, Chemical Communications*, 1202–1203.

Bensasson, C. S., Cornforth, J., Du, M.-H., and Hanson, J. R. (1997). Unusual generation of methoxy groups in Barton deoxygenations of alcohols. *Chemical Communications*, 1509–1510.

Boger, D. L. and Mathvink, R. J. (1990). Tandem free radical ring expansion and 5-*exo-dig* 5-hexynyl radical cyclisation: a useful approach to fused bicyclic carbocycles. *Journal of Organic Chemistry*, **55**, 5442–5444.

Boger, D. L. and McKie, J. A. (1995). An efficient synthesis of 1,2,9,9a-tetrahydrocyclopropa[c]benz[e]indol-4-one (CBI): an enhanced and simplified analog of the CC-1065 and duocarmycin alkylation subunit. *Journal of Organic Chemistry*, **60**, 1271–1275.

Boivin, J., Fouquet, E., and Zard, S. Z. (1990). Cyclisation of iminyl radicals derived from sulphenilimines: a simple access to $\Delta^1$-pyrrolines. *Tetrahedron Letters*, **31**, 85–88.

Boivin, J., Callier-Dublanchet, A.-C., Quiclet-Sire, B., Schiano, A.-M., and Zard, S. Z. (1995). Iminyl, amidyl, and carbamyl radicals from *O*-benzoyl oximes and *O*-benzoyl hydroxamic acid derivatives. *Tetrahedron*, **51**, 6517–6528.

Bowman, W. R., Bridge, C. F., and Brookes (2000). Synthesis of heterocycles by radical cyclisation. *Journal of the Chemical Society, Perkin Transactions, 1*, 1–14.

Bowman, W. R., Clark, D. N., and Marmon, R. J. (1994). Synthesis of pyrrolizidines using aminyl radicals generated from sulfenamide precursors. *Tetrahedron*, **50**, 1295–1310.

Breslow, R., Barrett, E., and Mohacsi, E. (1962). Free radical additions to squalenes. *Tetrahedron Letters*, **3** 1207–1211.

Broeker, J. L. and Houk, K. N. (1991). MM2 model for the intramolecular addition of acyl-substituted radicals to alkenes. *Journal of Organic Chemistry*, **56**, 3651–3655.

Büchi, G. and Wüest, H. (1979). New synthesis of β-agarofuran and of dihydroagarofuran. *Journal of Organic Chemistry*, **44**, 646–649.

Callier-Dublanchet, A.-C., Quiclet-Sire, B., and Zard, S. Z. (1995). A new source of nitrogen centered radicals. *Tetrahedron Letters*, **36**, 8791–8794.

Chateauneuf, J., Lusztyk, J., and Ingold, K. U. (1988). Absolute rate constants for the reaction of some carbon-centered radicals with 2,2,6,6-tetramethylpiperidine-*N*-oxyl. *Journal of Organic Chemistry*, **53**, 1629–1632.

Chatgilialoglu, C. and Lucarini, M. (1995). Rate constants for the reaction of acyl radicals with Bu$_3$SnH and (TMS)$_3$SiH. *Tetrahedron Letters*, **36**, 1299–1302.

Chatgilialoglu, C. and Newcomb, M. (1999). Hydrogen donor abilities of the group 14 hydrides. *Advances in Organometallic Chemistry*, **44**, 67–112.

Chen, Y.-J., Chen, C.-M., and Lin, W.-Y. (1993). A new approach to the formal synthesis of (±)-cedrene. *Tetrahedron Letters*, **34**, 2961–2962.

Chen, L., Gill, G. B., and Pattenden, G. (1994). New radical mediated polyolefin cyclisations directed towards steroid ring synthesis. *Tetrahedron Letters*, **35**, 2593–2596.

Clive, D. L. J. and Yang, W. (1995). A nitrogen containing stannane for free radical chemistry. *Journal of Organic Chemistry*, **60**, 2607–2609.

Corey, E. J. and Suggs, J. W. (1975). A method for catalytic dehalogenations via trialkyltin hydrides. *Journal of Organic Chemistry*, **40**, 2554–2555.

Crich, D. and Quintero, L. (1989). Radical chemistry associated with the thiocarbonyl group. *Chemical Reviews*, **89**, 1413–1432.

Crich, D. and Recupero, F. (1998). Steric acceleration in the reaction of aryl bromides with tributylstannyl radicals. *Journal of the Chemical Society, Chemical Communications*, 189–190.

Curran, D. P. (1998). Strategy-level separation in organic synthesis: from planning to practice. *Angewandte Chemie International Edition in English*, **37**, 1174–1196.

Curran, D. P. and Chang, C.-T. (1989). Atom transfer cyclisation reactions of α-iodo esters, ketones, and malonates: examples of selective 5-*exo*, 6-*endo*, 6-*exo*, and 7-*endo* ring closures. *Journal of Organic Chemistry*, **54**, 3140–3157.

Curran, D. P. and Hadida, S. (1996). Tris(2-(perfluorohexyl)ethyl)tin hydride: a new fluorous reagent for use in traditional organic synthesis and liquid phase combinatorial synthesis. *Journal of the American Chemical Society*, **118**, 2531–2532.

Curran, D. P. and Rakiewicz, D. M. (1985). Radical-initiated polyolefinic cyclisations in linear triquinane synthesis. Model studies and total synthesis of (±) hirsutene. *Tetrahedron*, **41**, 3943–3958.

Curran, D. P., Jasperse, C. P., and Totleben, M. J. (1991). Approximate absolute rate constants for the reaction of tributyltin radicals with aryl and vinyl halides. *Journal of Organic Chemistry*, **56**, 7169–7162.

Curran, D. P., Ko, S.-B., and Josien, H. (1995). Cascade radical reactions of isonitriles: a second- generation synthesis of (20S)-camptothecin, topotecan, irinotecan, and GI-147211C. *Angewandte Chemie International Edition in English*, **34**, 2683–2684.

Dolan, S. C. and MacMillan, J. (1985). A new method for the deoxygenation of secondary and tertiary alcohols. *Journal of the Chemical Society, Chemical Communications*, 1588–1589.

Dumartin, G., Ruel, G., Kharbouti, J., Delmond, B., Connil, M.-F., Jousseaume, B., and Pereyre, M. (1994). Straightforward synthesis and reactivity of polymer-supported organotin hydrides. *Synlett*, 952–954.

Engel, P. S. (1980). Mechanism of the thermal and photochemical decomposition of azo-alkanes. *Chemical Reviews*, **80**, 99–150.

Esker, J. L. and Newcomb, M. (1993*a*). Amidyl radicals from *N*-(phenylthio)amides. *Tetrahedron Letters*, **34**, 6877–6880.

Esker, J. L. and Newcomb, M. (1993*b*). The generation of nitrogen radicals and their cyclisations for the construction of the pyrrolidine nucleus. *Advances in Heterocyclic Chemistry*, **58**, 1–45.

Fallis, A. G. and Brinza, I. M. (1997). Free radical cyclisations involving nitrogen. *Tetrahedron*, **53**, 17543–17594.

Fischer, H. and Paul, H. (1987). Rate constants for some prototype radical reactions in liquids by kinetic electron spin resonance. *Accounts of Chemical Research*, **20**, 200–206.

Fischer, H. and Radom, L. (2001). Factors controlling the addition of carbon-centered radicals to alkenes—an experimental and theoretical perspective. *Angewandte Chemie International Edition in English*, **40**, 1340–1375–206.

Galli, C. and Pau, T. (1998). The dehalogenation reactions of organic halides by tributyltin radicals: the energy of activations the BDE of the C–X bond. *Tetrahedron*, **54**, 2893–2904.

Garden, S. J., Avila, D. V., Beckwith, A. L. J., Bowry, V. W., Ingold, K. U., and Lusztyk, J. (1996). Absolute rate constant for the reaction of aryl radicals with tri-*n*-butyltin hydride. *Journal of Organic Chemistry*, **61**, 805–809.

Gastdaldi, S. and Stien, D. (2002). PAH-supported tin hydride: a new tin reagent easily removable from reaction mixtures. *Tetrahedron Letters*, **43**, 4309–4311.

Giese, B. (1983). Formation of CC bonds by addition of free radicals to alkenes. *Angewandte Chemie International Edition in English*, **22**, 753–764.

Giese B. (1986). *Radicals in organic synthesis: formation of carbon–carbon bonds*, p. 64. Pergamon Press, Oxford.

Giese, B. and Dupuis, J. (1983). Diastereoselective synthesis of *C*-glycopyranosides. *Angewandte Chemie International Edition in English*, **22**, 622–623.

Giese, B. and Lachhein, S. (1982). Addition of alkyl radicals to alkynes: distinction between radical and ionic nucleophiles. *Angewandte Chemie International Edition in English*, **21**, 768–769.

Giese, B. and Witzel, T. (1986). Synthesis of 'C-disaccharides' by radical C-C bond formation. *Angewandte Chemie International Edition in English*, **25**, 450–451.

Giese, B., González-Gómez, J. A., and Witzel, T. (1984). The scope of the radical C–C-coupling by the 'tin method'. *Angewandte Chemie International Edition in English*, **23**, 69–70.

Giese, B., Kopping, B., Göbel, T., Dickhaut, J., Thoma, G., Kulicke, K. J., and Trach, F. (1996). Radical cyclisation reactions. *Organic Reactions*, **48**, 301–856.

Grady, G. L. and Kuivila, H. G. (1969). A simple technique for performing reactions with organotin hydrides. *Journal of Organic Chemistry*, **34**, 2014–2016.

Griller, D. and Ingold, K. U. (1980). Free radical clocks. *Accounts of Chemical Research*, **13**, 317–323.

Harendza, M., Leßmann, K., and Neumann, W. P. (1993). A polymer-supported distannane as photochemical, regenerable source of stannyl radicals for organic synthesis. *Synlett*, 283–285.

Hartung, J., Gottwald, T., and Spehar, K. (2002). Selectivity in the chemistry of oxygen-centered radicals—the formation of carbon–oxygen bonds. *Synthesis*, 1469–1498.

Hartwig, W. (1983). Modern methods for the radical deoxygenation of alcohols. *Tetrahedron*, **39**, 2609–2645.

Hayashi, K., Iyoda, J., and Shiihara, I. (1967). Reaction of organotin oxides, alkoxides and acyloxides with organosilicon hydrides. New preparative method of organotin hydride. *Journal of Organometallic Chemistry*, **10**, 81–94.

Héberger, K. and Lopata, A. (1998). Assessment of nucleophilicity and electrophilicity of radicals, and of polar and enthalpy effects on radical addition reactions. *Journal of Organic Chemistry*, **63**, 8646–8653.

Hitchcock, S. A. and Pattenden, G. (1992). Total synthesis of the mycotoxin ($-$)-zearalenone based on macrocyclisation using a cinnamyl radical intermediate. *Journal of the Chemical Society, Perkin Transactions,* **1**, 1323–1328.

Ivor John, D., Thomas, E. J., and Tyrrell, N. D. (1979). Reduction of $6\alpha$-alkyl-$6\beta$-isocyano penicillanates by tri-*n*-butyltin hydride. A stereoselective synthesis of $6\beta$-alkylpenicillanates. *Journal of the Chemical Society, Chemical Communications*, 345–347.

Johnston, L. J., Lusztyk, J., Wayner, D. D. M., Abeywickreyma, A. N., Beckwith, A. L. J., Scaiano, J., and Ingold, K. U. (1985). Absolute rate constants for reaction of phenyl, 2,2-dimethylvinyl, cyclopropyl, and neopentyl radicals with tri-*n*-butylstannane. Comparison of the radical trapping abilities of tri-*n*-butylstannane and -germane. *Journal of the American Chemical Society*, **107**, 4594–4596.

Keck, G. E., McHardy, S. F., and Murry, J. A. (1995). Total synthesis of ($+$)-7-pancratistatin: a radical cyclisation approach. *Journal of the American Chemical Society*, **117**, 7289–7290.

Keusenkothen, P. F. and Smith, M. B. (1992). Asymmetric radical cyclisation: the synthesis of 6-alkyl pyrrolizidines-2-ones. *Tetrahedron*, **48**, 2977–2992.

Khan, F. A. and Prabhudas, B. (1999). A simple and preparatively useful tributylstannane mediated selective reduction and bridgehead functionalization of tetrahalonorbornene derivatives. *Tetrahedron Letters*, **40**, 9289–9292.

Khoo, L., E. and Lee, H. H. (1968). Hydrostannolysis reactions. Part 1. Reduction of esters. *Tetrahedron Letters*, **9**, 4351–4354.

Knapp, S., Gibson, F. S., and Choe, Y. H. (1990). Radical based annulation of iodolactams. *Tetrahedron Letters*, **31**, 5397–5400.

Ladlow, M. and Pattenden, G. (1984). Intramolecular free radical cyclisation onto vinyl ethers. A method for the synthesis of $\beta$-oxo-$\gamma$-butyrolactones. *Tetrahedron Letters*, **25**, 4317–4320.

Le Tadic-Biadatti, M.-H., Callier-Dublanchet, A.-C., Horner, J. H., Quiclet-Sire, B., Zard, S. Z., and Newcomb, M. (1997). Absolute rate constants for iminyl radical reactions. *Journal of Organic Chemistry*, **62**, 559–577.

Lee, E., Lim, J. W., Yoon, C. H., Sung, Y.-S., and Kim, Y. K. (1997). Total synthesis of ($+$) cladantholide and ($-$) estafiatin: 5-*exo*, 7-*endo* radical cyclisation strategy for the construction of guainolide skeleton. *Journal of the American Chemical Society*, **119**, 8391–8392.

Leibner, J. E. and Jacobus, J. (1979). Facile product isolation from organostannane reductions of organic halides. *Journal of Organic Chemistry*, **44**, 449–450.

Lopez, R. M., Hays, D. S., and Fu, G. C. (1997). Bu$_3$SnH-catalyzed Barton-McCombie deoxygenation of of alcohols. *Journal of the American Chemical Society*, **119**, 6949–6950.

Mampuya Bimwala, R. and Vogel, P. (1992). Synthesis of $\alpha$-(1→2)-, $\alpha$-(1→3)-, $\alpha$-(1→4)-, and $\alpha$-(1→5)-*C*-linked disaccharides through 2,3,4,6-tetra-*O*-acetylglucopyranosyl radical additions to 3-methylene-7-oxabicyclo[2.2.1]heptan-2-one derivatives. *Journal of Organic Chemistry*, **57**, 2076–2083.

Milstein, D. and Stille, J. K. (1978). A general, selective, and facile method for ketone synthesis from acid chlorides and organotin compounds catalyzed by palladium. *Journal of the American Chemical Society*, **100**, 3636–3638.

Nagahara, K., Ryu, I., Kambe, N., Komatsu, M., and Sonoda, N. (1995). Rate constants for the addition of a primary alkyl radical to carbon monoxide. *Journal of Organic Chemistry*, **60**, 7384–7385.

Neumann, W. P. (1987). Tri-*n*-butyltin hydride as reagent in organic synthesis. *Synthesis*, 665–683.

Newcomb, M. (1993). Competition methods and scales for alkyl radical reaction kinetics. *Tetrahedron*, **49**, 1151–1176.

Nicolau, K. C., Sato, M., Theodorakis, E. A., and Miller, N. D. (1995). Conversion of thionoesters and thionolactones to ethers; a general and efficient radical desulfurisation. *Journal of the Chemical Society, Chemical Communications*, 1583–1585.

Nozaki, K., Oshima, K., and Utimoto, K. (1987). Et₃B-Induced radical addition of R3SnH to acetylenes and its application to cyclisation reactions. *Journal of the American Chemical Society*, **109**, 2547–2549.

Ono, N., Miyake, H., Kamimura, A., Hamamoto, I., Tamura, R., and Kaji, A. (1985). Denitrohydrogenation of aliphatic nitro citro compounds and a new use of aliphatic nitro compounds as radical precursors. *Tetrahedron*, **41**, 4013–4023.

Overberger, C. G., Hale, W. F., Berenbaum, M. B., and Finestone, A. B. (1954). Azo-bis nitriles. XI. Decomposition of azo compounds. Steric effects. *Journal of the American Chemical Society*, **76**, 6185–6187.

Parnes, H. and Pease, J. (1979). Simple method for the reductive dehalogenation of 9α-bromosteroids. *Journal of Organic Chemistry*, **44**, 151–152.

Pfenninger, J., Heuberger, C., and Graf, W. (1980). The radical induced stannane reduction of selenoesters and selenocarbonates: a new method for the degradation of carboxylic acids to nor-alkanes and for desoxygenation of alcohols to alkanes. *Helvetica Chimica Acta*, **63**, 2328–2337.

Porter, N. A., Magnin, D. R., and Wright, B. T. (1986). Free radical macrocyclization. *Journal of the American Chemical Society*, **108**, 2787–2790.

Porter, N. A., Chang, V. H.-T., Magnin, D. R., and Wright, B. T. (1988). Free-radical macrocyclisation–transannular cyclisation. *Journal of the American Chemical Society*, **110**, 3554–3560.

Porter, N. A., Lacher, B., Chang, V. H.-T., and Magnin, D. R. (1989). Regioselectivity and diasterioselectivity in radical macrocyclisation. *Journal of the American Chemical Society*, **111**, 8309–8310.

RajanBabu, T. V. (1991). Stereochemistry of intramolecular free radical cyclisation reactions. *Accounts of Chemical Research*, **24**, 139–145.

Renaud, P. and Gerster, M. (1998). Use of Lewis acids in free radical reactions. *Angewandte Chemie International Edition in English*, **37**, 2562–2579.

Renaud, P., Lacôte, E., and Quaranta, L. (1998). Alternative and mild procedures for the removal of organotin residues from the reaction mixtures. *Tetrahedron Letters*, **39**, 2123–2126.

Rhéault, T. R. and Sibi, M. P. (2003). Radical-mediated annulation reactions. *Synthesis*, 803–819.

Rhee, J. U., Bliss, B. I., and RajanBabu, T. V. (2003). A new reaction manifold for the Barton radical intermediates: synthesis of *N*-heterocyclic furanosides and pyranosides via the formation of the C1–C2 bond. *Journal of the American Chemical Society*, **125**, 1492–1493.

Rodriguez, G., Castedo, L., Dominguez, D., and Saá, C. (1998). Transannular cyclisations of 10-membered lactams: an easy route to isoquinoline alkaloids. *Tetrahedron Letters*, **39**, 6651–6654.

Russell, G. A. (1973). Reactivity, selectivity, and polar effects in hydrogen atom transfer reactions. In *Free radicals*, Vol 1 (ed. J. K. Kochi), pp. 275–331. Wiley Interscience, New York.

Russell, G. A. and Tashtoush, H. (1983). Free-radical chain substitution reactions of alkylmercury halides. *Journal of the American Chemical Society*, **105**, 1398–1399.

Ryu, I., Sonoda, N., and Curran, D. P. (1996). Tandem radical reactions of carbon monoxide, isonitriles, and other reagent equivalents of the geminal radical acceptor/radical precursor synthon. *Chemical Reviews*, **96**, 177–194.

Ryu, I., Niguma, T., Minakata, S., Komatsu, M., Hadida, S., and Curran, D. P. (1997). Hydroxymethylation of organic halides. Evaluation of a catalytic system involving a fluorous tin hydride reagent for radical carbonylation. *Tetrahedron Letters*, **38**, 7883–7886.

Saegusa, T., Kobayashi, S., Ito, Y., and Yasuda, N. (1968). Radical reaction of isocyanide with organotin hydride. *Journal of the American Chemical Society*, **90**, 4182.

Sawamura, M., Kawagushi, Y., and Nakamura, E. (1997). Conversion of alkyl halides into alcohols using near stoichiometric amount of molecular oxygen: an efficient route to $^{18}O$- and $^{17}O$-labeled alcohols. *Synlett*, 801–802.

Singh, A. K., Bakshi, R. K., and Corey, E. J. (1987). Total synthesis of ($\pm$) atractyligenin. *Journal of the American Chemical Society*, **109**, 6187–6189.

Spellmeyer, D. C. and Houk, K. N. (1987). A force field model for intramolecular radical additions. *Journal of Organic Chemistry*, **52**, 959–974.

Stork, G., Mook, R. Jr., Biller, S. A., and Rychnovsky, S. D. (1983). Free radical cyclization of bromoacetals. Use in the construction of bicyclic acetals and lactones. *Journal of the American Chemical Society*, **105**, 3741–3742.

Stork, G. (1990). A survey of the radical mediated cyclization of $\alpha$-haloacetals of cyclic allyl alcohols as a general route for the control of vicinal regio- and stereochemistry. *Bulletin de la Société Chimique de France*, 675–680.

Stork, G. and Kahn, M. (1985). Control of ring junction via radical cyclisation reactions. *Journal of the American Chemical Society*, **107**, 500–501.

Stork, G. and Sher, P. M. (1986). A catalytic tin system for trapping of radicals from cyclization reactions. Regio- and stereocontrolled formation of two adjacent chiral centres. *Journal of the American Chemical Society*, **108**, 303–304.

Stork, G. and Sofia, M. J. (1986). Stereospecific reductive methylation via a radical cyclization–desilylation process. *Journal of the American Chemical Society*, **108**, 6826–6828.

Stork, G., Sher, P. M., and Chen, H.-L. (1986). Radical cyclization-trapping in the synthesis of natural products. A simple stereocontrolled route to prostaglandin $F_{2\alpha}$. *Journal of the American Chemical Society*, **108**, 6384–6385.

Stork, G., Suh, H. S., and Kim, G. (1991). The temporary silicon connection method in the control of regio- and stereochemistry. Application to radical-mediated reactions. The stereospecific synthesis of *C*-glycosides. *Journal of the American Chemical Society*, **113**, 7054–7056.

Suga, S., Manabe, T., and Yoshida, J. (1999). Control of the free radical reaction by dynamic coordination: unique reactivity of of pyridylethyl-substituted tin hydride. *Journal of Chemical Society, Chemical Communications*, 1237–1238.

Takano, S., Nishizawa, S., Akiyama, M., and Ogasawara, K. (1984). A. new synthesis of *cis*-2,2-dimethyl-3-hydroxymethylcyclopropanecarboxylic acid. *Synthesis*, 949–950.

Tanner, D. D., Blackburn, E. V., and Diaz, G. E. (1981). A free radical chain reaction involving electron transfer. The replacement of a nitro group by hydrogen using trialkyltin hydride, a variation of the Kornblum reaction. *Journal of the American Chemical Society*, **103**, 1557–1559.

Tauh, P. and Fallis, A. G. (1999). Rate constants for 5-*exo* secondary alkyl radical cyclizations onto hydrazones and oxime ethers via intramolecular competition experiments. *Journal of Organic Chemistry*, **64**, 6960–6968.

Tedder, J. M. and Walton, J. C. (1976). The kinetics and orientation of free-radical addition to olefins. *Accounts of Chemical Research*, **9**, 183–191.

Ueno, Y., Moriya, O., Chino, K., Watanabe, M., and Okawara, M. (1982). Homolytic carbocyclisation by use of heterogeneous supported organotin catalyst. A new synthetic route to 2-alkoxytetrahydrofurans and γ-butyrolactones. *Journal of the American Chemical Society*, **104**, 5564–5566.

Ueno, Y., Moriya, O., Chino, K., Watanabe, M., and Okawara, M. (1986). General synthetic route to γ-butyrolactones via stereoselective radical cyclisation by organotin species. *Journal of the Chemical Society, Pertkin Transactions, 1*, 1351–1356.

Villar, F. and Renaud, P. (1998). Diastereoselective radical cyclisation of bromoacetals (Ueno–Stork reaction) controlled by the acetal center. *Tetrahedron Letters*, **39**, 8655–8658.

Walbiner, M., Wu, J. Q., and Fischer, H. (1995). Absolute rate constants for the addition of benzyl and cumyl radicals to alkenes in solution. *Helvetica Chimica Acta*, **78**, 910–924.

Walling, C. and Huyser, E. S. (1963). Free radical additions to olefins to form carbon–carbon bonds. *Organic Reactions*, **13**, 91–149.

Wayner, D. D. M., McPhee, D. J., and Griller, D. (1988). Oxidation and reduction potentials of transient free radicals. *Journal of the American Chemical Society*, **110**, 132–137.

Weinschenken, N. M., Crosby, G. A., and Wong, J. Y. (1975). Polymeric reagents. IV. Synthesis and utilization of an insoluble polymeric organotin dihydride reagent. *Journal of Organic Chemistry*, **40**, 1966–1971.

Wu, M. and Begley, T. P. (2000). β-Scission of the N–O bond in alkyl hydroxamate radicals: a fast radical trap. *Organic Letters*, **2**, 1345–1348.

Yet, L. (1999). Free radicals in the synthesis of medium sized rings. *Tetrahedron*, **55**, 9349–9402.

Yorimitsu, H., Nakamura, T., Shinokubo, H., and Oshima, K. (1998). Triethylborane-mediated atom transfer radical cyclization reaction in water. *Journal of Organic Chemistry*, **63**, 8604–8605.

Zard, S. Z. (1996). Iminyl radicals: a fresh look at a forgotten species (and some of its relatives). *Synlett*, 1148–1155.

Zhang, X.-M. (1998). Homolytic bond dissociation enthalpies of the C–H bonds adjacent to radicals. *Journal of Organic Chemistry*, **63**, 1872–1877.

Zytowski, T. and Fischer, H. (1996). Absolute rate constants for the addition of methyl radicals to alkenes in solution: new evidence for polar interactions. *Journal of the American Chemical Society*, **118**, 437–439.

# 3 Further chain reactions of stannanes

## 3.1 Radical rearrangements

Except for Sections 3.8 and 3.9, where we shall examine the synthetic potential of generating stannyl radicals by $\beta$-scission (in contrast to hydrogen abstraction from a triorganotin hydride), most of this chapter will be devoted to radical rearrangements. A radical cyclization may be considered as a special case of a rearrangement (or isomerization) process. The radical site thus moves along the molecule from one position to another, with one or more new $\sigma$-bonds being formed at the expense of a corresponding number of $\pi$-bonds. In favourable cases, it is possible to shift the position of the radical in the reverse manner through ring opening: now $\pi$-bonds can be created by breaking $\sigma$-bonds. Another way to modify the location of the radical within a molecule is by intramolecular hydrogen abstraction or, more generally, through atom or group migration, a topic that will be discussed in Section 3.7.

If the potential reversibility of the various steps in a ring-closure/ring-opening process is taken into consideration, then the reaction manifold becomes quite complex. This manifold can be accessed from any point by generating a given radical from a suitable precursor. From a synthetic standpoint, it is of course essential to bias the various reaction parameters so as to channel the process towards one specific product and, as we have seen, many factors have to be taken into consideration. The size of the ring and the electronic and steric parameters of the substituents influence enormously the rates of ring closure or ring opening. The temperature (higher temperature favours open forms with greater entropy) and the concentration of the reactants also play a crucial role (e.g. $Bu_3SnH$: at high concentrations, quenching of the radicals might be too fast to allow equilibration between closed and open structures). The relative importance of these and other factors is best underscored by analysing specific examples.

## 3.2 Opening and closure of 3- and 4-membered rings

The cyclopropylmethyl radical undergoes a very fast ring opening to a butenyl radical (Scheme 3.1), the rate being of the order of $1.2 \times 10^8$ s$^{-1}$ at 37 °C and $4 \times 10^7$ s$^{-1}$ at 25 °C (Bowry et al. 1991; Horner et al. 1998; Furxhi et al. 1999). The rate is even higher when substituents are present on the ring due to a combination of electronic ('polar') and steric weakening of the cyclopropyl bond. Almost any substituent will weaken the bond: electron-withdrawing groups remove electrons from the bonding orbitals of the bond in question whereas electron-donating groups pump electrons into the anti-bonding orbitals. The presence of a substituent will thus affect the energy level of the HOMO and LUMO of the bond to be broken, resulting in a

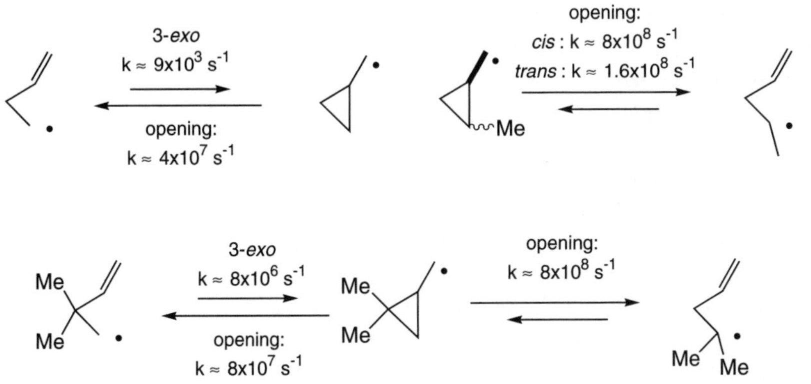

**Scheme 3.1** Absolute rate constants for the opening of cyclopropyl-carbinyl radicals

modification of the energy gap with the SOMO of the radical; as a consequence, the rate of opening will be altered in a way that reflects the nature of the dominant interaction. Steric repulsion causes a lengthening, and hence weakening, of the bond. This is seen by comparing the rate of opening of the *cis*- and *trans*- methylcyclopropylmethyl radicals. Steric repulsion is greater in the former resulting in a larger rate of ring opening ($8 \times 10^8 \, s^{-1}$ and $1.6 \times 10^8 \, s^{-1}$ respectively at 37 °C). The reverse process, that is 3-*exo* cyclization of the butenyl radical, is also fairly fast, with an estimated rate of $9 \times 10^3 \, s^{-1}$ at 25°C; this reaction is strongly affected by the Thorpe–Ingold accelerating effect: 2,2-dimethyl-3-butenyl radicals for example cyclize nearly 1000 times faster ($8 \times 10^6 \, s^{-1}$ at 25 °C) than the parent radical.

It is useful to mention in this context the opening of radicals related to the cyclopropyl system, namely the oxiranylmethyl and aziridinylmethyl radicals (Scheme 3.2). Unless the R-group is especially stabilizing (e.g. R = phenyl), then opening occurs by scission of the C–O (Barton *et al.* 1981) or C–N bonds (Dickinson and Murphy 1992; Schwan and Refvik 1993; De Kimpe *et al.* 1994; De Smaele *et al.* 1998), even though these are normally 'stronger' than the C–C bond in the 3-membered ring (Grossi and Strazzari 1997; Smith *et al.* 1998). The rate of opening of an oxiranyl-carbinyl radical has been estimated to be of the order of $10^{10} \, s^{-1}$ (Krishnamurthy and

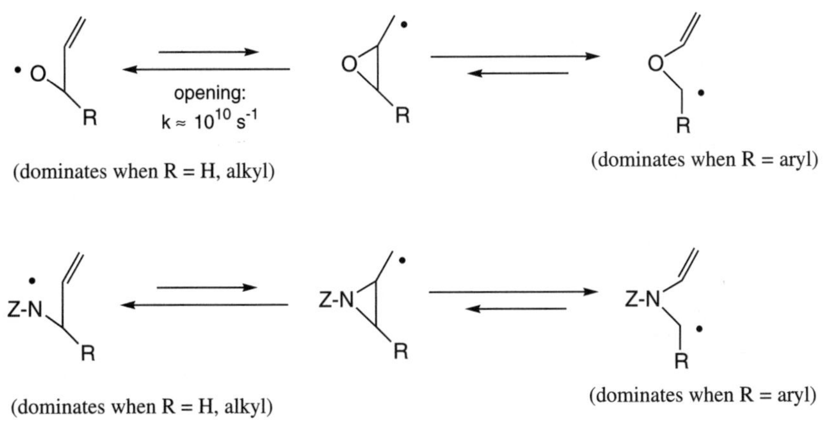

(dominates when R = H, alkyl)

(dominates when R = aryl)

**Scheme 3.2** Opening of oxiranylalkyl and aziridinylalkyl radicals

Rawal 1997). This extremely high rate appears to be due to the 'polar' factors briefly discussed above: a better matching between the SOMO of the radical and LUMO of the bond to be broken, even though the alkoxy radical produced is highly energetic. The cleavage of both the C–O and C–C bonds in oxiranylcarbinyl systems is reversible (Ziegler and Petersen 1995; Marples *et al.* 1997).

Reaction rates for the cyclobutylmethyl system are much lower (Scheme 3.3). For the parent structure, a rate of $4.3 \times 10^3$ s$^{-1}$ at 60 °C has been estimated for the ring opening; if a geminal dimethyl moiety is placed in a vicinal position, the rate increases 300-fold, whereas substitution by geminal diphenyl groups enhances the opening by 5 orders of magnitude (Choi *et al.* 2000). A geminal dimethyl group *distal* to the radical bearing carbon slows the ring opening by about 40% (a 'reverse' Thorpe–Ingold effect; Beckwith and Moad 1980). The rate constant for cyclization of the 4-pentenyl radical has not been measured but has been estimated to be of the order of $0.1$–$1$ s$^{-1}$ at 60 °C. This is very slow to be synthetically useful but, by associating a Thorpe–Ingold type acceleration and activation of the olefin by a cyano group, the rate of 4-*exo* cyclization can be increased to nearly $2 \times 10^4$ s$^{-1}$ (Park *et al.* 1990).

One final but fundamental point is that, like any elementary reaction, the opening of 3- and 4-membered rings is subject to the principle of maximum orbital overlap (stereoelectronic control). In a rigid system where rotation around the bonds is restricted, overlap of the orbital containing the single electron might be easier to accomplish with only one of the bonds of the ring: it is this particular bond that will break the fastest. This stereoelectronic factor, which usually overrides the other factors discussed above, can lead to a kinetic control of the ring cleavage with important consequences on the nature and stereochemistry of the product. This is illustrated by a classical study on cyclopropyl radicals derived from *i*-steroids summarized in Scheme 3.4 (Beckwith and Phillipou 1976). The $3\alpha$-5-cyclocholestan-6-yl radical opens to the secondary cholest-5-en-3-yl radical whereas the isomeric $3\beta$-5-cyclocholestan-6-yl radical rearranges stereospecifically to the less stable cyclopentylmethyl radical. In each case, efficient orbital overlap can occur with only one of the two bonds of the cyclopropane ring adjoining the radical. A similar kinetic control was applied on a cyclobutane system to introduce a methyl group with

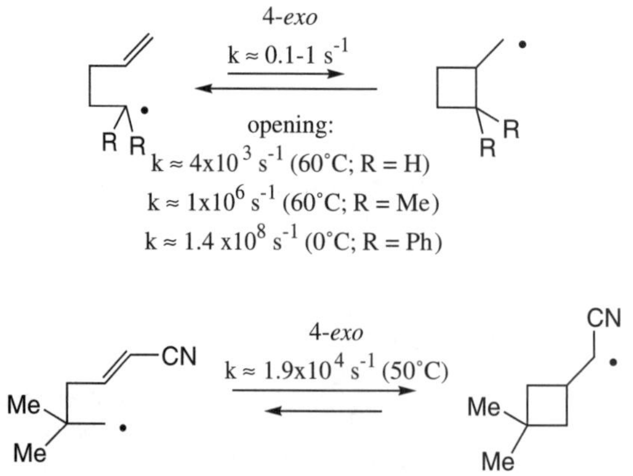

Scheme 3.3  Opening of cyclobutylcarbinyl radicals

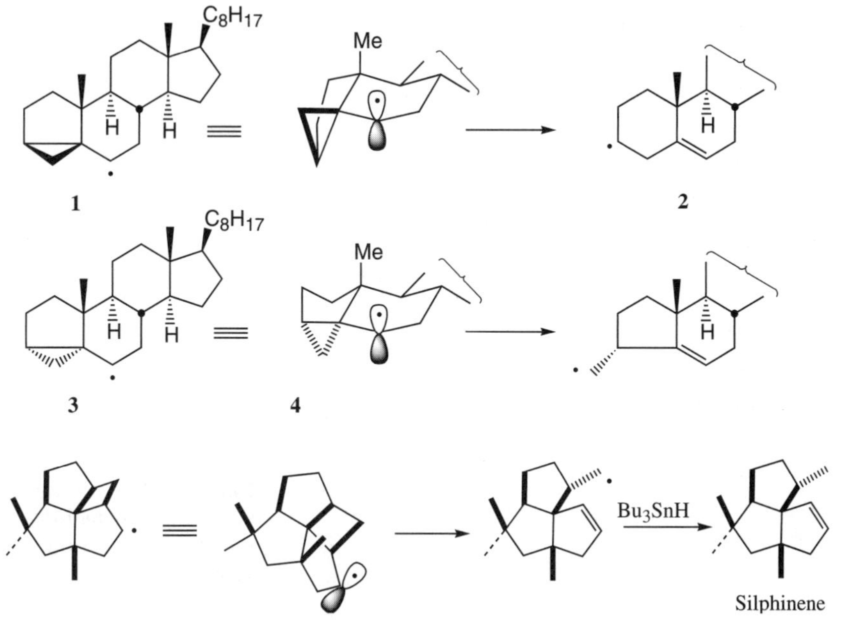

**Scheme 3.4** Stereo-electronic control in the rupture of small rings

a defined stereochemistry in an elegant approach to silphinene (Crimmins and Mascarella 1987). Again, because of better orbital overlap, the cyclobutane ring (generated by a photochemical [2 + 2] cycloaddition reaction) opens selectively to give the less stable primary radical, which then becomes a methyl group following hydrogen abstraction from tributylstannane.

The exceedingly fast ring opening of a cyclopropylcarbinyl radical has been extensively used as a mechanistic probe in chemical and enzymatic reactions, and as a radical 'clock' against which other fast radical processes are compared. As for organic synthesis, this step can be readily coupled to other radical transformations, often in spectacular cascades (Nonhebel 1993); some synthetic applications will be presented in the following section.

## 3.3   Opening of small rings: some synthetic applications

Cyclopropylcarbinols are easily available in a stereoselective manner from allylic alcohols by exploiting the directing effect of the hydroxy group in the Simmons–Smith reaction (Scheme 3.5). The hydroxy group can then be converted into a radical precursor via the corresponding selenide or by using a Barton–McCombie type precursor. This strategy allows the stereoselective introduction of an angular methyl group as illustrated by the first example in Scheme 3.5 (Clive and Daigneault 1991). In the two other examples, opening of the cyclopropane rings leads to the less stable primary radicals, again as a result of better orbital overlap. Capture by the internal alkyne finally gives two epimeric spiro compounds (Batey *et al.* 1992). Since the hydroxy group directs the formation of the cyclopropane ring,

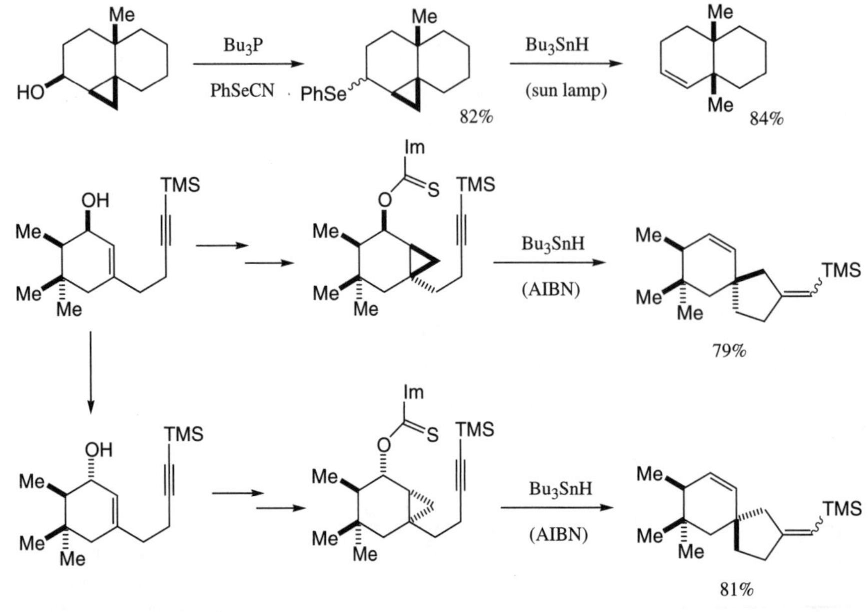

**Scheme 3.5** Some synthetic applications of the cyclopropyl ring opening

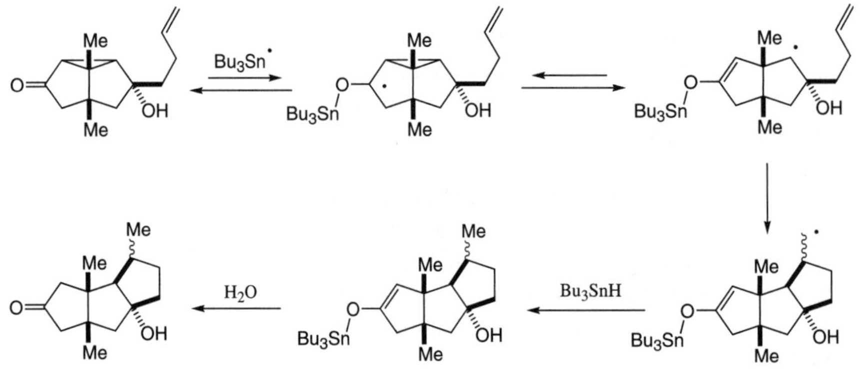

**Scheme 3.6** Tandem reactions of cyclopropyl ketones

its stereochemistry ultimately controls the relative configuration of the spiro junction.

Ketones and ordinary esters are not normally affected by stannyl radicals: the addition is relatively slow and highly reversible and, macroscopically, it seems as if nothing has happened. However, in the case of a cyclopropyl ketone, the unfavourable equilibrium leading to an *O*-stannyl ketyl radical is drained by the subsequent fast opening of the cyclopropane ring. One interesting application is the synthesis of a linear triquinane skeleton by combining the ring opening with a 5-*exo* cyclization as shown in Scheme 3.6 (Enholm and Jia 1996). Stereoelectronic constraints cause yet again the faster formation of the thermodynamically less stable secondary radical.

Because ring opening of the cyclobutane ring is generally sluggish, its use in synthesis has not been so extensive (Crimmins 1988). Problems are often encountered in obtaining complete $\beta$-scission, especially when the opening leads to a relatively high-energy primary radical. In the example related to the synthesis of silphinene displayed in Scheme 3.4, total opening of the cyclobutane ring could only be achieved by using a syringe pump addition of tributyltin hydride to the iodide precursor. When the reaction was carried out at 0.01 M concentration of stannane, a 1:1 mixture of the desired ring-opened structure (silphinene) and the product of direct reduction was obtained (Crimmins and Mascarella 1987).

A key step in the synthesis of (−)-dendrobine involves the rupture of the cyclobutane ring present in *trans*-verbenol (Cassayre and Zard 1999). The geminal dimethyl groups facilitate the fragmentation and no difficulty in terms of premature reduction was encountered in this case, in contrast to the synthesis of silphinene discussed above. To ease the removal of tin residues, the initially produced cyclic carbamate was not purified but directly cleaved with alkali to give the corresponding amino alcohol, which could then be extracted by an acid wash. The radical sequence triggered by the amidyl radical thus allows the control of three contiguous asymmetric centres present in dendrobine.

Cyclobutyliminyl radicals cleave more readily than simple cyclobutylmethyl radicals, the estimated rate for the parent cyclobutyliminyl being greater than $10^3$ s$^{-1}$ at −73 °C (Roberts and Winter 1979). In the example outlined in Scheme 3.8, reduction in the absence of an external trap leads almost exclusively to one of the two possible nitriles resulting from the ring-opening mode leading to the more stable secondary radical (Boivin *et al.* 1991). The intermediate carbon radical may be captured with methyl acrylate, which reacts as expected from the least hindered face, to give the *trans*-adduct. This compound is a potential precursor for *trans*-hydrindane structures through a base induced Dieckmann-type cyclization.

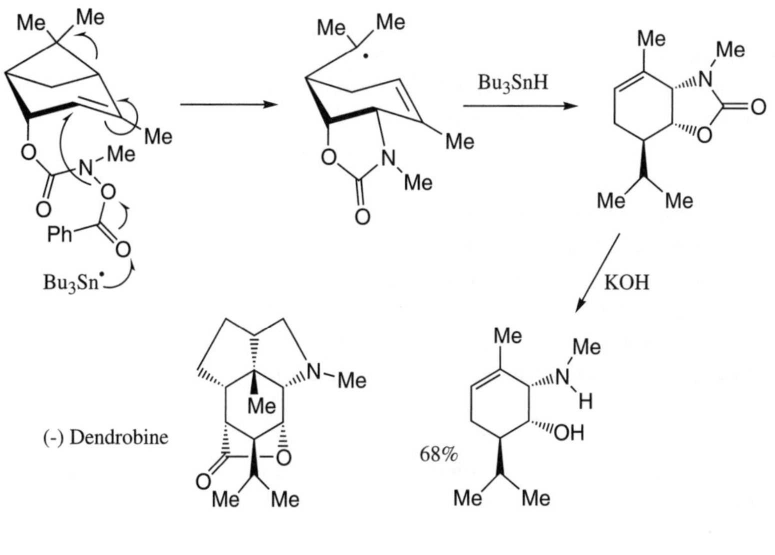

**Scheme 3.7** Cyclobutyl ring opening in the synthesis of (−)-dendrobine

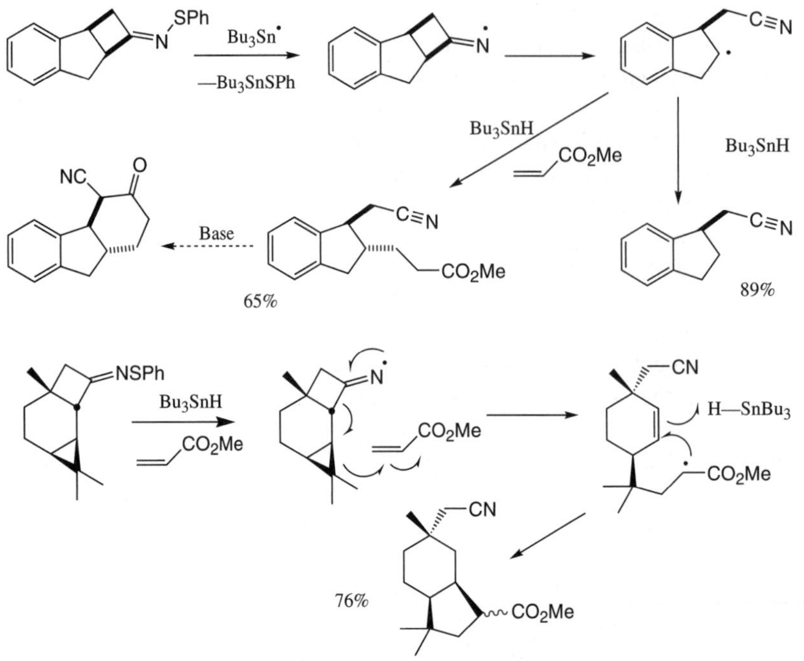

**Scheme 3.8**  Ring opening of cyclobutyliminyl radicals

A more complex sequence is observed with the sulfenimine derived from $\Delta^2$-(+)-carene displayed in the lower part of Scheme 3.8. A cyclobutane then a cyclopropane are successively broken and the tertiary homoallylic radical thus created undergoes an annelation with methyl acrylate to give a *cis*-hydrindane (Boivin *et al.* 1991). The stereochemistry of the product is dictated by that of the starting carene.

## 3.4   Formation of small rings

The fact that ring opening is faster than ring closure complicates the construction of 3- and 4-membered rings by radical processes. This difficulty can be overcome in two ways: (a) by favouring the small ring structure either through some special geometric constraints or by strongly stabilizing the closed radical, or (b) by combining the cyclization step with one or more fast and preferably irreversible radical or non-radical steps. These contrivances are illustrated by the examples set out in Schemes 3.9 and 3.10. In the first reaction in Scheme 3.9, it turns out that the special compact shape of the molecule makes the cyclopropylcarbinyl radical less strained than the corresponding homoallyl radical (Denis *et al.* 1993). The special stability of the allyl radical favours the cyclized form in the second example (Weng and Luh 1993). A similar cyclization sequence is involved in the last example, but now the cyclopropane formation is made essentially irreversible by the departure of the phenylthiyl radical (Denis and Gravel 1994). No hydrogen atom source is needed in this case and tributyltin radicals were generated using a combination of hexabutylditin and irradiation with a sun lamp.

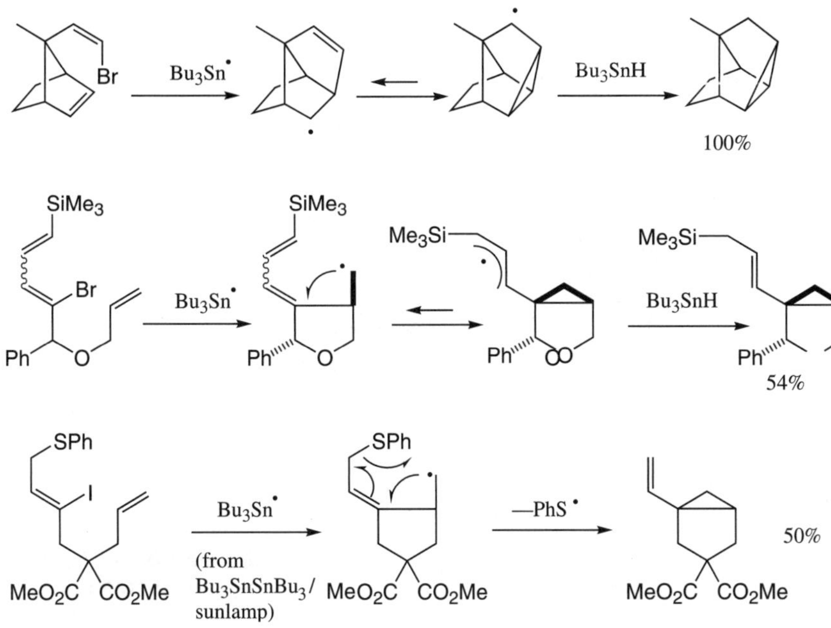

**Scheme 3.9** Formation of cyclopropyl rings

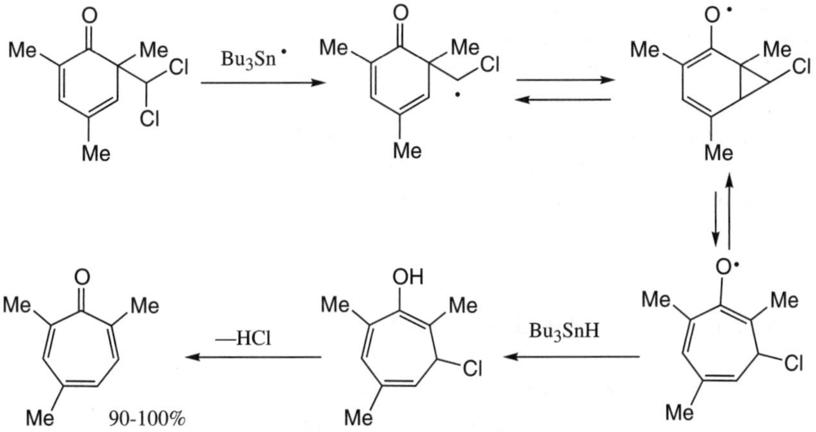

**Scheme 3.10** Synthesis of tropones

The sequence outlined in Scheme 3.10 represents an interesting and remarkably efficient route to tropones where a reversible electrocyclic ring-opening step within the radical sequence and elimination of HCl ultimately drive the process forward (Barbier *et al.* 1984). The cyclohexadienone precursors for this transformation are readily obtained by reaction of appropriately substituted phenols with dichlorocarbene.

The construction of 4-membered rings by a radical cyclization may be illustrated by the transformations in Scheme 3.11. The two β-lactam syntheses (Fremont *et al.* 1991; Ishibashi *et al.* 1993) require the presence of phenyl or phenylthio groups both to drive the equilibrium towards ring closure by stabilizing the desired radical and to

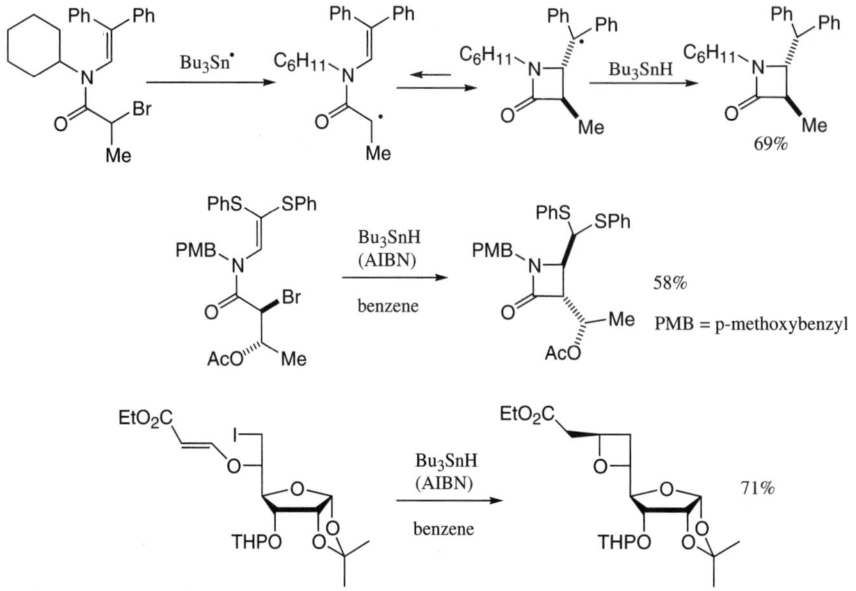

**Scheme 3.11**  Synthesis of 4-membered rings

accelerate an otherwise sluggish cyclization step which, it must be remembered, is always in competition with premature hydrogen abstraction by the stannane. The more stable *trans*-geometry observed in the product may be the sole result of kinetic control through a less crowded and therefore lower energy transition-state but could also be dictated by the reversibility of the cyclization step which introduces an element of thermodynamic control. In the last example, the ester group activates the olefin and stabilizes the cyclized radical, the cyclization being further improved by a compressing effect due to lone pair repulsion on the oxygen of the enol ether (Araki *et al.* 1989).

## 3.5   1,2-shift of unsaturated groupings and related rearrangements

1,2-Migration of hydrogen or simple alkyl groups does not normally occur with radicals (Brooks and Scott 1999). This is in stark contrast to cationic chemistry where Wagner–Meerwein rearrangements (also defined as suprafacial [1,2]-sigmatropic shifts) leading to the most stable cation are quite common. A formal 1,2 shift can however be accomplished with some unsaturated groupings through the transient formation and cleavage of a 3-membered ring. Thus, as shown in Scheme 3.12, it is possible to go indirectly to a more stable radical by a formal 1,2 shift of groups such as vinyl, aryl, cyano, aldehyde, etc. in a two-step process.

Such a cyclization/fragmentation was invoked by both Beckwith and O'Shea (1986) and Stork and Mook (1986) to explain why the cyclopentane/cyclohexane ratio in the cyclization of a vinyl radicals varies considerably with the concentration of tributyltin hydride. In the example displayed in Scheme 3.13, the ratio of **7:9** increased from 3 : 1 at 0.02 M to greater than 97 : 3 at 1.7 M (i.e. neat tributylstan-

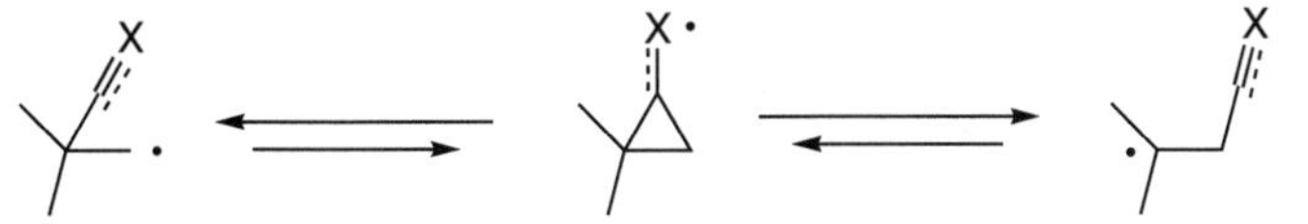

**Scheme 3.12** Stepwise 1,2-migration of unsaturated groups

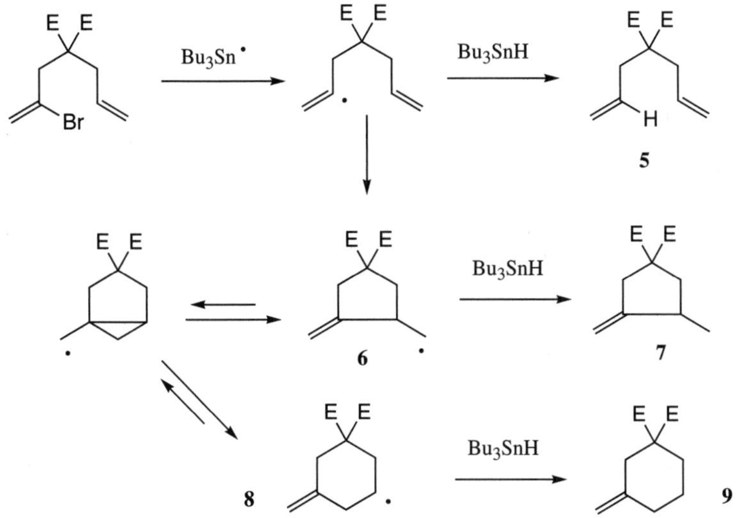

**Scheme 3.13** Stepwise formation of cyclohexane structures

nane; the 5-*exo* cyclization of vinyl radicals is so fast that it still dominates—to about 4:1—the direct reduction leading to **5** even under these highly concentrated conditions!). If both rings were derived via two competing simple unimolecular 5-*exo*- and 6-*endo*- closures, then the ratio should have been independent of the concentration in stannane.

The formation of the methylenecyclohexane isomer involves in fact a more complex route which gains in importance as the concentration of stannane is lowered: radical **6** acquires sufficient lifetime to undergo the ring closure–ring opening sequence. In other words, diminishing the concentration of stannane increases the thermodynamic control over the system by allowing the reversibility of the intermediate steps to manifest itself. The ring expansion studied by the groups of Beckwith and Stork is really a well-disguised 1,2-shift of a vinyl group from a less stable primary radical in **6** to a more stable secondary radical **8**. It represents indeed a prototype for a large number of transformations which can be rationalized by a similar mechanistic manifold. The analogous but generally slower 1,2-shift of aromatic rings (estimated rate $9 \times 10^2$ s$^{-1}$ at 25 °C; Franz *et al.* 1984), also termed a neophyl rearrangement, is sometimes observed and can be used to advantage in organic synthesis. Two examples are shown in Scheme 3.14. The first concerns an approach to mesembrine-type alkaloids which was in fact complicated by the unexpected migration of the aryl group (Ishibashi *et al.* 1991); the second is a deliberate ring expansion which exploits the large difference in stability

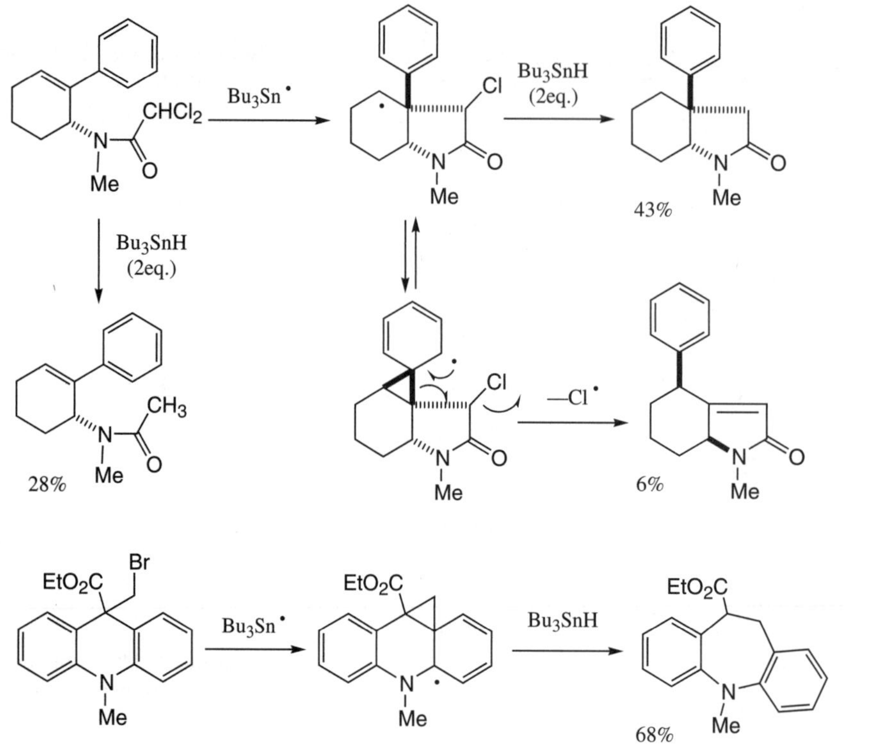

**Scheme 3.14**  Aryl group migration

between the initial and final carbon radicals involved in the neophilic rearrangement (Zheng and Dowd 1993). A naphthyl migrates faster than a phenyl and, as would be expected, electron-withdrawing groups on the aromatic ring increase the rate of rearrangement (Abeywickrema *et al.* 1987).

Ring enlargements occurring through the intermediacy of alkoxy radicals generated by 3-*exo* additions to carbonyl groups are of greater synthetic interest, and various clever combinations have been reported (Dowd and Zhang 1993). Two examples, pictured in Scheme 3.15, will suffice to demonstrate the power of this strategy. The first illustrates the passage from a 7- to an 8-membered ring in an approach to fusicoplagin D (Mehta *et al.* 1991) while the second is a tandem ring expansion/ 5-*exo-dig* ring closure (Boger and Mathvink 1990).

The one-carbon ring expansion can be combined with the stereoelectronically controlled ring opening of a cyclobutane, as illustrated by the key step in an elegant synthesis of lubiminol, a spirovetivane phytoalexin (Scheme 3.16; Crimmins *et al.* 1996). The rate constant for 3-*exo* addition to the carbonyl group in a simple cyclopentanone has been estimated to be of the order of $5.2 \times 10^4$ s$^{-1}$ at 25 °C (Chatgilialoglu *et al.* 1998).

Migration of a nitrogen atom to an adjacent carbon by a radical process is very rare. The biosynthesis of β-lysine from *L*-lysine, mediated in nature by lysine 2,3-aminomutase (Frey 1990; Wetmore *et al.* 2000), appears to involve the intermediacy of an aziridinecarbinyl radical as shown by the model transformation in Scheme 3.17,

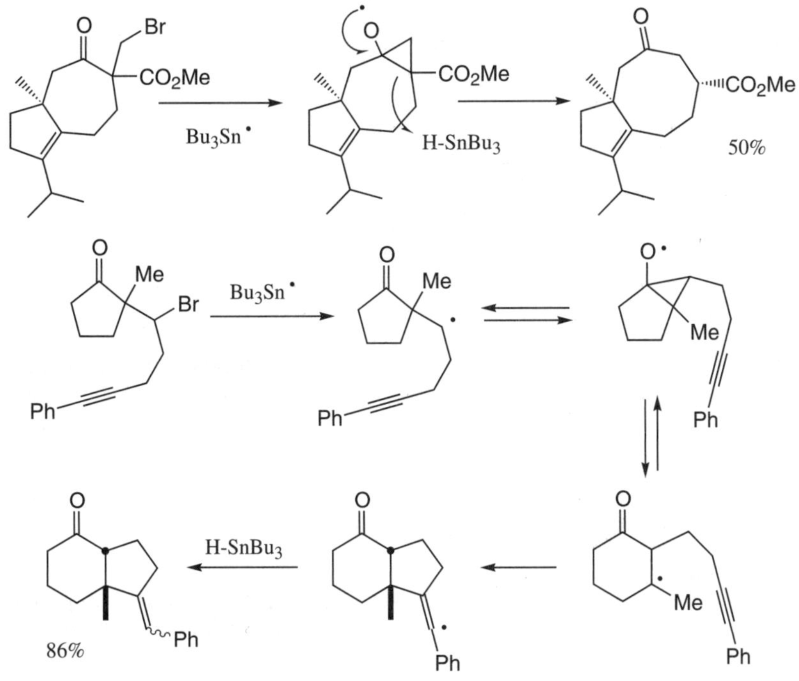

**Scheme 3.15** One-carbon ring expansion

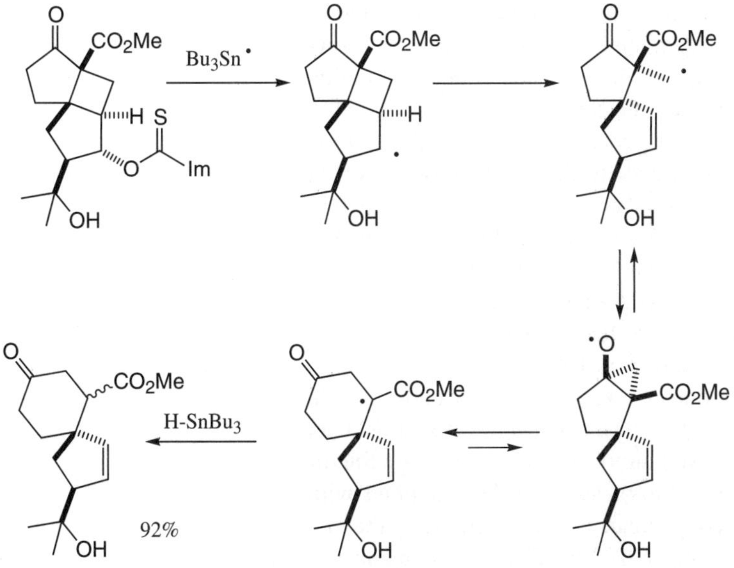

**Scheme 3.16** Sequential opening of a cyclobutane and ring expansion

where the benzaldehyde-derived Schiff base acts as the radical acceptor (Han and Frey 1990; Wetmore *et al.* 2000). In the actual enzymatic system, which also requires *S*-adenosyl methionine (SAM), the corresponding Schiff base intermediate is generated with pyridoxal, another essential component.

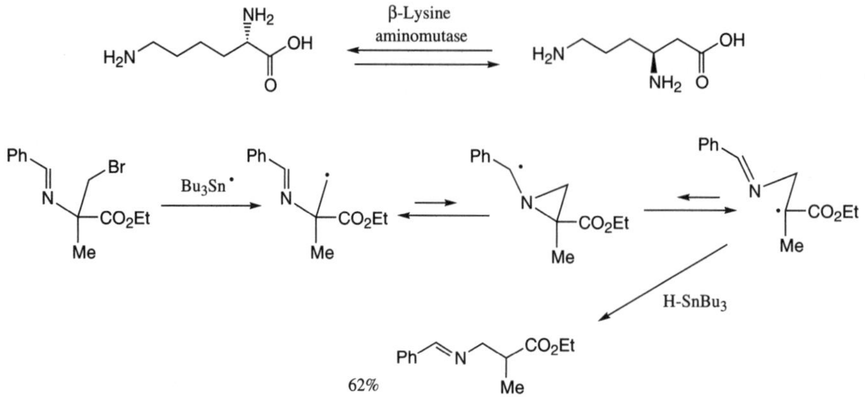

**Scheme 3.17** Stepwise 1,2-migration of nitrogen

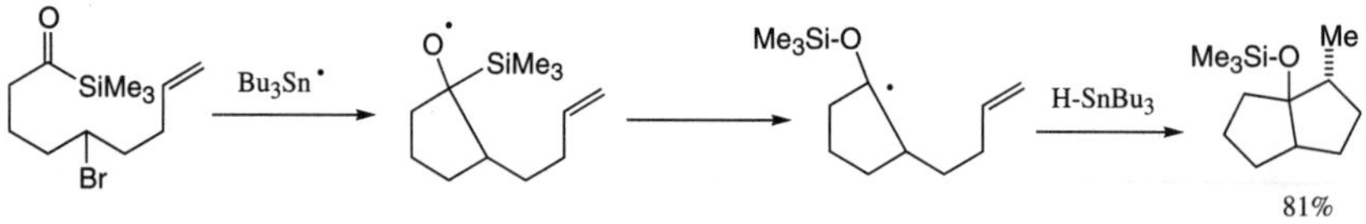

**Scheme 3.18** Migration of a trimethylsilyl group

The vast majority of 1,2-shifts in radical chemistry involve unsaturated groupings which can undergo a fast 3-*exo*-cyclization–fragmentation process. There are however some related migrations which do not hinge on the presence of an unsaturation such as the 1,2-shifting of silicon groups (Studer and Steen 1999; Sugimoto *et al.* 1999). The key step in the example displayed in Scheme 3.18 is a radical Brook rearrangement, involving perhaps a transient pentavalent silicon whereby the silicon migrates to the oxygen atom leaving behind an $\alpha$-siloxy carbon radical which is intercepted by the internal olefin (Curran *et al.* 1993; Tsai *et al.* 1993). It is interesting to note that the carbonyl carbon in the acyl silane group has acted successively as a radical acceptor and then as a radical donor. Migration of a silicon group from one carbon atom to another adjacent carbon has also been reported but the mechanism has not been unambiguously clarified (Shuto *et al.* 1997).

The 1,2-shift of acyloxy and phosphate groups are known to occur under certain relatively rare circumstances (Beckwith *et al.* 1997). The unexpected migration of an acetate, albeit in modest yield, was first observed almost simultaneously by Surzur and Tessier (1970) and by Tanner and Law (1969), and was later applied with remarkable efficiency in the synthesis of 2-deoxy-sugars as demonstrated by the first example in Scheme 3.19 (Giese and Gröninger 1990). The transformation is intramolecular and the acyloxy group remains on the same side of the molecule as in a suprafacial rearrangement. Five- (and even three-) membered ring transition-states with some polar character have been proposed but the question has not been completely and unambiguously settled so far (Beckwith *et al.* 1997). The second example is a more recent extension to nucleosides where the pivalate group shifts away from the anomeric centre (Itoh *et al.* 1995).

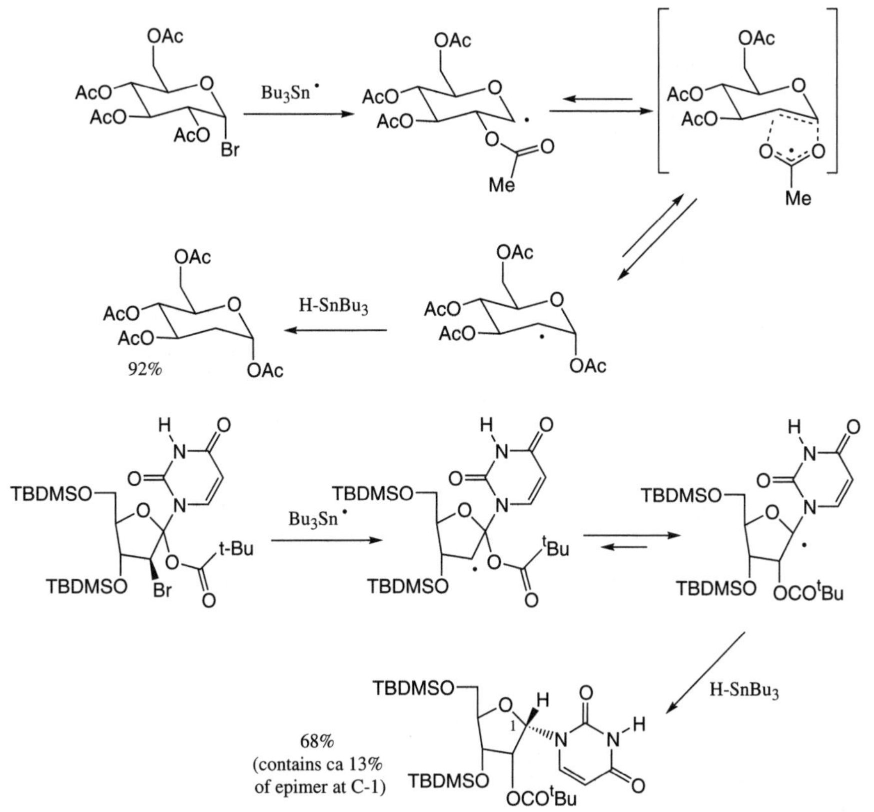

**Scheme 3.19**
Migration of ester groups

## 3.6   Ring opening of 5-membered and larger rings

Cleavage of 5-membered and larger rings with little inherent strain is harder to accomplish. An increase in the temperature favours in principle the open form with higher entropy content; however, with most compounds, the temperature needed to induce ring opening is too high to be practical. Placing substituents which strongly stabilize the opened radical sometimes results in a reversible cyclization under milder conditions. The same can be achieved by having a relatively weak linkage (e.g. a C–S bond) in the ring which undergoes β-scission easily. Introducing reversibility leads to a thermodynamic control of the cyclization process: a kinetically slower 6-*endo-trig* closure can thus prevail over a faster 5-*exo-trig* if the former leads to a more stable radical. Some examples of cyclizations under thermodynamic control will be discussed in Chapter 6 but the principle of incorporating a weak bond in a 5-membered ring to promote its rupture has been exploited in what may be considered as the radical analogue of the Smiles rearrangement (Scheme 3.20). It involves *ipso*-substitution on an aromatic ring and subsequent extrusion of sulfur dioxide (Motherwell and Pennell 1991); the other competing reaction is a direct ring closure which of course cannot be avoided by slow addition of the stannane. The rearomatization embodies sufficient driving force to induce the

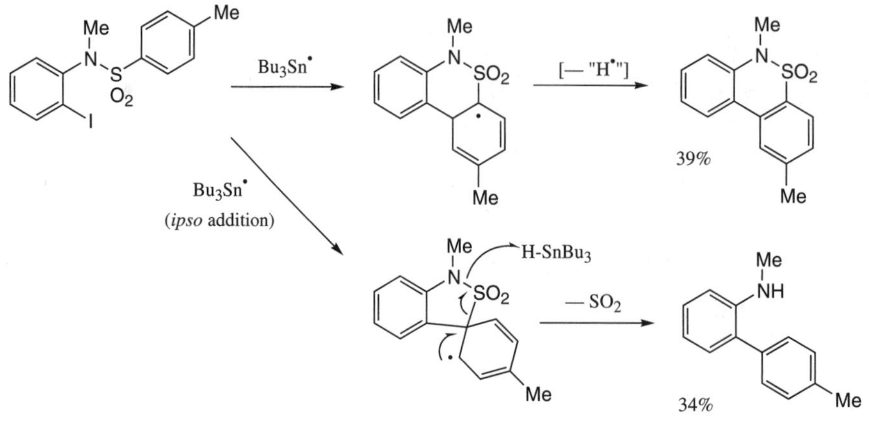

**Scheme 3.20** *Ipso* substitution on arylsulfonamides

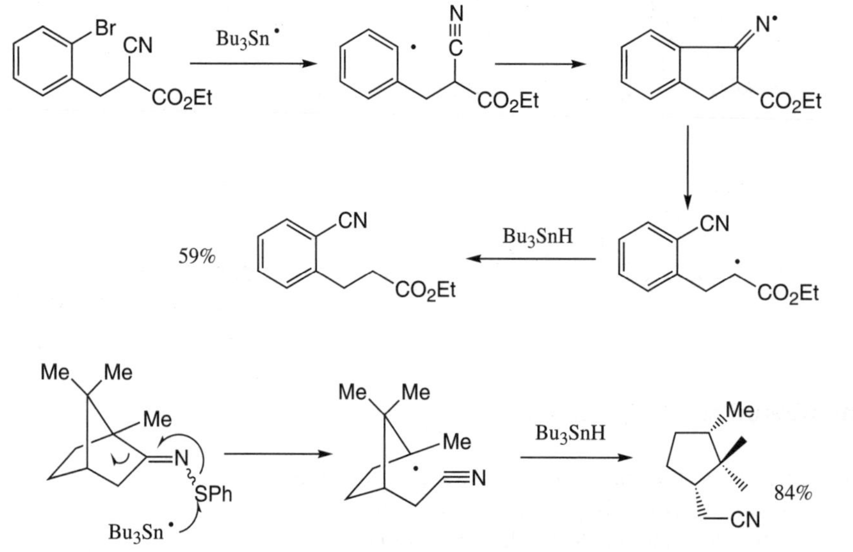

**Scheme 3.21** Opening of cyclopentyliminyl radicals

rupture of strong C–C, C–Si, C–N, or C–O bonds which would not normally break under other circumstances (Lee *et al.* 1995; Baguley and Walton 1998).

Because of geometrical constraints, some 5-membered rings contain enough strain to cause them to rupture. Cyclopentyliminyl radicals fused to aromatic rings are of this type and, since they can be generated by addition of an aromatic radical to a nitrile group, then an overall 1,5-transfer of the nitrile takes place as shown by the first example in Scheme 3.21 (Beckwith *et al.* 1988). Iminyls with a bicyclo[2.21]heptane structure also undergo ring opening as illustrated by the transformation camphor-derived sulfenimine (Boivin *et al.* 1994). Hydrogen delivery occurs from the least hindered side to furnish the *cis*-substituted cyclopentane.

One interesting synthetic application of the opening of strained cyclopentyliminyl radicals is the inversion of the quaternary C-13 centre in 17-ketosteroids. The

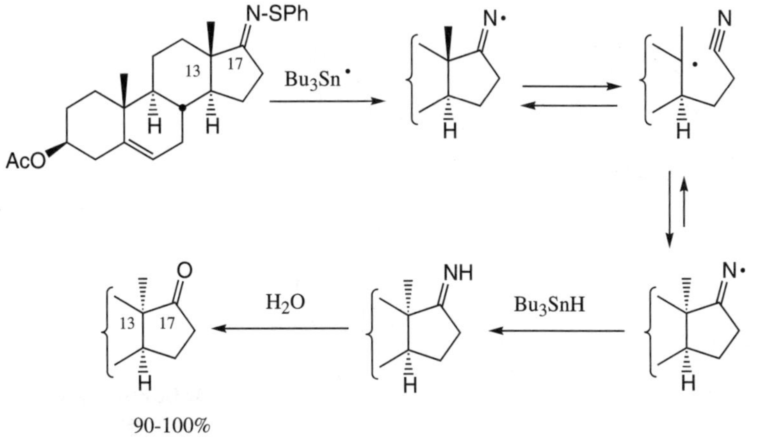

**Scheme 3.22** Synthesis of 13-episteroids

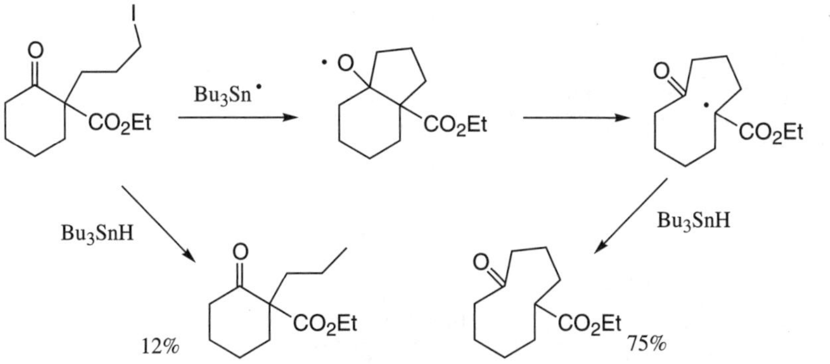

90-100%

**Scheme 3.23** Fragmentation of alkoxy radicals

C- and D-rings in the natural series have the thermodynamically less stable *trans*-hydrindane structure. If, as shown in Scheme 3.22, a reversible rupture of the C-13–C-17 bond can be implemented, then inversion to give the more stable, but unnatural, 13α-epimer will occur (Boivin *et al.* 1994). The same transformation can also be accomplished by a Norrish type I photochemical cleavage–ring closure directly from the 17-ketone.

The last, but most general, parameter that can be used to induce ring opening is to proceed via a high-energy cycloalkoxy radical. The instability of the oxygen centred radical and the strength of the carbonyl group represent a powerful driving force capable of tearing apart rings of any size, whether strained or not. This approach has a tremendous synthetic potential. Thus, by analogy with the one-carbon enlargement illustrated in Scheme 3.15, three- and four-carbon ring expansions are conceivable, allowing an expedient access to medium and large rings of various sizes and substitution patterns. The example shown in Scheme 3.23 has been selected from the extensive and pioneering study of Dowd and Zhang (1993). The intermediate alkoxy radical rapidly fragments to the more stable carbon radical which is then quenched by the stannane. Some premature reduction by the stannane was also observed.

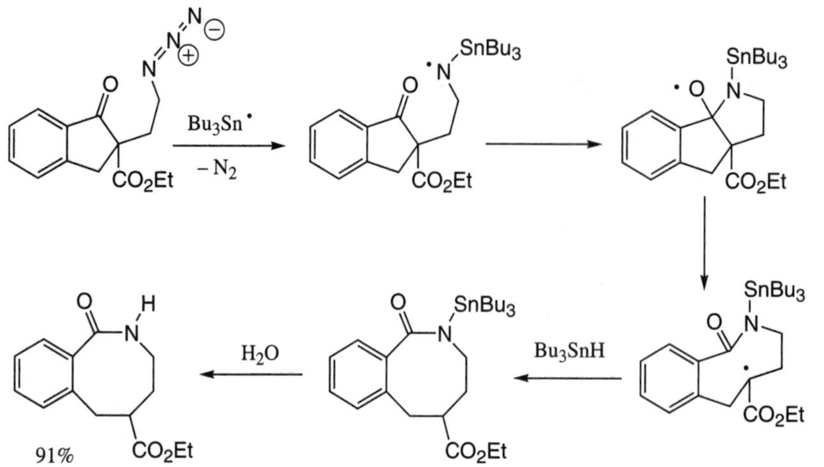

**Scheme 3.24**
Synthesis of medium-sized lactams

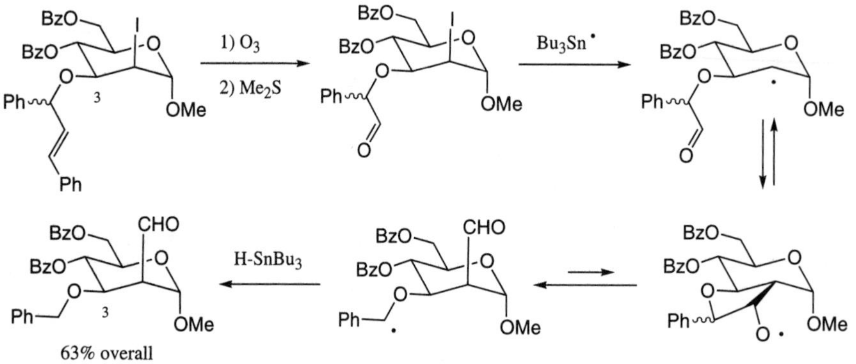

**Scheme 3.25** Migration of an aldehyde group

This ring-expansion sequence can be incorporated in complicated radical cascades. But an interesting variation concerns the synthesis of macrolactams by a variation of the above approach, whereby an aminyl radical is used to attack the carbonyl group. The transformation portrayed in Scheme 3.24 (Kim *et al.* 1993) shows the synthesis of an 8-membered lactam and highlights a simple and efficient way to generate aminyl radicals from azides (for a possible mechanism for the reaction of stannyl radicals with azides, see Scheme 3.32).

Another ingenious application of the addition fragmentation to a carbonyl group is outlined in Scheme 3.25 (Jarreton *et al.* 1996). The aldehyde precursor was obtained by ozonolysis of the styryl olefin and used directly without purification in the radical step. The fast fragmentation to the stabilized benzylic radical gives an aldehyde with a well-defined stereochemistry, and the alcohol on C-3 remains protected as an easily cleavable benzyl ether. Intramolecular radical additions to aldehydes within a sugar framework have been extensively studied by Fraser-Reid and his group (Walton and Fraser-Reid 1991).

## 3.7 Radical translocation by intramolecular hydrogen transfer

The other major isomerization process leading to a translocation of the radical centre is the intramolecular transfer of a hydrogen atom. This reaction can be a serious source of trouble, especially when the desired step is comparatively slow (cf. Scheme 2.26). Unless deuterium labelling is used, the unruly behaviour of the radical may go unnoticed because the stannane reduces the various rearranged radicals to the same end product. In some cases, the mischief is revealed by an unexpected modification in the stereochemistry at a remote position in the molecule. One such instance is depicted in Scheme 3.26 where partial epimerization at C-5 of a carbo-hydrate was observed following cyclization of the anomeric radical onto an ally-loxy group on C-2 (De Mesmaeker *et al.* 1992). Initially, the rapid 5-*exo* cyclization leads to two epimeric radicals **10** and **11** in an approximate ratio of 85 : 15 in favour of the *endo* isomer **10.** Under high dilution conditions (0.001 M), this radical has enough lifetime to undergo partial 1,5-hydrogen atom migration to give radical **12** which reacts from the least hindered *exo* face to furnish c-5 epimer **13**. At the higher concentration in stannane (0.2 M), this secondary path is suppressed. Note that the ratio of (**13**+**14**): **15** does not vary with the concentration of the stannane and simply reflects the initial ratio of **10 : 11.** This elegant experiment embodies much of what should be understood about radical processes and how a feel for the relative rates can guide the organic chemist to devise a simple solution to what otherwise would have been a vexing problem.

But far from being just a nuisance, intramolecular hydrogen transfer can be deliberately included in synthetic planning with spectacular effect. Radicals, carbenes, and some special transition metal complexes or enzymatic systems containing transition metals (e.g. cytochrome P450) are perhaps the only 'reagents' reactive enough to allow the modification of an unactivated alkane portion of a molecule. The functionalization of hydrocarbons certainly remains one of the most difficult problems in organic chemistry and the use of a well-designed radical-based scheme is one way to tackle it. Several examples will appear throughout this book, but it is useful at this stage to outline some of the factors that must be taken into account when considering this mode of radical translocation.

(a): [Bu₃SnH] = 0.001M
(b): [Bu₃SnH] = 0.2M
(combined yield under both dilutions: 86%)

**10**

**11**

**13**
(a): 43%; (b): 0%

**12**

**14**
(a): 42%; (b): 86%

**15**
(a): 15%; (b): 14%

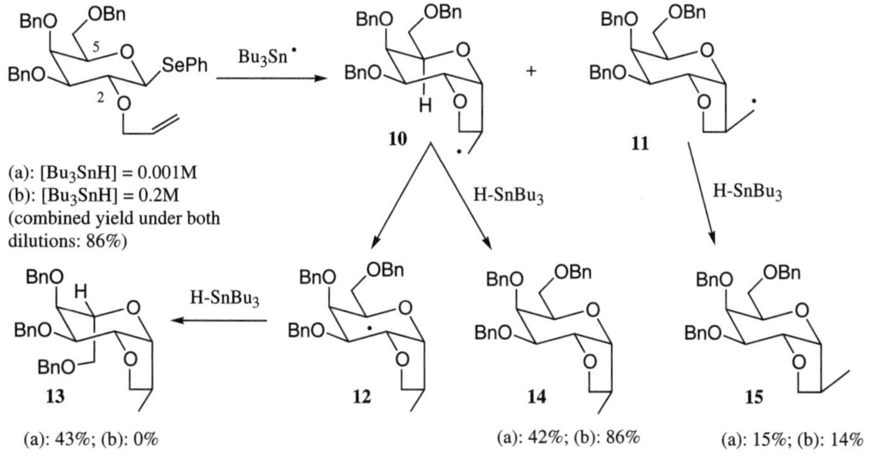

**Scheme 3.26**
Intramolecular hydrogen abstraction and remote inversion of configuration

Angle of approach and proximity are prime factors. *Ab initio* calculations on the model reaction of hydroxyl radicals with methane indicate a nearly linear transition state but a distortion of up to 20–30° has only a small effect on the activation energy (Dorigo and Houk 1988). For intramolecular abstractions of hydrogen, the strain incurred in order to attain a favourable disposition and the necessary decrease in entropy content translate into a strong preference for (1,5) shifts, as shown by comparing the estimated activation energy of various (1, *n*) H-atom migrations given in Fig. 3.1 (Viskolcz *et al.* 1996). Thus, a 6-membered transition-state is the most favourable but (1,4) and (1,6), and even (1,7) and higher shifts are not completely out of reach. In the case of larger rings, although the linear arrangement can be readily attained, conformational strain and an unfavourable entropy term slow down the reaction (Brunton *et al.* 1976; Huang and Dannenberg 1991). It must be remembered that theoretical calculations are usually performed for simple flexible models; the presence of rigidifying motifs and other geometric and steric constraints in the structure (bulky groups, unsaturations aromatic or alicyclic rings, etc.) may especially favour an otherwise rare transposition.

The movement of a hydrogen from a saturated carbon to another saturated carbon of a similar type is essentially thermoneutral and hence reversible in principle. The equilibrium may be shifted in a given direction by adjoining a fast irreversible step such as a cyclization, an elimination, or the fragmentation of a small ring. In the transformation outlined in Scheme 3.27, the structure and rigidity of the pinene

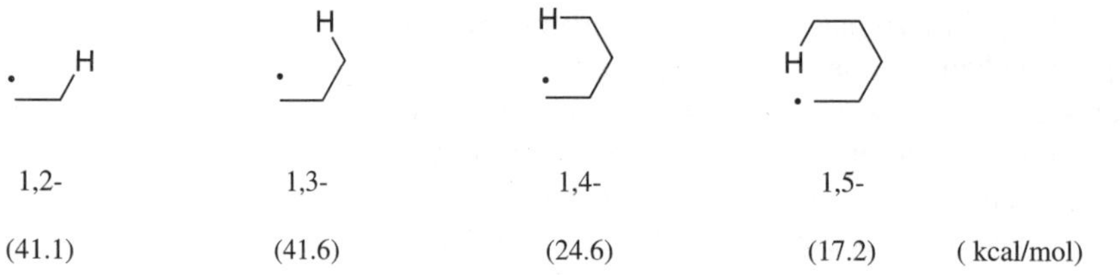

| 1,2- | 1,3- | 1,4- | 1,5- | |
| (41.1) | (41.6) | (24.6) | (17.2) | ( kcal/mol) |

**Fig. 3.1** Activation energy for intramolecular hydrogen transfer

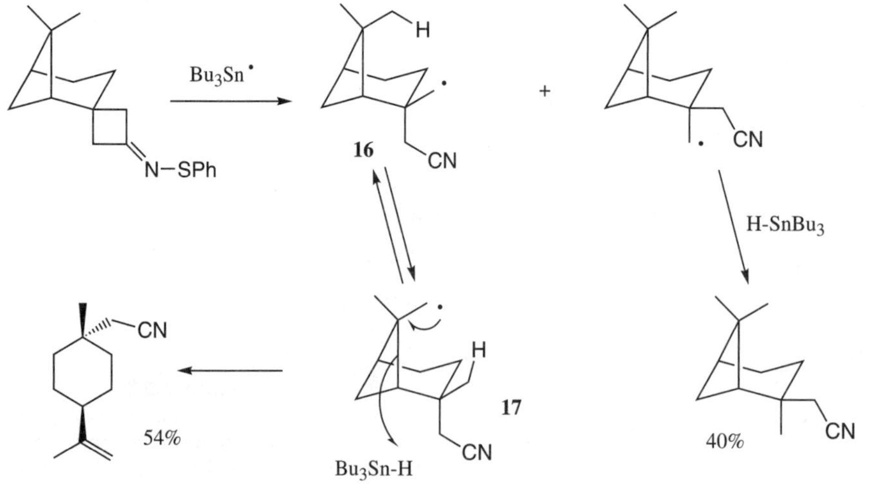

**Scheme 3.27** Intramolecular hydrogen abstraction and regioselective ring opening

skeleton forces into close proximity the primary radical in **16** and one of the geminal methyl groups, allowing a fast—but thermoneutral—exchange of a hydrogen
atom (Boivin *et al.* 1991). This equilibrium between the two primary radicals **16** and
**17**, which are of similar stability, is drained by the irreversible rupture of the weak-
est (because of steric repulsion) of the two adjacent cyclobutane bonds.

In most applications, however, the migration of the hydrogen atom is made
exothermic, and therefore essentially irreversible, by generating at one point a high-
energy primary, vinylic, aromatic, or an oxygen- or nitrogen-centred radical and terminating with a more stable entity. Two examples of 1,5-intramolecular hydrogen
abstraction by a vinylic radical are displayed in Scheme 3.28. The ensuing stabilized
tertiary radical is in both cases captured by an activated internal olefin to give on one
hand a cyclopentane derivative (Curran *et al.* 1988; Curran and Shen 1993) and, on
the other, a spiroketal structure (Brown *et al.* 1993; Simpkins and Clinch 1993).

A conceptually similar transformation can be envisaged starting with an aromatic
radical. In the sequence shown in Scheme 3.29, the hydrogen abstraction is followed
by ring closure onto the aromatic ring to provide an oxindole (Beckwith and Storey
1995). Like many of the previous additions to aromatic rings, the exact way rearomatization occurs is not clear. Nevertheless, high temperature favours the cyclization
step and the reaction is best conducted in refluxing *t*-butyl benzene with di-*t*-butyl
peroxide (TBPO) as the initiator. TBPO has a long enough half-life (0.5–1 h) at the

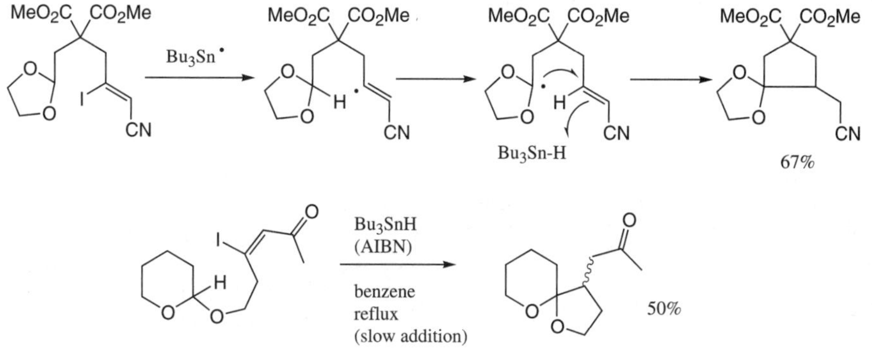

**Scheme 3.28**
Sequential intramolecular
hydrogen abstraction
and cyclization

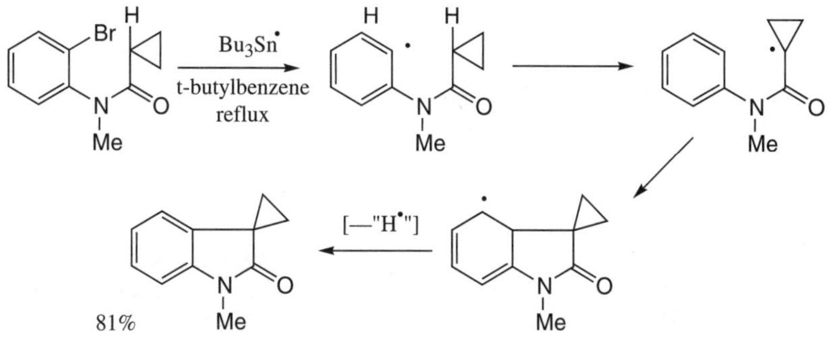

**Scheme 3.29** Oxindole
synthesis

boiling point of *t*-butyl benzene (160 °C) to behave as an efficient initiator under these conditions.

Radical translocation involving aromatic radicals are becoming more and more common (Snieckus *et al.* 1990; Denenmark *et al.* 1992; Renaud and Gerster 1995). Their applications can be quite unusual: in the transformation shown in Scheme 3.30, the hydrogen abstraction is followed by a fragmentation to give a stabilized trityl radical (Curran and Yu 1992); thus, overall, an oxidation is performed on a protected alcohol using a reducing agent! Because the bromotrityl group can be introduced selectively on the less hindered primary alcohol, the aldehyde is obtained despite the presence of the normally easier to oxidize secondary allylic hydroxy group.

The highly reactive alkoxy radicals are even more powerful in their hydrogen abstraction ability. As we shall see in Chapter 7, dramatic use can be made of this property to functionalize angular methyl groups in steroids and related structures. One example involving an alkoxy radical intermediate produced through the opening of an oxirane ring is outlined in Scheme 3.31 (Rawal *et al.* 1990).

Electrophilic nitrogen centered radicals such as aminium (e.g. protonated aminyl radicals) or amidyl type radicals are also efficient hydrogen abstractors, but these species have more often been generated in the past by methods not based on stannanes. This is part due, at least in the case of aminium radicals, to the incompatibility of tin hydrides with an acidic medium. Hydrogen abstractions by neutral aminyl radicals are rare but an example where a relatively labile C–H bond was broken has recently been reported and is displayed in Scheme 3.32 (Kim *et al.* 1997). Deuterium labelling was employed to establish the reaction course. The stannylamine end

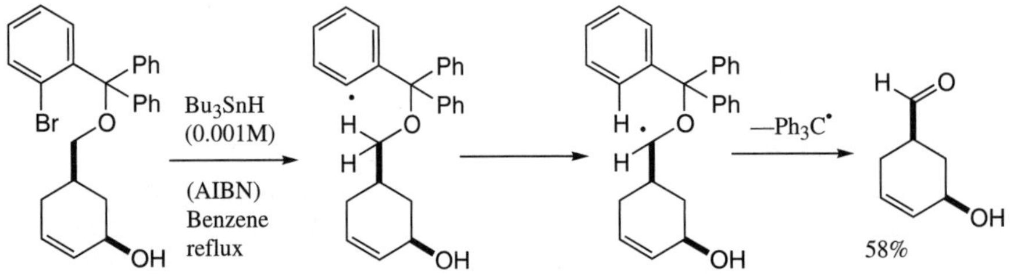

**Scheme 3.30** Formation of an aldehyde by hydrogen abstraction and fragmentation

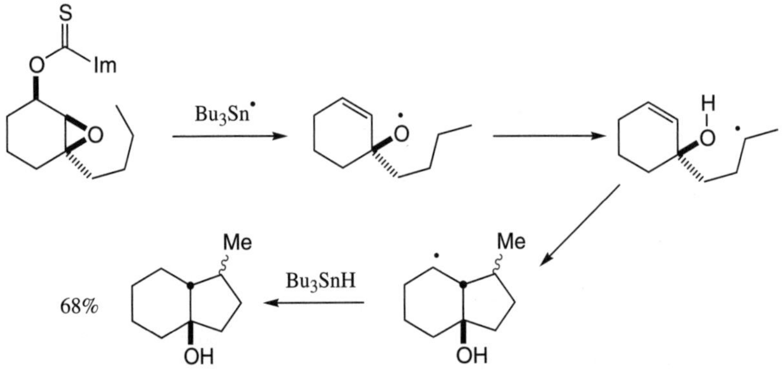

**Scheme 3.31** Hydrogen abstraction by an alkoxy radical

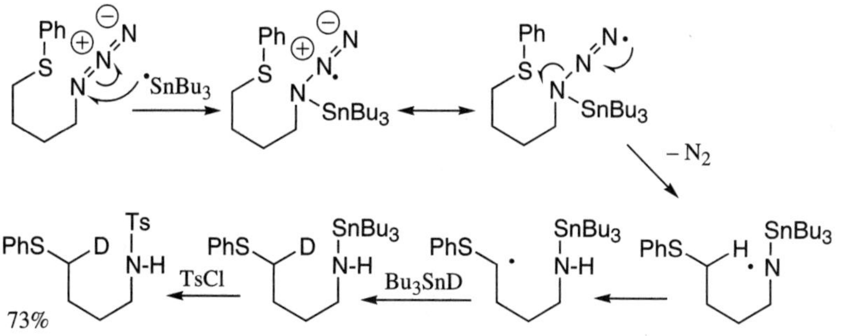

**Scheme 3.32**
Hydrogen abstraction by
an aminyl radical

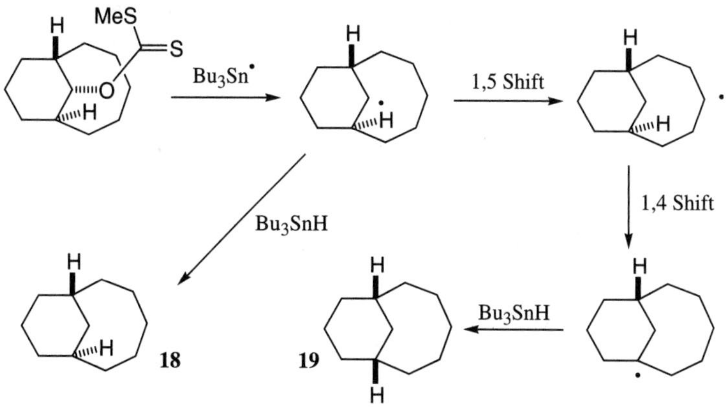

**Scheme 3.33**  Example
of a 1,4-hydrogen
migration

product is very easily hydrolysed and is best characterized in this case as the toluene-sulfonamide derivative of the parent amine.

Radical translocations by other than 1,5 migrations are far less common. 1,4 Hydrogen shifts are especially rare but can occur when steric compression for example forces the C–H bond into close proximity to the radical centre, and with the correct alignment to permit hydrogen exchange (Crich *et al.* 1996; Boiteau *et al.* 1998). Isomerizations involving 1,6 and, but to a lesser extent, higher order hydrogen atom transfers are encountered more frequently. In the example displayed in Scheme 3.33, the radical generated via the Barton–McCombie reaction undergoes first a 1,5- then a 1,4-shift to give ultimately the alkane with a bicyclic structure isomeric to that of the starting material (Winkler *et al.* 1989). The ratio of isomers **18** and **19** depends of course on the concentration of the stannane, a higher concentration favouring the unrearranged alkane **18**. This is another instance where a change in the stereochemistry signals the occurrence of a radical translocation.

In another example (Scheme 3.34), an unexpected participation of the benzyl group was observed in an interesting approach to fredericamycin A (Bennett and Clive 1986). The desired 5-*exo* cyclization was followed by a 1,6-migration of hydrogen then a 6-*endo* ring closure to give finally a pentacyclic product.

The 1,6-hydrogen shifts in both of the above examples are favoured by the rigidity of the structures involved which reduces the negative change in entropy needed

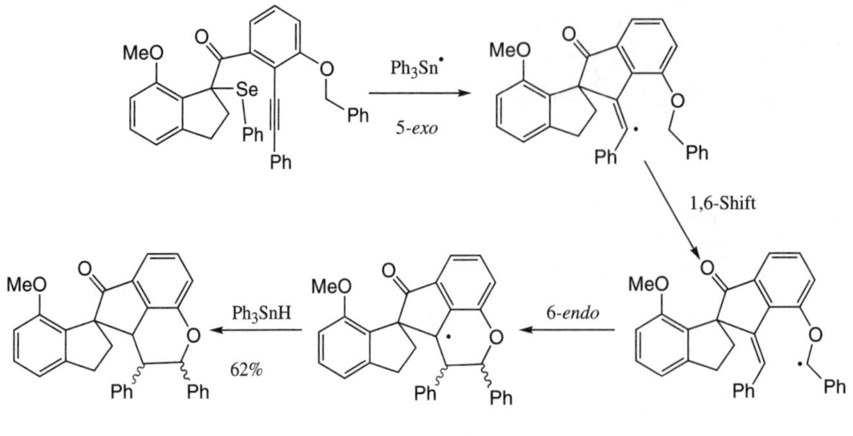

**Scheme 3.34** Example of a 1,6-hydrogen migration

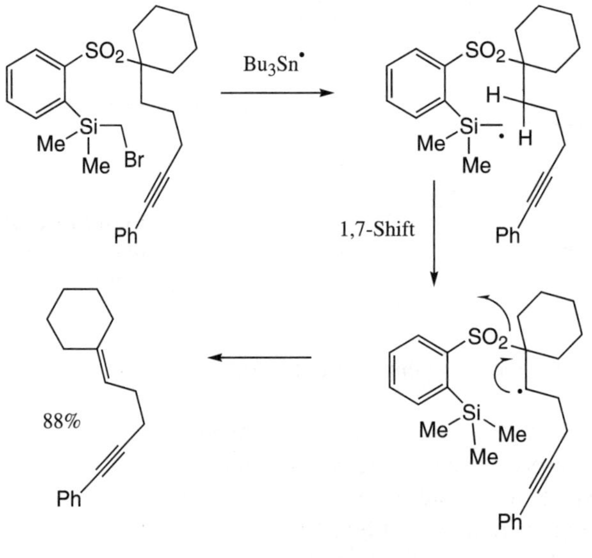

**Scheme 3.35** Example of a 1,7-hydrogen migration

to reach the more organized transition state. This is perhaps even more true for the amazingly efficient 1,7-translocation depicted in Scheme 3.35, where the number of rotational degrees of freedom is curtailed by the presence of the aromatic ring and the bulkiness of the substituents around the silicon and sulfone groups (Van Dort and Fuchs 1997).

## 3.8   Stannyl radical generation by fragmentation

The $\beta$-elimination of the sulfonyl radical portrayed in Scheme 3.35 is a typical fragmentation process which occurs every time a radical is located next to a relatively weak bond. Carbon–tin bonds also readily undergo $\beta$-scission to give stannyl radicals and these may be incorporated in chain reactions similar to those we have so far seen

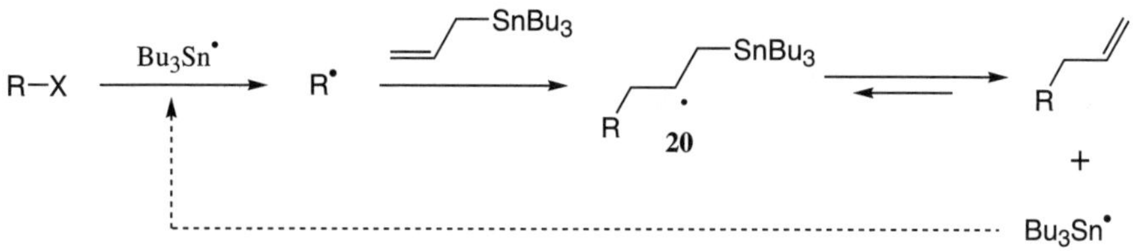

**Scheme 3.36**
β-Elimination of stannyl radicals

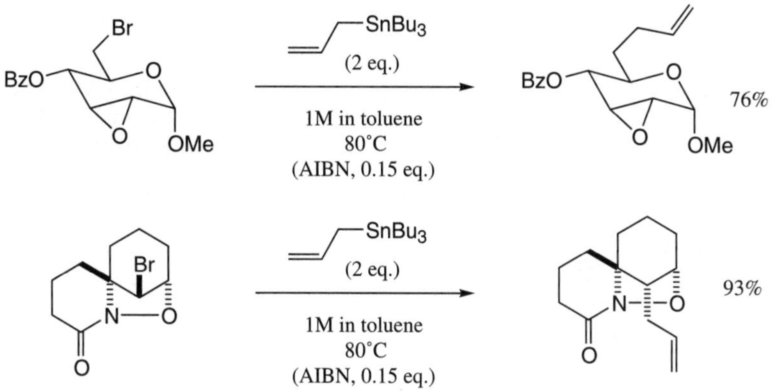

**Scheme 3.37**　Radical allylations with allyl tri-*n*-butyltin

with triorganotin hydrides. It is thus possible to envisage a chain reaction whereby the stannyl radical is not generated from the corresponding hydride. Addition of a radical to an allyltin is perhaps the most common way of creating a radical centre vicinal to the carbon–tin bond. Allyl triorganotin derivatives are indeed excellent allylating agents, operating through the general mechanism depicted in Scheme 3.36.

X represents, as usual, any functionality capable of reacting with tin radicals. Overall, the sequence may be viewed as a transfer of an allyl group via an $S_H2'$ process. Group transfers in general will be discussed in more detail in Chapter 6 but the key feature of this system is that it is possible to perform an *intermolecular* addition of a carbon radical on a relatively unactivated olefin. The estimated rate for the addition of an alkyl radical to allyl tributyltin is $1-6 \times 10^4 \text{ M}^{-1} \text{ s}^{-1}$ at 50–80 °C, a value somewhat greater than for the addition to a simple terminal olefin but much smaller than rates of additions to electrophilic alkenes (Curran *et al.* 1990). The slight activation by the triorganotin group ensures that by using a modest excess of the reagent (e.g. 2 equivalents), the radicals will mostly attack the allyltin rather than the olefin in the product. Moreover, the rapid unimolecular fragmentation of the intermediate adduct **20** means that this species is too short lived to participate in undesirable side reactions. Because of the absence of a fast hydrogen atom transfer as in the tin hydride system, it is not necessary to add slowly the tin reagent: on the contrary, operating under high concentration becomes an advantage because the slow step is the bimolecular addition to the allylstannane.

Radical allylations with allylstannanes were discovered simultaneously by Grignon and Pereyre (1973) and by Kosugi and his group (1973) and later improved and extended to complex molecules by Keck and Yates (1982). The two examples shown in Scheme 3.37 (Keck and Yates 1982) give an idea of the power of this

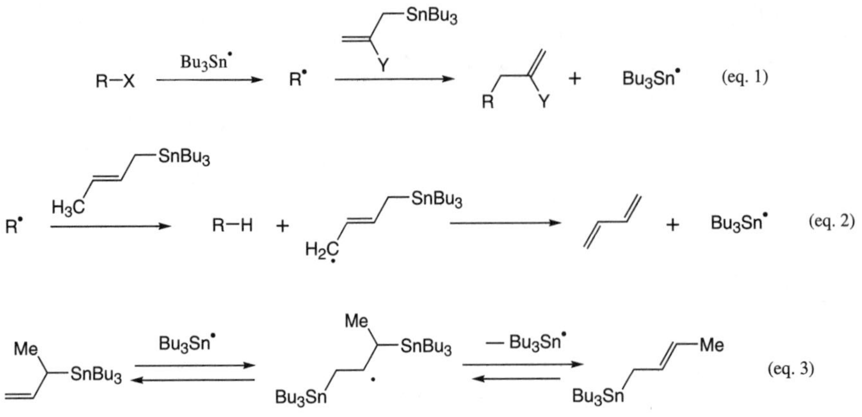

Scheme 3.38 Effect of substituents on the allylating agent

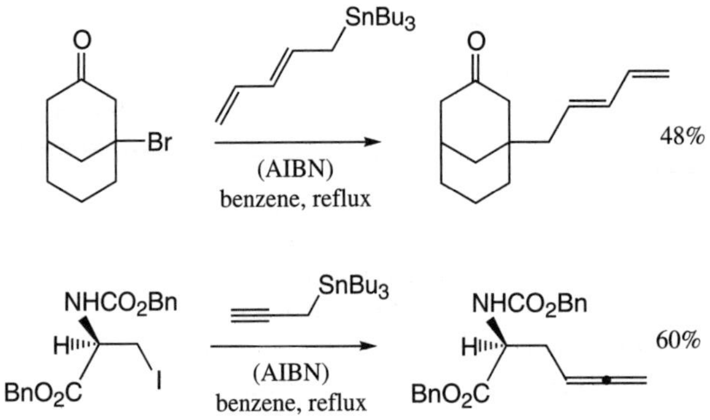

Scheme 3.39 Introduction of diene and allene groups

approach for introducing new carbon–carbon bonds under mild, neutral conditions. The reaction is performed by simply heating all the ingredients in the degassed solvent. Notice, however, the relatively high concentration (1 M) of the medium.

The method is applicable to substituted allyl groups, as long as the substituent is in the 2-position. Methyl (Keck and Yates 1982); ester, nitrile, or chloride (Baldwin *et al.* 1988); various silicon containing groups (Lee *et al.* 1988; Clive *et al.* 1991); and even complex structures (Toru *et al.* 1988) have thus been introduced. Attempts to extend the method to crotyl tributyltin proved problematic however: mostly reduction by abstraction of the allylic hydrogen was observed (Keck and Yates 1983). Normal addition is slowed down by the presence of the methyl group at the terminus of the olefin. On the other hand, reaction with (1-buten-3-yl)-tributylstannane is frustrated by its competing conversion into the useless crotyl isomer by a chain addition–fragmentation of a tributyltin radical. These various considerations are summarized in Scheme 3.38: a substituent on position-2 of the allyl system (eqn 1, Scheme 3.38) is acceptable, but can cause difficulties when it is on positions-1 or -3 (eqns 2 and 3, Scheme 3.38). Some solutions to these general problems have been devised based on a combination of allyl sulfides and hexabutylditin. The process is not a chain and is normally performed by irradiation (Keck and Byers 1985; Yanagisawa *et al.* 1988).

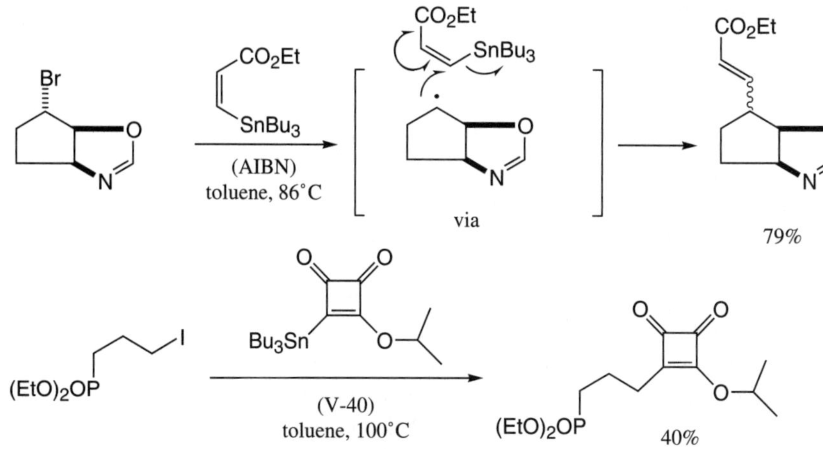

Scheme 3.40   Radical vinylation reactions

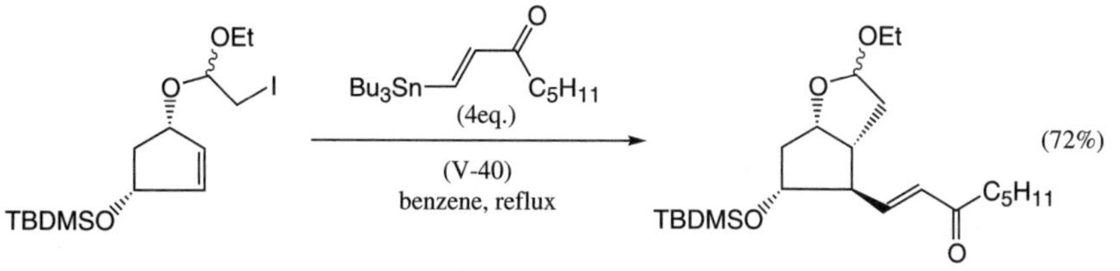

V-40 is 1,1'-azobis(cyclohexanenitrile)

**Scheme 3.41** Sequential cyclization and vinylation in the synthesis of prostaglandins

One interesting extension concerns the introduction of a pentadiene (Kraus *et al.* 1993) or an allene unit (Baldwin *et al.* 1984) by using the corresponding stannane, as shown by the two examples in Scheme 3.39; the yield is only moderate but these groups are somewhat more subject to side reactions in comparison with a simple allyl. Other variations have been described by Russell and Herold (1985).

Vinylations are also possible, as long as an activating group (aryl, ester, ketone, sulfone, etc.) is present on the vinylstannanes to compensate for the steric hindrance due to the tin substituent (Baldwin and Kelly 1985; Keck *et al.* 1988). Moreover, easily abstractable allylic hydrogens must preferably be absent from the reagent. In this way a simple vinylcarboxylate (Baldwin and Kelly 1985) or a more complex cyclobutanedione (Kinney 1993) were efficiently introduced starting with a bromide and an iodide, respectively (Scheme 3.40).

This radical-based vinylation technology is exceedingly powerful and nicely complements palladium- and other transition-metal-induced couplings which are not generally applicable to aliphatic systems. It can be combined with other radical processes allowing the design and implementation of highly convergent synthetic strategies. The concise approach to prostaglandins outlined in Scheme 3.41 is one such example (Keck and Burnett 1987). Again, there is no need for high dilution because the vinyl transfer is too slow to compete with the cyclization step, in stark contrast with the many similar transformations we have explored involving organotin hydrides.

## 3.9   Variations on the fragmentation theme

The principle of generating tin-centred radicals through a fragmentation step can appear under various guises. The ring expansion depicted in Scheme 3.42 is one instance whereby a small amount of stannyl radicals, generated from tributyltin hydride and AIBN, serves to trigger a cascade where the last step is an elimination giving back the stannyl radical (Baldwin *et al.* 1989). It is important to have a quaternary carbon next to the carbonyl group (R≠H) to avoid an undesirable intramolecular hydrogen abstraction which is quite efficient in this case.

Another cascade where a hydrogen abstraction was deliberately used is shown in Scheme 3.43 (Chambournier *et al.* 1997). Although in principle the sequence is catalytic in tributylstannyl radicals, in practice it was found that one equivalent each of tributylstannane and AIBN, added together through a syringe pump, were necessary to get the reaction to completion. The reason presumably is that the addition of the stannyl radicals to the carbonyl group is rather inefficient and only a fraction actually participates in the desired transformation with, as a consequence, a significant wastage of the reagent.

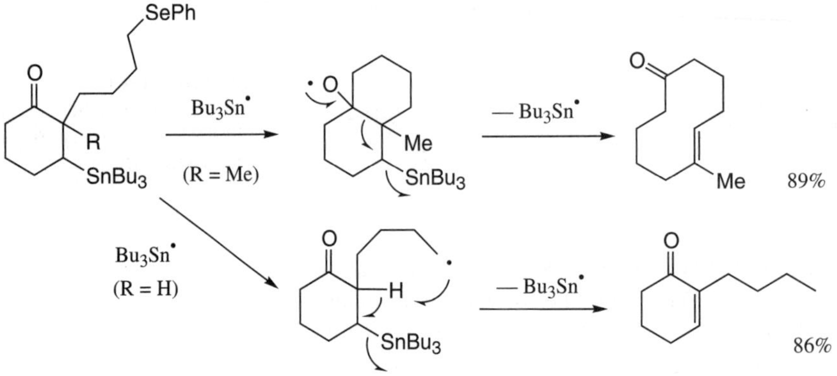

**Scheme 3.42** Synthesis of a medium-sized ring by a double fragmentation

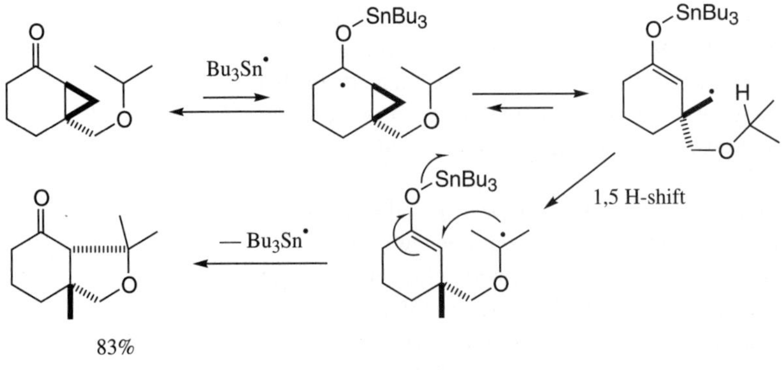

Conditions: Slow addition of Bu₃SnH (1 eq.) and AIBN (1eq.)
benzene (0.2M), reflux

**Scheme 3.43** Cascade reaction of a cyclopropyl-ketone-induced by stannyl radicals

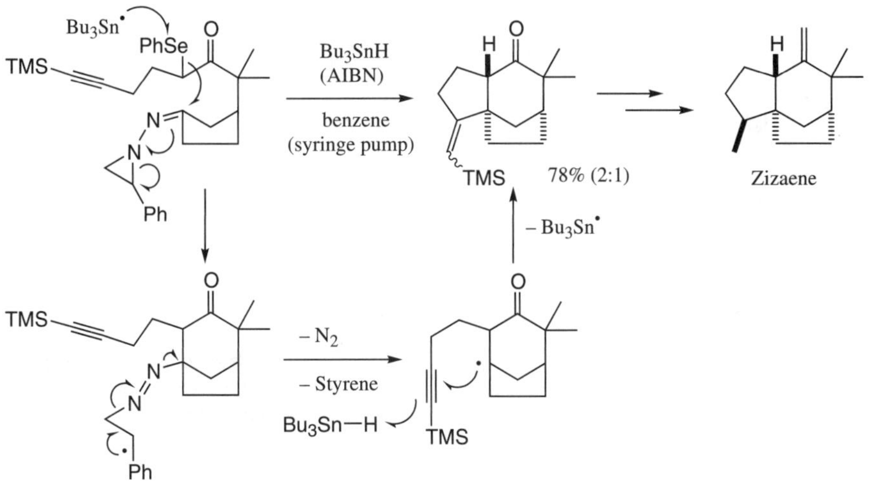

**Scheme 3.44** Key radical sequence in the synthesis of zizaene

It is appropriate to conclude the section with a brilliant approach to zizaene embodying many of the transformations discussed in this and the previous chapters (Kim and Cheong 1997). The sequence of cyclizations, ring opening, and β-eliminations leading to essentially the complete zizaene structure is outlined in a summarized form in Scheme 3.44. The first 6-*exo* ring-closure onto the hydrazone is followed by ring opening of the aziridine ring and elimination of styrene and nitrogen, bringing the radical centre into the correct position for the final 5-*exo* cyclization. The hydrazone carbon thus acts both as a radical 'acceptor' and as a radical 'donor' in this beautiful cascade.

## 3.10 Some aspects of the stereochemistry of radical reactions

The radical allylation with allylstannane has often been used as a tool to study the stereochemistry of intermolecular carbon–carbon bond formation, especially in the acyclic series. It is thus appropriate, before concluding this chapter, to consider some further aspects of the important problem of stereochemical control in radical reactions (Beckwith 1993; Curran *et al.* 1996; Sibi and Porter 1999).

In ring closures, the structure of the transition-state is more rigid than that of the substrate and the substituents tend to occupy regions of space as far apart as possible, unless some special attractive force (dipole–dipole interaction or hydrogen bonding) is operating. The radical then reacts from the least–hindered face. Because of this inherent rigidity, the use of a permanent or temporary cyclic template is the easiest way to control stereochemistry. Attaining high selectivities with the more floppy acyclic derivatives is more problematic. The difference in energy between the various possible conformations is often too low to be useful. The difficulty may in some cases be lifted by incorporating substituents which, often because of allylic strain, strongly favour one particular rotamer and force the attack to take place from mostly one side (Giese *et al.* 1992). This is shown by the example in Scheme 3.45. To minimize allylic strain, the smallest group (hydrogen) is the one that staggers the

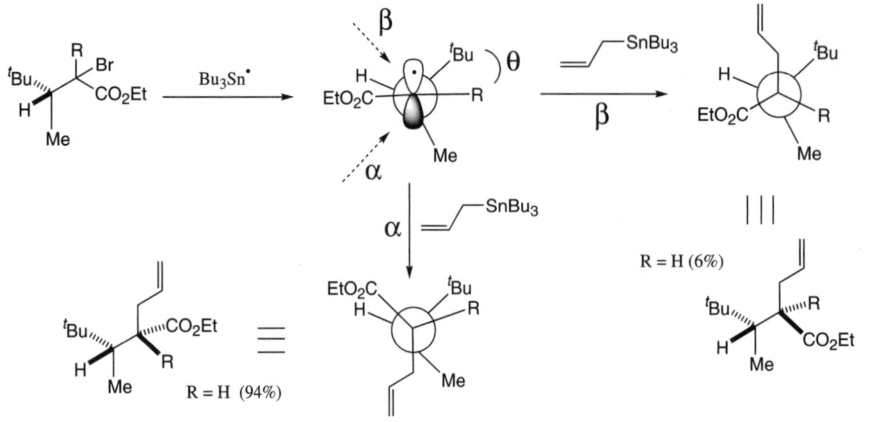

**Scheme 3.45** Stereo-chemistry of radical allylation

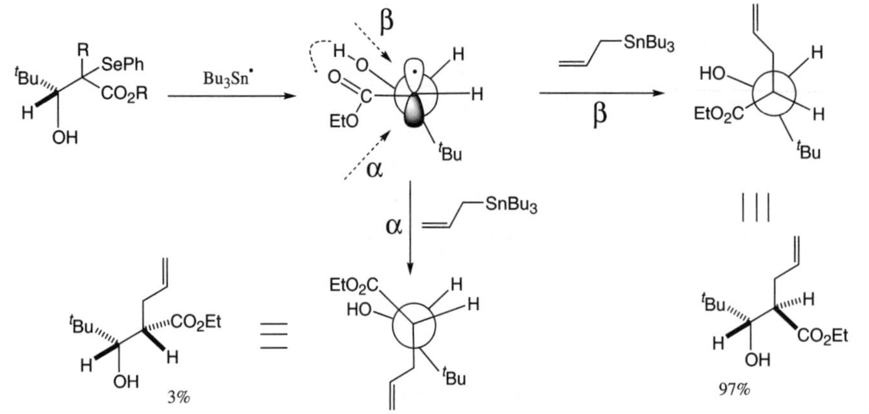

**Scheme 3.46** Effect of hydrogen bonding on the stereochemistry of radical allylation

ester on the adjacent carbon, leaving the bulky *t*-Bu- to block the top (or β-) face of the intermediate radical. The exact position of the *t*-Bu- will depend on the size of substituent R: for R = H, the value of angle $\theta$ has been estimated to be 40°, whereas for R = Me, $\theta$ increases to about 80°. The shielding of the top side will be more effective in the latter case. The ratio of $\alpha : \beta$ attack with tributyl tin deuteride thus improves from 87 : 13 to 98 : 2 on going from R = H to R = Me. With the slower reacting and bulkier allyl tributyltin reagent, the selectivity even with R = H is already an impressive 94 : 6.

If the methyl in the starting material is replaced by a hydroxyl group, then an attractive hydrogen bonding with the ester carbonyl modifies the preferred confor-mation (Scheme 3.46). The *t*-Bu- group moves to the lower part of the molecule, leaving the top side open to attack by the allylating reagent. The selectivity is now opposite to that with the methyl group, the $\alpha : \beta$ ratio being 3 : 97 (yield, 90%; Hart and Krishnamurthy 1992).

Complexation with Lewis acids, transition metals, or lanthanide salts can also bring about a modification of the preferred conformation and can thus influence strongly the selectivity (Renaud and Gerster 1995). Hydrogen bonding and complexation by a Lewis acid are really just an expedient way of creating a temporary ring (and therefore rigidity) to distinguish between the two sides of an otherwise flexible molecule.

Another way to bring into play allylic strain is to exploit the relatively slow rotation around an amide bond, as illustrated by the example in Scheme 3.47 (Porter *et al.* 1992). Steric repulsion between the more or less bulky R group in the oxazolidine and the alkane chain on the acid side forces the intermediate radical to adopt a conformation where the R group is nearly perpendicular to the nodal plane of the p-orbital containing the single electron. Attack therefore occurs mostly from the other side, the extent depending on the size of R: about 4:1 in favour of $\alpha$-side approach for R = $^i$Pr and greater than 20:1 for R = $^t$Bu. Various other amide-type structures have been employed (Curran *et al.* 1996).

## 3.11 Some concluding remarks

Now that we have surveyed the main reactions based on the radical chemistry of organotin derivatives, it is perhaps useful to make a brief summary of some of the salient points. A radical is a defined entity which is the same, whatever the method used to generate it. In most cases, it is a reactive species which, if its concentration is not kept very low, will rapidly react with another radical to give dimers or disproportionation products. The rate of decomposition of the initiator at the reaction temperature (or the intensity of the light in the case of photochemical initiation) determines the amount of radicals pumped into the system at a given time. It is thus important to select the appropriate initiator or initiation method in accord with the temperature and with the expected rate of the subsequent desired transformations. The faster these steps are, the greater the rate of radical production the system tolerates without interference from radical–radical interactions, and the less initiator one needs if one is dealing with a chain process. In this respect, the choice of the radical precursor (an iodide instead of a bromide or a selenide in place of a less reactive sulfide), the purity and absence of inhibitors, the effect of temperature if unimolecular fragmentation steps entailing a large increase in entropy are present, the concentration and mode of addition of the reagents are all important factors that must be carefully considered.

It must also be remembered that the intermediate radicals can always react with the solvent. The selection of the solvent is thus a very important consideration, since the rate of reaction with a solvent will impose a lower limit on the rate of the desired pathway below which it will be difficult to capture the radicals in a useful manner. Unwanted hydrogen abstraction from within the substrate can also occur and which may not be obvious upon examination of the reaction products but which may be revealed by deuteration experiments. Inspection of models might help in locating potentially troublesome hydrogens and incorporation of temporary activating or blocking groups can sometimes help in channelling the radical process in the desired direction by speeding up some steps, slowing down, or completely eliminating others.

The examples that have been discussed give an idea of the synthetic possibilities and show how some of the problems have been solved. In the following chapters we shall build upon these foundations by describing other methods for generating radicals which may or may not involve chain processes. But, even if the way by which the initial radical is produced and the sequence terminated differ, the steps in between (additions, cyclizations, fragmentations, etc.) will be essentially the same to the ones described in this chapter.

## References

Abeywickrema, A. N., Beckwith, A. L. J., and Gerba, S. (1987). Consecutive ring closure and neophyl rearrangement of some alkenyl radicals. *Journal of Organic Chemistry*, **52**, 4072–4078.

Araki, Y., Endo, T., Arai, Y., Tanji, M., and Ishido, Y. (1989). Stereoselective radical cyclisation for the synthesis of bicyclic higher-carbon sugars. Synthesis of the sugar moiety of octosyl acids. *Tetrahedron Letters*, **30**, 2829–2832.

Baguley, P. A. and Walton, J. C. (1998). Flight from the tyranny of tin: the quest for practical radical sources free from metal encumbrances. *Angewandte Chemie International Edition in English*, **37**, 3072–3082.

Baldwin, J. E. and Kelly, D. R. (1985). Applications of consecutive radical addition-elimination reactions in synthesis. *Journal of the Chemical Society, Chemical Communications*, 682–684.

Baldwin, J. E., Adlington, R. M., and Basak, A. (1984). Allene transfer reactions. A new synthesis of terminal allenes. *Journal of the Chemical Society, Chemical Communications*, 1284–1285.

Baldwin, J. E., Adlington, R. M., and Robertson, J. (1989). Carbocyclic ring expansion reactions via radical chain processes. *Tetrahedron*, **45**, 909–922.

Baldwin, J. E., Adlington, R. M., Lowe, C., O'Neil, I. A., Sanders, G. L., Schofield, C. I., and Sweeney, J. B. (1988). Reactions of a glycidyl radical equivalent with 2-functionalised allyl stannanes. *Journal of the Chemical Society, Chemical Communications*, 1030–1031.

Barbier, M., Barton, D. H. R., Devys, M., and Topgi, R. S. (1984). A simple synthesis of tropones and related compounds. *Tetrahedron*, **43**, 5031–5038.

Barton, D. H. R., Hay Motherwell, R. S., and Motherwell, W. B. (1981). Radical-induced ring opening of epoxides. A convenient alternative to the Wharton rearrangement. *Journal of the Chemical Society, Perkin Transactions 1*, 2363–2367.

Batey, R. A., Grice, P., Harling, J. D., Motherwell, W. B., and Rzepa, H. S. (1992). Origins of the regioselectivity of cyclopropylcarbinyl ring opening reaction in bicyclo[n.1.0] systems. *Journal of the Chemical Society, Chemical Communications*, 942–943.

Beckwith, A. L. J. (1993). The pursuit of selectivity in radical reactions. *Chemical Society Reviews*, 143–151.

Beckwith, A. L. J. and Moad, G. (1980). The kinetics and mechanism of ring opening of radicals containing the cyclobutylcarbinyl system. *Journal of the Chemical Society, Perkin Transactions 1*, 1083–1082.

Beckwith, A. L. J. and Phillipou, G. (1976). Specific $\beta$-scission of 3$\beta$,5-cyclocholestan-6-yl radical. *Australian Journal of Chemistry*, **29**, 123–131.

Beckwith, A. L. J. and O'Shea, D. M. (1986). Kinetics and mechanism of some vinyl radical cyclisations. *Tetrahedron Letters*, **27**, 4525–4528.

Beckwith, A. L. J. and Storey, J. M. D. (1995). Tandem radical translocation and homolytic aromatic substitution: a convenient and efficient route to oxindoles. *Journal of the Chemical Society, Chemical Communications*, 977–978.

Beckwith, A. L. J., O'Shea, D. M., and Westwood, S. W. (1988). Rearrangement of suitably constituted aryl, alkyl, or vinyl radicals by acyl or cyano group migration. *Journal of the American Chemical Society*, **110**, 2565–2575.

Beckwith, A. L. J., Crich, D., Duggan, P. J., and Yao, Q. (1997). Chemistry of $\beta$-(acyloxy)alkyl and $\beta$-phosphatoxy)alkyl radicals and related species: radical and radical ionic migrations and fragmentations of carbon-oxygen bonds. *Chemical Reviews*, **97**, 3273–3312.

Bennett, S. M. and Clive, D. L. J. (1986). Synthesis of spiro-compounds related to frederi-camycin A. *Journal of the Chemical Society, Chemical Communications*, 878–880.

Boger, D. L. and Mathvink, R. J. (1988). Acyl radicals: functionalized free radicals for intramolecular cyclisation reactions. *Journal of Organic Chemistry*, **53**, 3379–3381.

Boiteau, L., Boivin, J., Quiclet-Sire, B., Saunier, J.-B., and Zard, S. Z. (1998). Synthetic routes to $\beta$-lactams. Some unexpected hydrogen atom transfer reactions. *Tetrahedron*, **54**, 2087–2098.

Boivin, J., Fouquet, E., and Zard, S. Z. (1991). Ring opening induced by iminyl radicals derived from cyclobutanones: new aspects of tin hydride cleavage of *S*-phenyl sulphenylimines. *Journal of the American Chemical Society*, **113**, 1055–1057.

Boivin, J., Fouquet, E., and Zard, S. Z. (1994). Iminyl radicals: part II. Ring opening of cyclobutyl- and cyclopentyliminyl radicals.*Tetrahedron*, **50**, 1757–1768.

Bowry, V., Lusztyk, J., and Ingold, K. U. (1991). Calibration of a new horlogery of fast radical 'clocks'. Ring-opening rates for ring- and $\alpha$-alkyl-substituted cyclopropylcarbinyl radicals and for the bicyclo[2.2.1]pent-2-yl radical. *Journal of the American Chemical Society*, **113**, 5687–5698.

Brooks, M. A. and Scott, L. T. (1999). 1,2-Shifts of hydrogen atoms in aryl radicals. *Journal of the American Chemical Society*, **121**, 5444–5449.

Brown, C. D. S., Simpkins, N. S., and Clinch, K. (1993). A route to spiroketals using radical translocation. *Tetrahedron Letters*, **34**, 131–132.

Brunton, G., Griller, D., Barclay, L. R. C., and Ingold, K. U. (1976). Kinetic application of electron paramagnetic resonance spectroscopy. 26. Quantum mechanical tunneling in the isomerisation of sterically hindered aryl radicals. *Journal of the American Chemical Society*, **98**, 6803–6811.

Cassayre, J. and Zard, S. Z. (1999). A short synthesis of (−)-dendrobine. *Journal of the American Chemical Society*, **121**, 6072–6073.

Chambournier, G., Krishnamurthy, V., and Rawal, V. H. (1997). Radical induced cyclopropane fragmentation-H abstraction-cyclisation cascade: synthesis of carbocyclic systems containing bridgehead methyl groups. *Tetrahedron Letters*, **38**, 6313–6316.

Chatgilialoglu, C., Timokhin, V. I., and Ballestri, M. (1998). One carbon ring expansion in cyclopentanones as a free-radical clock. *Journal of Organic Chemistry*, **63**, 1327–1329.

Choi, S.-Y., Horner, J. H., and Newcomb, M. (2000). Laser flash photolysis measurements of the kinetics of ring opening of the 2,2-diphenylcyclobutylcarbinyl radical. *Journal of Organic Chemistry*, **65**, 4447–4449.

Clive, D. L. J. and Daigneault, S. (1991). Use of radical ring opening for introduction of alkyl and substituted alkyl groups with stereochemical control: a synthetic application of cyclopropylcarbinyl radicals. *Journal of Organic Chemistry*, **56**, 3801–3814.

Clive, D. L. J., Paul, C. C., and Wang, Z. (1991). Radical allylations with trimethyl [2-[(tributyl-stannyl)methyl]-2-propenyl]silane or trimethyl [2-[(triphenylstannyl)methyl]-2-propenyl] silane. *Journal of Organic Chemistry*, **62**, 7028–7032.

Crich, D., Sun, S., and Brunckova, J. (1996). Chemistry of 1-alkoxy-1-glycosyl radicals: the manno- and rhamnopyranosyl series. Inversion of $\alpha$- to $\beta$-pyranosides and the fragmentation of anomeric radicals. *Journal of Organic Chemistry*, **61**, 605–615.

Crimmins, M. T. (1988). Synthetic applications of intramolecular enone–olefin photocycloadditions. *Chemical Reviews*, **88**, 1453–1473.

Crimmins, M. T. and Mascarella, S. W. (1987). Radical cleavage of cyclobutanes: alternative routes to (±)-silphinene. *Tetrahedron letters*, **28**, 5063–5066.

Crimmins, M. T., Wang, Z., and McKerlie, L. A. (1996). Rearrangement of cyclobutyl carbinyl radicals: total synthesis of the spirovetivane phytoalexin (±) lubiminol. *Tetrahedron Letters*, **37**, 8703–8706.

Curran, D. P. and Shen, W. (1993). Radical translocation reactions of vinyl radicals: substituent effects on 1,5-hydrogen atom transfer reactions. *Journal of the American Chemical Society*, **115**, 6051–6059.

Curran, D. P. and Yu, H. (1992). New application of 1,5-hydrogen atom transfer reactions: self-oxidizing protecting groups. *Synthesis*, 123–127.

Curran, D. P., Porter, N. A., and Giese, B. (1996). *Stereochemistry of radical reactions*. VCH, Weinheim.

Curran, D. P., Jiaang, W.-T., Palovich, M., and Tsai, Y.-M. (1993). Acylsilanes as geminal radical acceptors/radical donors equivalents in tandem cyclisation/addition reactions. *Synlett*, 403–404.

Curran, D. P., Kim, D., Liu, H. T., and Shen, W. (1988). Translocation of radical sites by 1,5-hydrogen atom transfer. *Journal of the American Chemical Society*, **110**, 5900–5902.

Curran, D. P., van Elburg, P. A., Giese, B., and Gilges, S. (1990). Approximate rate constants for the addition of alkyl radicals to allylstannanes. *Tetrahedron Letters*, **31**, 2861–2864.

De Kimpe, N., De Smaele, D., and Bogaert, P. (1994). Radical-induced ring opening of 2-(bromomethyl)aziridines. *Synlett*, 287–288.

De Mesmaeker, A., Waldner, A., Hoffmann, P., Hug, P., and Winkler, T. (1992). C(5)-Epimerisation in glycopyranosides during the cyclisation of anomeric radicals: a comparison with glycofuranosides. *Synlett*, 285–290.

De Smaele, D., Bogaert, P., and De Kimpe, N. (1998). Synthesis of pyrrolizidines by cascade reactions of *N*-alkenylaziridinylmethyl radicals. *Tetrahedron Letters*, **39**, 9797–9800.

Denenmark, D., Winkler, T., Waldner, A., and De Mesmaeker, A. (1992). Competeing radical translocation reactions of tertiary *N*-(2-bromobenzyl)- and *N*-(8-bromonaphthyl)-acetamides. *Tetrahedron Letters*, **33**, 3613–3616.

Denis, R. C. and Gravel, D. (1994). New access to fused vinylcyclopropanes by radical cyclisation. *Tetrahedron Letters*, **35**, 4531–4534.

Denis, R. C., Rancourt, J., Ghiro, E., Boutonnet, F., and Gravel, D. (1993). Cyclopropanations in vinyl radical cyclisations: the importance of structural constraints. *Tetrahedron Letters*, **34**, 2091–2094.

Dickinson, J. M. and Murphy, J. A. (1992). Pyrrolidines and allylic amines from radical-induced cleavage of aziridines. *Tetrahedron*, **48**, 1317–1326.

Dorigo, A. E. and Houk, K. N. (1988). On the relationship between proximity and reactivity. An *ab initio* study of the OH$^{\bullet+}$ CH$_4$ hydrogen abstraction transition state and a force-field model for the transition states of intramolecular hydrogen abstractions. *Journal of Organic Chemistry*, **53**, 1650–1664.

Dowd, P. and Zhang, W. (1993). Free radical-mediated ring expansion and related annulations. *Chemical Reviews*, **93**, 2091–2115.

Enholm, E. J. and Jia, Z. J. (1996). Construction of linear triquinane skeleton by an *O*-stannyl ketyl radical rearrangement. *Journal of the Chemical Society, Chemical Communications*, 1567–1568.

Franz, J. A., Barrows, R. D., and Camaioni, D. M. (1984). Arrhenius parameters for rearrangements of the neophyl, 1-indanylmethyl, 2-allylbenzyl, and 2-(2-vinylphenyl)ethyl radicals relative to hydrogen abstraction from tributylstannane. *Journal of the American Chemical Society*, **106**, 3964–3967.

Fremont, S. L., Belletire, J. L., and Ho, D. M. (1991). Free radical cyclizations leading to four membered rings. 1. Beta-lactam production using tributyltin hydride. *Tetrahedron Letters*, **32**, 2335–2338.

Frey, P. A. (1990). Importance of organic radicals in enzymatic cleavage of unactivated C–H bonds. *Chemical Reviews*, **90**, 1343–1357.

Furxhi, E., Horner, J. H., and Newcomb, M. (1999). Direct measurements of the kinetics of 3-*exo* radical cyclisations using radical reporter groups. *Journal of Organic Chemistry*, **64**, 4064–4068.

Giese, B. and Gröninger, K. S. (1990). 1,3,4,6-Tetra-*O*-acetyl-2-deoxy-*α*-*D*-glucopyranose. *Organic Synthesis*, **69**, 66–71.

Giese, B., Damm, W., Wetterich, F., and Zeitz, H.-G. (1992). 1,2-Stereoinduction in acyclic radicals: allylic strain effects. *Tetrahedron Letters*, **33**, 1863–1866.

Grignon, J. and Pereyre, M. (1973). Mécanisme de la substitution des dérivés halogénés par les organostanniques allyliques. *Journal of Organometallic Chemistry*, **61**, C33–C35.

Grossi, L. and Strazzari, S. (1997). Oxiranylmethyl radicals: EPR detection by spin trapping. *Journal of the Chemical Society, Chemical Communications*, 917–918.

Han, O. and Frey, P. A. (1990). Chemical model for the pyridone 5′-phosphate dependent lysine aminomutases. *Journal of the American Chemical Society*, **112**, 8982–8983.

Hart, D. J. and Krishnamurthy, R. (1992). Investigation of a model for 1,2-asymmetric induction in reactions of *α*-carbalkoxy radicals: a stereochemical comparison of reactions of *α*-carbalkoxy radicals and ester enolates. *Journal of Organic Chemistry*, **57**, 4457–4470.

Horner, J. H., Tanaka, N., and Newcomb, M. (1998). Kinetics of cyclopropylcarbinyl radical ring opening measured directly by laser flash photolysis. *Journal of the American Chemical Society*, **120**, 10379–10390.

Huang, X. L. and Dannenberg, J. J. (1991). Molecular orbital estimation of the activation enthalpies for intramolecular hydrogen transfer as functions of size of the cyclic transition state and the C–H–C angle. *Journal of Organic Chemistry*, **56**, 5421–5424.

Ishibashi, H., So, T. S., Okochi, K., Sato, T., Nakamura, N., and Ikeda, M. (1991). Radical cyclisation of *N*-(cyclohex-2-enyl)-*α*,*α*-dichloroacetamides. Stereoselective syntheses of (±)-mesembranol and (±)-elwesine. *Journal of Organic Chemistry*, **56**, 95–102.

Ishibashi, H., Kameoka, C., Yoshikawa, A., Ueda, R., Kodama, K., Sato, T., and Ikeda, M. (1993). Synthesis or *β*-lactams by means of sulfur-controlled regioselective radical cyclisations of *N*-ethenyl-*α*-bromoalkanamides. *Synlett*, 649–650.

Itoh, Y., Haraguchi, K., Tanaka, H., Matsumoto, K., Nakamura, K. T., and Miyasaka, T. (1995). Radical-initiated 1,2-acyloxy migration which generates a nucleoside anomeric radical. *Tetrahedron Letters*, **36**, 3867–3870.

Jarreton, O., Skrydstrup, T., and Beau, J.-M. (1996). The stereospecific synthesis of methyl *α-L*-mannobioside: a potential inhibitor of M. *tuberculosis* binding to human macrophages. *Journal of the Chemical Society, Chemical Communications*, 1661–1662.

Keck, G. E. and Yates, J. B. (1982). Carbon–carbon bond formation via the reaction of trialkylallystannanes with organic halides. *Journal of the American Chemical Society*, **104**, 5829–5831.

Keck, G. E. and Yates, J. B. (1983). Hydrogen transfer from crotyltri-*n*-butylstannane to carbon radicals. *Journal of Organometallic Chemistry*, **248**, C21–C25.

Keck, G. E. and Byers, J. H. (1985). A new 'one electron' carbon–carbon bond forming reactions: separation of the chain propagating steps in free radical allylations. *Journal of Organic Chemistry*, **50**, 5442–5444.

Keck, G. E. and Burnett, D. A. (1987). $β$-Stannyl enones as radical traps: a very direct route to PGF$_{2α}$. *Journal of Organic Chemistry*, **52**, 2958–2960.

Keck, G. E., Byers, J. H., and Tafesh, A. M. (1988). A free radical addition–fragmentation reaction for the preparation of vinyl sulfones and phosphine oxides. *Journal of Organic Chemistry*, **53**, 1127–1128.

Kim, S and Cheong, J. H. (1997). Efficient synthesis of *dl*-zizaene sesquiterpenes via tandem radical cyclization of *N*-aziridinylimines. *Synlett*, 947–949.

Kim, S., Joe, G. H., and Do, J. Y. (1993). Highly efficient intramolecular addition of aminyl radicals to carbonyl groups: a new ring expansion reaction leading to lactams. *Journal of the American Chemical Society*, **115**, 3328–3329.

Kim, S., Yeon, K. M., and Yoon, K. S. (1997). 1,5-Hydrogen transfers from carbon to *N*-tributyltin substituted nitrogen. *Tetrahedron Letters*, **38**, 3919–3922.

Kinney, W. A. (1993). Synthesis of alkyl substituted cyclobutanediones by free radical chemistry. Carbon for nitrogen replacement in the $α$-amino acid bioisostere—3,4-diamino-3-cyclobutene-1,2-dione. *Tetrahedron Letters*, **34**, 2715–2718.

Kosugi, M., Kurino, K., Takayama, K., and Migita, T. (1973). The reaction of organic halides with allyl trimethyltin. *Journal of Organometallic Chemistry*, **56**, C11–C13.

Kraus, G., Andersh, B., Su, Q., and Shi, J. (1993). Bridgehead radicals in organic chemistry. An efficient construction of the ABDE ring system of the lycoctonine alkaloids. *Tetrahedron Letters*, **34**, 1741–1744.

Krishnamurthy, V. and Rawal, V. H. (1997). Kinetics of the oxiranylcarbinyl radical rearrangement. *Journal of Organic Chemistry*, **62**, 1572–1573.

Lee, E., Whang, H. S., and Chung, C. K. (1995). Radical isomerization via intramolecular ipso substitution of *N*-arylamides: aryl translocation from nitrogen to carbon. *Tetrahedron Letters*, **36**, 913–914.

Lee, E., Yu, S.-G., Hur, C.-U., and Yang, S.-M. (1988). Radical reaction of (2-trimethylsilyl-allyl)triphenyl stannane with alkyl halides: a neutral acetone enolate equivalent. *Tetrahedron Letters*, **29**, 6969–6970.

Marples, B. A., Rudderham, J. A., Slawin, A. M. Z., Edwards, A. J., and Hird, N. W. (1997). Demonstration of reversible C–C bond cleavage in oxiranylcarbinyl radicals. *Tetrahedron Letters*, **38**, 3599–3602.

Mehta, G., Krishnamurthy, N., and Karra, S. R. (1991). Terpenoids to terpenoids: enantioselective construction of 5,6-, 5,7-, and 5,8- fused bicyclic systems. Application to the total synthesis of isodaucane sesquiterpenes and dolastane diterpenes. *Journal of the American Chemical Society*, **113**, 5765–5775.

Motherwell, W. B. and Pennell, A. M. K. (1991). A novel route to biaryls via intramolecular free radical *ipso* substitution reactions. *Journal of the Chemical Society, Chemical Communications*, 877–879.

Nonhebel, D. C. (1993). The chemistry of cyclopropylmethyl and related radicals. *Chemical Society Reviews*, 347–359.

Park, S.-U., Varick, T. R., and Newcomb, M. (1990). Acceleration of the 4-*exo* radical cyclization to a synthetically useful rate. Cyclization of the 2,2-dimethyl-5-cyano-4-pentenyl radical. *Tetrahedron Letters*, **21**, 2975–2978.

Porter, N. A., Rosenstein, I. J., Breyer, R. A., Bruhnke, J. D., Wu, W.-X., and McPhail, A. T. (1992). Origins of stereoselectivity in radical additions: reactions of alkenes and radicals bearing oxazolidine and thiazolidine amide groups. *Journal of the American Chemical Society*, **114**, 7664–7676.

Rawal, V. H., Newton, R. C., and Krishnamurthy, V. (1990). Synthesis of carbocyclic systems via radical-induced epoxide fragmentation. *Journal of Organic Chemistry*, **55**, 5181–5183.

Renaud, P. and Gerster, M. (1995). Stereoselectivity in reactions of 1,2-dioxy-substituted radicals: electronic versus chelation control. *Journal of the American Chemical Society*, **117**, 6607–6608.

Roberts, B. P. and Winter, J. N. (1979). Electron spin resonance studies of radicals derived from organic azides. *Journal of the Chemical Society, Perkin Transactions 2*, 1353–1361.

Russell, G. A. and Herold, L. L. (1985). Free-radical chain substitution reactions ($S_H2'$) of alkenyl-, alkynyl-, and alkenyloxy-stannanes. *Journal of Organic Chemistry*, **50**, 1037–1040.

Schwan, A. L. and Refvik, M. D. (1993). Substituent control over the regiochemistry of ring opening of 2-aziridinylmethyl radicals. *Tetrahedron Letters*, **34**, 4901–4904.

Shuto, S., Kanazaki, M., Ichikawa, S., and Matsuda, A. (1997). A novel ring enlargement reaction of (3-oxa-2-silacyclopentyl)methyl radicals. Stereoselective introduction of a hydroxyethyl group via unusual 6-*endo* cyclisation products derived from 3-oxa-4-silahexanyl radicals and its application to the synthesis of a 4'-α-branched nucleoside. *Journal of Organic Chemistry*, **62**, 5676–5677.

Sibi, M. P. and Porter, N. A. (1999). Enantioselective free radical reactions. *Accounts of Chemical Research*, **32**, 163–171.

Simpkins, N. S. and Clinch, K. (1993). A route to spiroketals using radical translocation. *Tetrahedron Letters*, **34**, 131–132.

Smith, D. M., Nicolaides, A., Golding, B. T., and Radom, L. (1998). Ring opening of the cyclopropylcarbinyl Radical and its *N*- and *O*-substituted analogues: a theoretical examination of very fast unimolecular reactions. *Journal of the American Chemical Society*, **120**, 10223–10233.

Snieckus, V., Cuevas, J.-C., Sloan, C. P., Liu, H., and Curran, D. P. (1990). Intramolecular α-amidyl to aryl 1,5-hydrogen atom transfer reactions. Heteroannulations and α-nitrogen functionalization by radical translocation. *Journal of the American Chemical Society*, **112**, 896–898.

Stork, G. and Mook, R. Jr. (1986). Five vs six membered ring formation in the vinyl radical cyclisation. *Tetrahedron Letters*, **27**, 4529–4532.

Studer, A. and Steen, H. (1999). The $S_Hi$ reaction at silicon—a new entry into cyclic alkoxysilanes. *Chemistry—A European Journal*, **5**, 759–773.

Sugimoto, I., Shuto, S., and Matsuda, A. (1999). Kinetics of a novel 1,2-rearrangement reaction of b-silyl radicals. The ring expansion of (3-oxa-2-silacyclopentyl)methyl radical into 4-oxa-3-silacyclohexyl radical is irreversible. *Synlett*, 1766–1768.

Surzur, J.-M. and Teissier, P. (1970). Réactions d'additions radicalaires sur les alcools non saturés. III—Additions sur les acétates éthyléniques: migration radicalaire 1,2-du groupement acétoxy. *Bulletin de la Société Chimique de France*, 3060–3070.

Tanner, D. D. and Law, F. C. P. (1969). Free radical acetoxy group migration. *Journal of the American Chemical Society*, **91**, 7535–7537.

Toru, T., Yamada, Y., Ueno, T., Maekawa, E., and Ueno, Y. (1988). A new route to the prostaglandin skeleton via radical alkylation. Synthesis of 6-oxo-prostaglandin $E_1$. *Journal of the American Chemical Society*, **110**, 4815–4817.

Tsai, Y.-M., Tang, K.-H., and Jiaang, W.-T. (1993). Radical cyclisation of bromoacylsilanes and intramolecular trapping of the rearranged $\alpha$-silyloxy radical. *Tetrahedron Letters*, **34**, 1303–1306.

Walton, R. and Fraser-Reid, B. (1991). Studies on the intramolecular competitive addition of carbon radicals to aldehydo and alkenyl groups. *Journal of the American Chemical Society*, **113**, 5791–5799.

Van Dort, P. C. and Fuchs, P. L. (1997). Free radical self immolative 1,2-elimination and reductive desulfonylation of aryl sulfones promoted by intramolecular reactions with *ortho*-attached carbon centered radicals. *Journal of Organic Chemistry*, **62**, 7142–7147.

Viskolcz, B., Lendvay, V. G., Körtvélyesi, T., and Seres, L. (1996). Intramolecular H atom transfer reaction in alkyl radicals and the ring strain energy in the transition structure. *Journal of the American Chemical Society*, **118**, 3006–3009.

Weng, W.-W. and Luh, T.-Y. (1993). Tandem radical [2 + 1] cycloaddition. Remarkable silyl substituent effect on the chemoselectivity of the radical cyclisation reactions. *Journal of Organic Chemistry*, **58**, 5574–5575.

Wetmore, S. D., Smith, D. M., and Radom, L. (2000). How $B_6$ helps $B_{12}$: the role of $B_6$, $B_{12}$, and the enzymes in aminomutase-catalyzed reactions. *Journal of the American Chemical Society*, **122**, 10208–10209.

Winkler, J. D., Sridar, V., Rubo, L., Hey, J. P., and Haddad, N. (1989). Inside–outside stereoisomerism. 4. An unusual rearrangement of the *trans*-bicyclo[5.3.1]undecan-11-yl radical. *Journal of Organic Chemistry*, **54**, 3004–3006.

Yanagisawa, A., Noritake, Y., and Yamamoto, H. (1988). Selective 1,5-diene synthesis. A radical approach. *Chemistry Letters*, 1899–1902.

Zheng, Z. B. and Dowd, P. (1993). A free radical route to to benzazepines and dibenzazepines. *Tetrahedron Letters*, **34**, 7709–7712.

Ziegler, F. E. and Petersen, A. K. (1995). Allyloxy radicals are formed reversibly from oxiranyl radicals: a kinetic study. *Journal of Organic Chemistry*, **60**, 2666–2667.

# 4 Organo-silicon, -germanium, and -mercury hydrides

## 4.1 Silicon and germanium derivatives: general considerations

Silicon and germanium are in the same periodic column as tin, and the radical chemistry associated with these elements is in many ways qualitatively similar. The increased strength of the bonding to hydrogen and carbon translates into slower kinetics and less efficient chain processes as compared with organotin derivatives (Chatgilialoglu and Newcomb 1999). For example, hydrogen abstraction from tri-*n*-butylgermane by a given carbon-centred radical is about 10–20 times slower than from tri-*n*-tributyl tin hydride (Johnston *et al.* 1985), and from triethylsilane and other simple silanes it becomes too slow to be generally useful. As we shall see a little later in this chapter, the silicon–hydrogen bond can be weakened by placing appropriate substituents around the silicon atom. Organogermanium and modified organosilicon hydrides are more expensive than triorganotin hydrides but this is somewhat compensated by a lesser toxicity and a greater ease of purification of the reaction mixtures.

## 4.2 Radical reactions with germanium

There is little advantage in using organogermanium hydrides to perform simple reductions or cyclizations that are inherently fast. Such reagents do become interesting when a desired radical transformation is too sluggish and cannot be accomplished efficiently using organotin technology (Pike *et al.* 1988). One such situation was encountered by Stork and Mah (1989) when the construction of a γ- and δ-lactams by 5- and 6-*exo* type cyclizations was attempted starting from bromo-acetamides. The cyclization step is compounded by the existence of two rotamers, only one of which can cyclize, and the rate of rotation around the amide bond is slow in comparison with hydrogen abstraction from the stannane (Musa *et al.* 1999), resulting in mostly reduction instead of cyclization. The situation could be greatly improved by placing a bulky substituent on the nitrogen to increase the proportion of the desired rotamer and, of course, by a slow addition of the stannane. These palliatives were not sufficiently effective in the case of δ-lactams (Scheme 4.1); much better results were obtained with the slower reducing triphenylgermane. Interestingly, only the *cis*-fused isomer is produced in this cyclization.

Another instance where the stronger bonding to germanium (and to silicon) manifests itself is in allylation reactions. Allyl germanium derivatives are poor radical allylating agents because the rate of β-elimination of a germyl radical is not high enough to ensure an efficient propagation of the chain (Light *et al.* 1987). The slow β-elimination step becomes an advantage in additions of germanes to terminal olefins, which are now more efficient than with the stannane analogues. For instance,

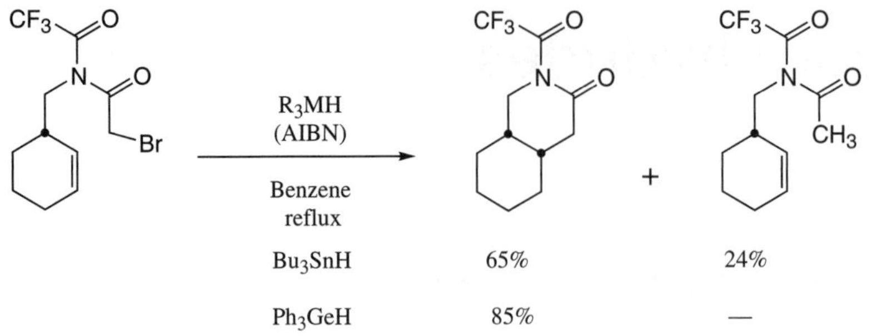

**Scheme 4.1**
Comparative reactions with triphenylstannane and triphenylgermane

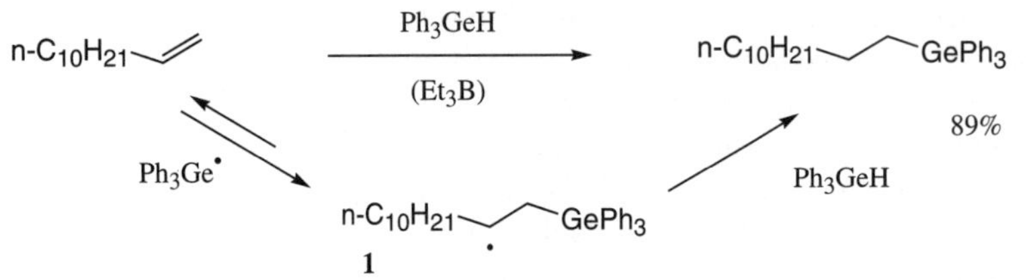

**Scheme 4.2**   Radical addition of triphenylgermane to 1-dodecene

the intermediate adduct **1** from the addition of a triphenylgermyl radical to 1-dodecene is sufficiently long lived to be able to capture a hydrogen to give the desired hydrogermylated product (Scheme 4.2); under the same conditions, no reaction occurs with triphenylstannane (Nozaki *et al.* 1990).

## 4.3   Radical reactions with silicon: tris(trimethylsilyl)silane

The Si–H bond is even stronger than the Ge–H bond. Bimolecular hydrogen atom transfer from alkyl and aryl silanes are generally too slow to propagate a chain reaction, and transformations analogous to those of tin and germanium are therefore more difficult to implement with simple silanes (Chatgilialoglu 1995). With triethylsilane for example, hydrogen abstraction from the ethyl groups themselves becomes a complicating factor (Chatgilialoglu *et al.* 1993). The efficacy of the hydrogen-abstraction step can be markedly improved by making it intramolecular, as illustrated by the sequence in Scheme 4.3 below (Curran *et al.* 1995). The slow bimolecular hydrogen abstraction (step **A**) allows the $\alpha$-carbonyl radical time to add to the unactivated olefin (step **B**) giving a radical that is now well positioned to pluck off the hydrogen from the silicon in an intramolecular fashion (step **C**). A small amount of hexabutyldistannane and irradiation are necessary to trigger the chain reaction.

Of more general utility is the fact that the Si–H bond can be weakened significantly by placing three trimethylsilyl groups around the central silicon atom, as in tris(trimethylsilyl)silane: $(Me_3Si)_3SiH$. The strength of the Si–H bond is now of the order of 79, or 11 kcal mole$^{-1}$ less than in triethylsilane (Kanabus-Kaminska *et al.* 1987); hydrogen abstraction from such a reagent is now only about one order of magnitude slower than from tributylstannane (Sn–H bond estimated to be 74 kcal mole$^{-1}$). Many of the typical tin-based reactions can thus be performed with

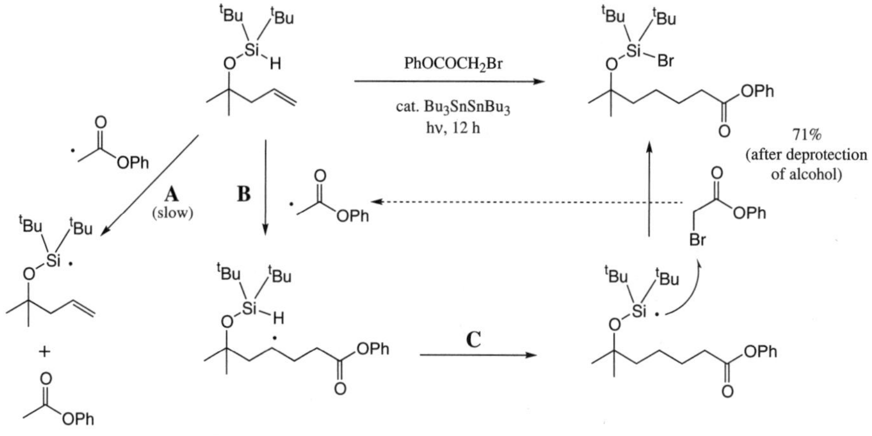
PhOCOCH2Br
cat. Bu3SnSnBu3
hv, 12 h
71%
(after deprotection
of alcohol)
A
(slow)
B
C

**Scheme 4.3**
Intermolecular silane-mediated radical addition to an unactivated alkene

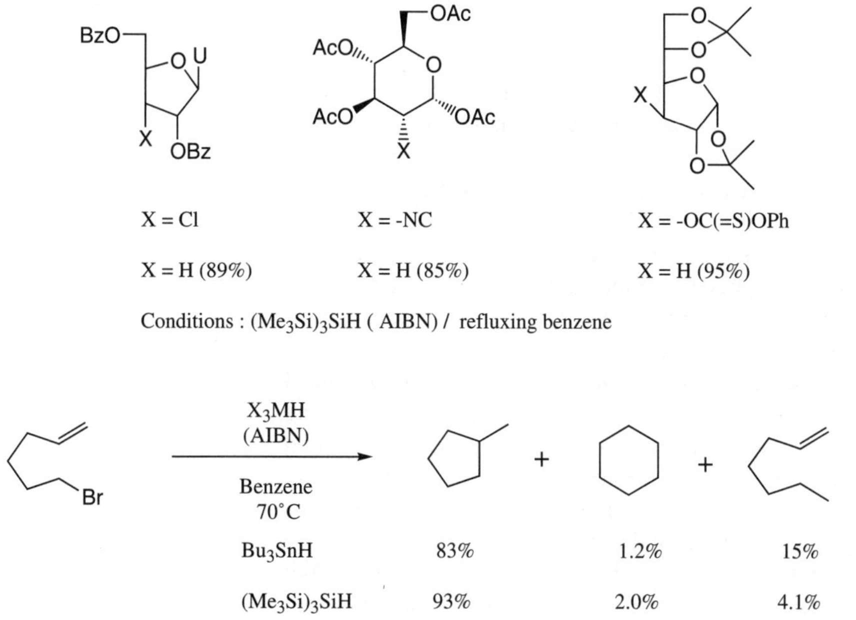
X = Cl
X = H (89%)
X = -NC
X = H (85%)
X = -OC(=S)OPh
X = H (95%)
Conditions : (Me3Si)3SiH ( AIBN) / refluxing benzene
X3MH
(AIBN)
Benzene
70°C
Bu3SnH 83% 1.2% 15%
(Me3Si)3SiH 93% 2.0% 4.1%

**Scheme 4.4**
Reductions with tris (trimethylsilyl) silane

this new silane, albeit with somewhat modified reaction rates (Chatgilialoglu *et al.* 1988; Chatgilialoglu 1992, 1995).

Tris(trimethylsilyl)silane is prepared by the action of trimethylsilyllithium on trichlorosilane in THF at −15 °C (Bürger and Kilian 1969). Halides, selenides, xanthates, isocyanides, and aliphatic nitro derivatives can be reduced in much the same way as with triorganotin hydrides. Some typical examples are displayed in Scheme 4.4 (Chatgilialoglu 1992). The consequences of the difference in the rate of hydrogen atom delivery are revealed by examining the reduction of 5-hexenyl bromide with tributylstannane and tris(trimethylsilyl)silane under the same reaction conditions. As expected, a better yield of cyclized material is obtained with the latter because of the longer lifetime of the intermediate radical.

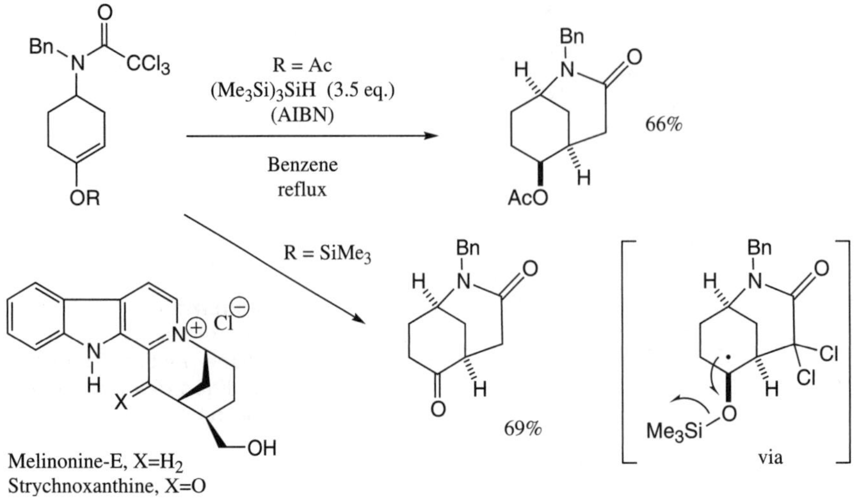

**Scheme 4.5** Cyclization mediated by tris (trimethylsilyl) silane

The slower transfer of hydrogen from tris(trimethysilyl)silane has been exploited for the construction of the azabicyclo[3.3.1]nonane subunit found in some alkaloids such as melinonine-E and strychnoxanthine (Quirante *et al.* 1997, 1998) by a 6-*exo*-cyclization as shown in Scheme 4.5. It is interesting to note that the end product depends on the nature of group R. The 'normal' derivative (only one isomer) is obtained with the enol acetate whereas the intermediate radical in the case of the trimethylsilyl enol ether apparently undergoes fragmentation to give the corresponding ketone. That the latter process indeed involves a homolytic cleavage of the strong Si–O bond remains to be demonstrated unambiguously.

As with stannane-based methodology, it is possible to capture the carbon radical in an intermolecular fashion. The same considerations apply in terms of competing side reactions and it is necessary to ensure the highest possible rates for the desired process by matching the polarities of the intermediate radical and the olefinic trap. This intermolecular variant, outlined in Scheme 4.6, has recently been used to elaborate the side chain of digitoxigenin (Almirante and Cerri 1997). The stereochemistry of the newly formed C–C bond in the adduct with maleic anhydride (which was not isolated but converted directly to the lactone) is dictated by the bowl-shaped steroid backbone containing the *cis*-CD ring junction commonly found in cardenolides.

Whereas bromine abstraction by tris(trimethysilyl)silyl and tributylstannyl radicals takes place with comparable rates, reaction with isocyanides appears to be significantly more efficient with the silicon reagent. Presumably the first step, that is, addition of the silicon-centred radical to the carbon end of the isonitrile group, is less reversible than the analogous stannyl case (cf. Scheme 2.11). This allows for example the reduction of primary isocyanides in refluxing benzene instead of refluxing xylene as with tributyltin hydride. Moreover, intermolecular C–C bond forming reactions by addition of radicals derived from isocyanides to activated olefins become possible, as illustrated by the example in Scheme 4.7.

Another important difference concerns the case of acid chlorides. These often react spontaneously with tributylstannane to give a number of products (aldehydes

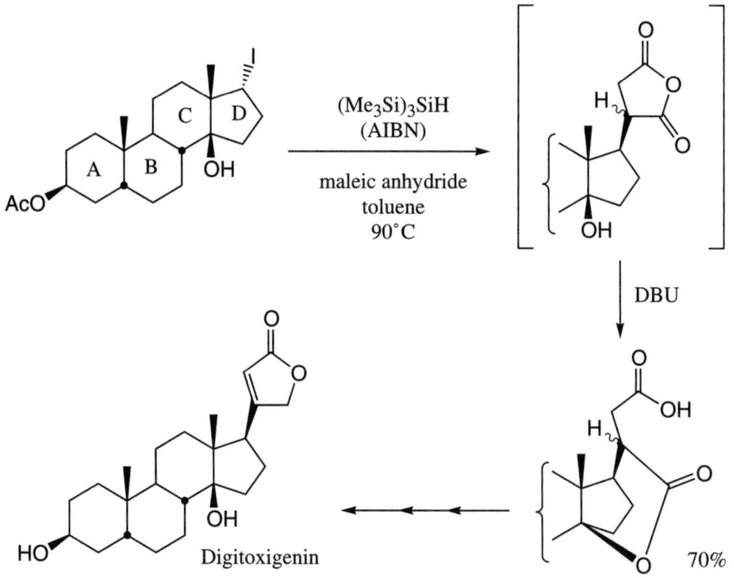

**Scheme 4.6**
Intermolecular radical addition mediated by tris (trimethylsilyl) silane

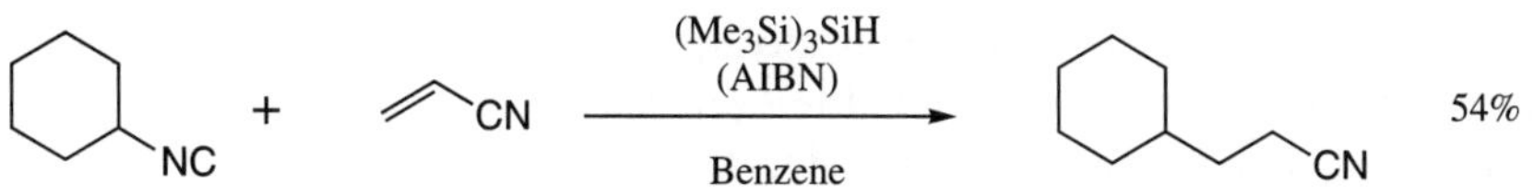

**Scheme 4.7**
Intermolecular radical addition using an isocyanide

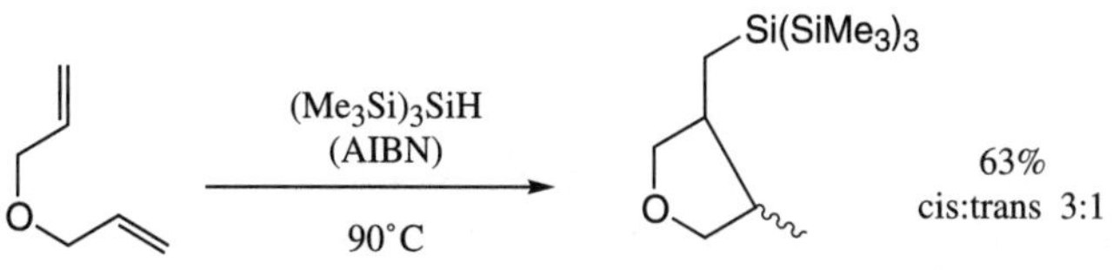

**Scheme 4.8**
Intermolecular addition–cyclization of tris (trimethylsilyl) silane

and esters mostly) arising from an ionic reduction pathway (Lusztyk *et al.* 1984). Hence the preferred use of acyl selenides in combination with organotin reagents to generate acyl radicals. Tris(trimethylsilyl)silane, in contrast, does not normally undergo spontaneous reaction with acid chlorides and can therefore be used to produce acyl radicals from acid chlorides (Ballestri *et al.* 1992). Because of the affinity of silicon for oxygen, silyl radicals do add to carbonyl groups, allowing the reduction of ketones and quinones under certain conditions (Alberti and Chatgilialoglu 1990).

In the same way as for germanium hydrides (Scheme 4.2), the stronger C–Si bond makes tris(trimethylsilyl)silane a good hydrosilylation reagent for alkenes and alkynes: the slower $\beta$-fragmentation step gives the intermediate adduct radical time either to abstract hydrogen or to undergo addition to an internal olefin. The latter variation is illustrated by the hydrosilylation–cyclization pictured in Scheme 4.8 (Kopping *et al.* 1992).

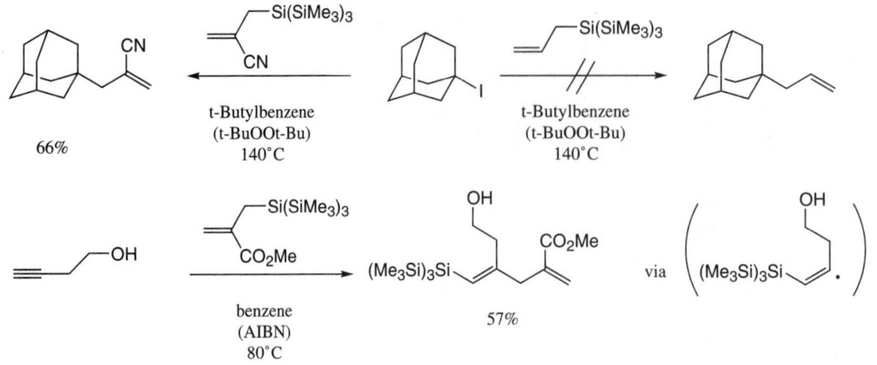

**Scheme 4.9** Addition–fragmentation to various allyl tris (trimethylsilyl) silanes

The reverse of the coin is that allylation reactions, where a fast $\beta$-scission step is an advantage, become less efficient. Thus, allyl tris(trimethylsilyl)silane is a poorer radical allylating reagent in comparison with allyl tributylstannane (Kosugi *et al.* 1991). More initiator and a higher reaction temperature are often required and curiously, the reaction seems to be more susceptible to polar factors. For instance, 'nucleophilic' adamantyl radicals do not add to the naked allyl derivative but will react with the more electrophilic cyano substituted analogue, as shown by the example in Scheme 4.9 (Chatgilialoglu *et al.* 1996). An example of silylation–allylation of an alkyne is also included in the same scheme (Miura *et al.* 1998). The *cis*-isomer is produced because the intermediate vinylic radical reacts from the side opposite to the very bulky silyl group.

## 4.4    Other silanes

Tris(alkylthio)silanes of formula (RS)3SiH, with R = methyl or isopropyl, have also been proposed as cheaper alternatives to tris(trimethylsilyl)silane (Chatgilialoglu *et al.* 1990). Unfortunately, they can liberate thiols on contact with moisture; these are not only obnoxious but can interfere with the radical process. Diphenylsilane has also been examined but does not appear to be sufficiently reactive to be generally useful (Barton *et al.* 1990). 1,9-Disilaanthracenes (Oba and Nishiyama 1994; Gimisis *et al.* 1995) and 1,1,2,2-tetraphenyldisilane (Yamazaki *et al.* 1998, 1999*a,b*; Togo *et al.* 2000) are recent, more promising additions to the family of silanes. The latter is an air stable crystalline solid, made by treating chlorodiphenylsilane with magnesium. It tolerates mildly acidic substances such as pyridinium salts and reactions can be run in refluxing alcohol (Scheme 4.10). Such conditions would usually destroy a stannane through protonation and generation of molecular hydrogen. In the addition to lepidinium trifluoroacetate, more than catalytic amounts of AIBN are needed because, in addition to its role as initiator, it serves to rearomatize the intermediate radical adduct into a pyridine (the overall process is not a chain in this case, cf. Scheme 2.47).

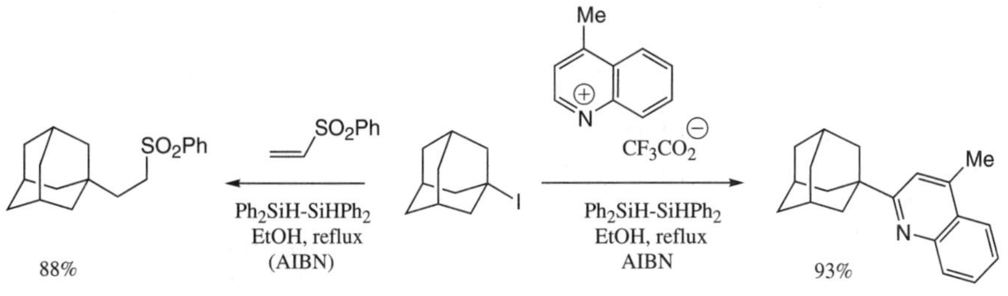

**Scheme 4.10** Silane-mediated intermolecular additions to electrophilic traps

$$R^\bullet \ + \ R'S\text{-}H \quad \longrightarrow \quad R\text{-}H \ + \ R'S^\bullet \quad (a)$$

$$R'S^\bullet \ + \ Et_3Si\text{-}H \quad \longrightarrow \quad R'S\text{-}H \ + \ Et_3Si^\bullet \quad (b)$$

$$Et_3Si^\bullet \ + \ R\text{-}X \quad \longrightarrow \quad Et_3Si\text{-}X \ + \ R^\bullet \quad (c)$$

**Scheme 4.11** Polarity reversal catalysis by a thiol

## 4.5  Polarity reversal catalysis

In addition to the inherent strength of the Si–H bond in simple silanes, the abstraction of the electron-rich hydrogen is compounded by the fact that alkyl radicals are themselves generally nucleophilic in character. The polarity mismatch that also obtains with stannanes is compensated by the weakness of the Sn–H bond. The S–H bond in aliphatic thiols is as strong as the Si–H bond in triethylsilane but has the reverse polarity: the rate of hydrogen atom transfer from a thiol to a nucleophilic alkyl radical becomes comparable to that from tributylstannane even though it is considerably less exothermic (Newcomb 1993). These considerations have resulted in an ingenious way to circumvent the slow hydrogen abstraction step from triethylsilane by replacing it with a cycle of two faster reactions (a) and (b) (Scheme 4.11), where a thiol acts as a relay and is not consumed overall (Cole *et al.* 1991; Roberts and Winter 1999).

The second step (b) is essentially thermoneutral but favoured by the electrophilicity of the thiyl radical and the nucleophilic character of the hydrogen attached to the silicon. This polarity reversal catalysis is based on earlier observations by Harris and Waters (1952) who found that small amounts of thiol accelerated the radical decarbonylation of aldehydes.

The efficient reduction of cholestanyl bromide or xanthate with triethylsilane and a catalytic amount of *t*-dodecanethiol (Cole *et al.* 1991) demonstrates the potential of the method (Scheme 4.12). No reduction occurs in the absence of the thiol, and cyclohexane or octane are better in this case than the more commonly used benzene because of the tendency of silyl (and but to a lesser extent germyl) radicals to add to the aromatic ring (Chatgilialoglu *et al.* 1983; Chatgilialoglu 1995). The choice of the initiator is dictated by its half-life at the reflux temperature of the solvent.

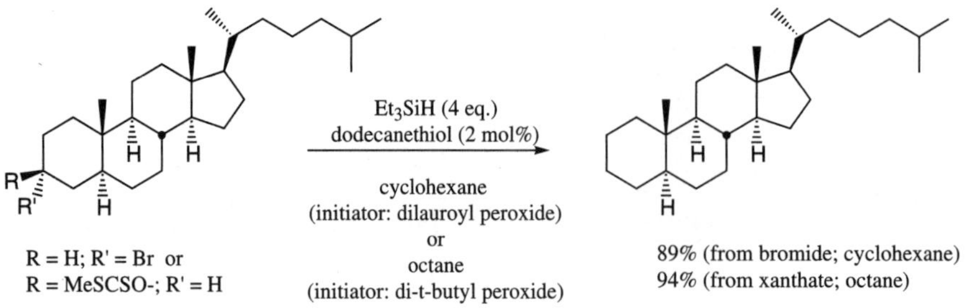
Et$_3$SiH (4 eq.)
dodecanethiol (2 mol%)
cyclohexane
(initiator: dilauroyl peroxide)
or
octane
(initiator: di-t-butyl peroxide)
R = H; R' = Br or
R = MeSCSO-; R' = H
89% (from bromide; cyclohexane)
94% (from xanthate; octane)

**Scheme 4.12** Reductions exploiting polarity reversal catalysis.

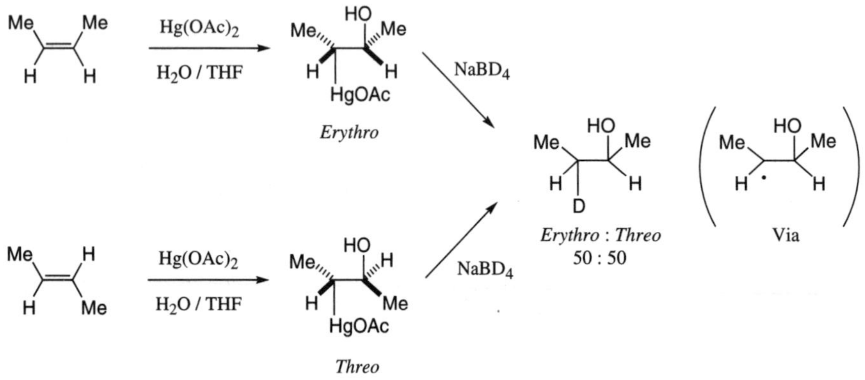
Hg(OAc)$_2$
H$_2$O / THF
HgOAc
Erythro
Threo
NaBD$_4$
Erythro : Threo
50 : 50
Via

**Scheme 4.13** Stereochemistry of the reductive demercuration reaction

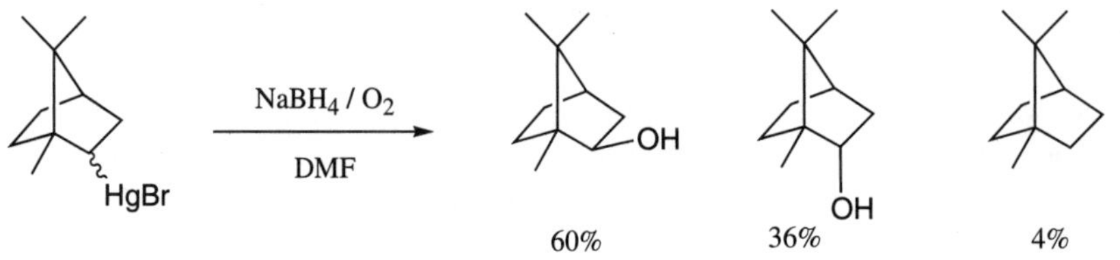
HgBr
NaBH$_4$ / O$_2$
DMF
OH
60%
36%
4%

**Scheme 4.14** Reductive demercuration in the presence of oxygen

## 4.6 Chain reactions of organomercury hydrides

Of the other metal or metalloid hydrides that could conceivably participate in radical processes, by far the most important from a synthetic standpoint are organomercury hydrides. Interest in the radical chemistry of organomercurials (Barluenga and Yus 1988; Russell 1990) blossomed when it was established that the classical demercuration reaction with sodium borohydride was in fact a radical chain reaction. Pasto and Gontarz (1969) found for example that both the *erythro*- and the *threo*-isomers from the acetoxy mercuration of *trans*- and *cis*-butene gave the same 50 : 50 mixture of *erythro*- and *threo*-3-deutero-2-butanol upon reduction with sodium borodeuteride (Scheme 4.13). The complete loss of stereochemical information could only be rationalized by invoking a carbon-centred radical intermediate.

Further evidence was adduced from the observation that alcohols were obtained when the reduction was performed under air (Whitesides and San Filippo 1970;

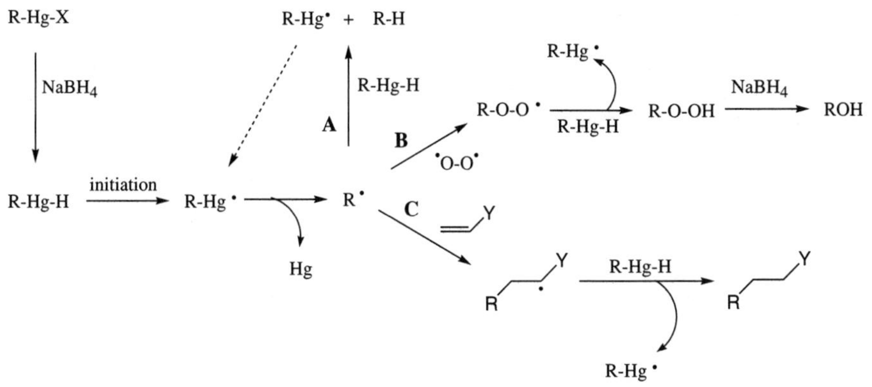

**Scheme 4.15** Reaction manifold of organomercury hydrides

Quirk 1972). Thus, isomeric bicyclomercuric bromides in Scheme 4.14 gave, upon exposure to sodium borohydride and oxygen in DMF, a high yield of a 3:2 mixture of *exo*- and *endo*-alcohols instead of the corresponding alkane, which was only a very minor component in the product mixture (Hill and Whitesides 1974).

As delineated in Scheme 4.15, the intervention of a labile organomercury hydride which gives rise to a carbon radical appears to be the most logical explanation. Aliphatic organomercury hydrides are so labile that there is usually no need to add any initiator: traces of oxygen (or perhaps some metallic impurities) are sufficient to trigger the radical chain. The organomercury radical extrudes mercury to produce a carbon radical which propagates the chain by abstracting a hydrogen atom from another molecule of hydride (path **A**). If the reaction is conducted in the deliberate presence of molecular oxygen, the carbon radical can be captured to give ultimately an alcohol by reduction of the intermediate hydroperoxide (path **B**).

The carbon radical may also be trapped in a variety of other ways such as with an external or internal olefin (path **C**). As with the tin- and silicon-based methods, the trapping process has to compete with the hydrogen atom transfer step of path **A**. The rate of the latter has been estimated to be at least as high as $10^7$ M$^{-1}$s$^{-1}$ (Giese and Kretzschmar 1984); the trapping agent must therefore be especially reactive in order to succeed. In practice, the organomercury hydride is not isolated but simply generated *in situ* at room temperature or below by reduction of an organomercuric halide or carboxylate, usually with a borohydride, but other reducing agents including stannanes, can be employed. The rate of addition of the reductant controls the concentration of the hydride; metallic mercury separates out and is easily removed by decantation. For small scale work, degassing of the solvent is especially important because of the relative solubility of oxygen in organic solvents at room temperature, unless of course the alcohol is the desired product.

## 4.7  Organomercury hydrides: synthetic applications

Organomercury hydrides constitute a convenient and synthetically highly useful source of carbon-centred radicals because of the diversity of methods allowing access to the precursors. Oxymercuration of olefins or cyclopropanes and mercuration of

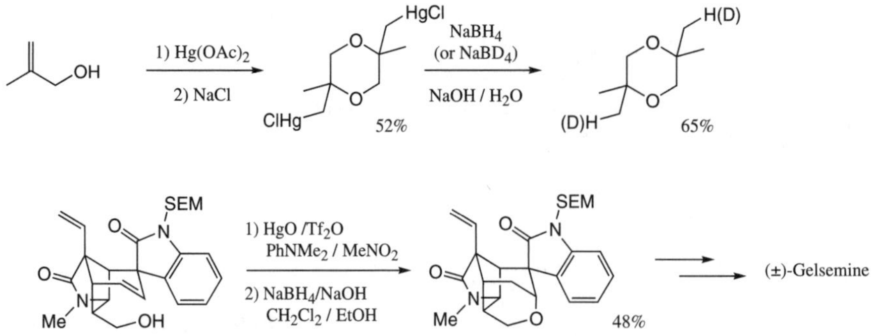

**Scheme 4.16** Examples of oxymercuration–demercuration

and 8α-isomer

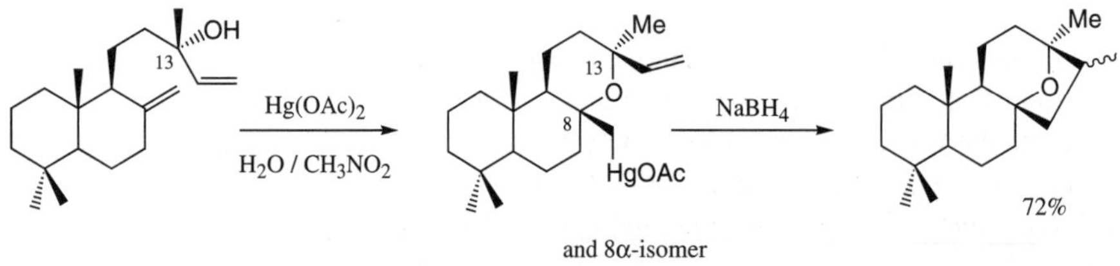

**Scheme 4.17** Regioselective oxymercuration–demercuration

organometallic reagents are especially powerful and versatile in this regard. The mercuric group may be simply replaced by hydrogen or deuterium as in the mechanistic study discussed in Scheme 4.13 above. Introducing deuterium or tritium via organomercury derivatives is especially easy. This is shown by the first sequence in Scheme 4.16 (Bordwell and Douglass 1966); the organomercuric precursor in this case was constructed by a double alkoxymercuration reaction. Addition of sodium chloride causes the formation of the organomercuric chloride; these are usually insoluble and are isolated by mere filtration. An intramolecular alkoxy-mercuration–demercuration sequence was used in the late stages of a recent synthesis of gelsemine (Newcombe *et al.* 1994).

The expedient construction of the strobane skeleton from 13-*epi*-manool (Scheme 4.17) provides an example of intramolecular capture of the intermediate carbon radical (Matsuki *et al.* 1979). The acetoxymercuration step leads to two epimeric derivatives at C-8 but only the β-isomer can undergo the desired cyclization. This occurs quite efficiently to give one isomer whose stereochemistry has not been ascertained.

Like with stannane chemistry, intermolecular capture of the carbon radical is more difficult to master because of the competitive fast hydrogen atom transfer from the organomercury hydride. Only highly reactive olefins such as methyl acrylate, methylvinyl ketone, or acrylonitrile are suitable, and they have to be used in excess (5–10-fold). Some examples are collected in Scheme 4.18. The variety lies really in the way the mercuration is conducted rather than in the radical reactions themselves.

The first shows how a glucal can be elaborated into a sugar with a new C–C bond in the 2-position (Giese and Gröninger 1984) whereas the second involves a nitrogen atom as the neighbouring group participant in the acetoxymercuration step

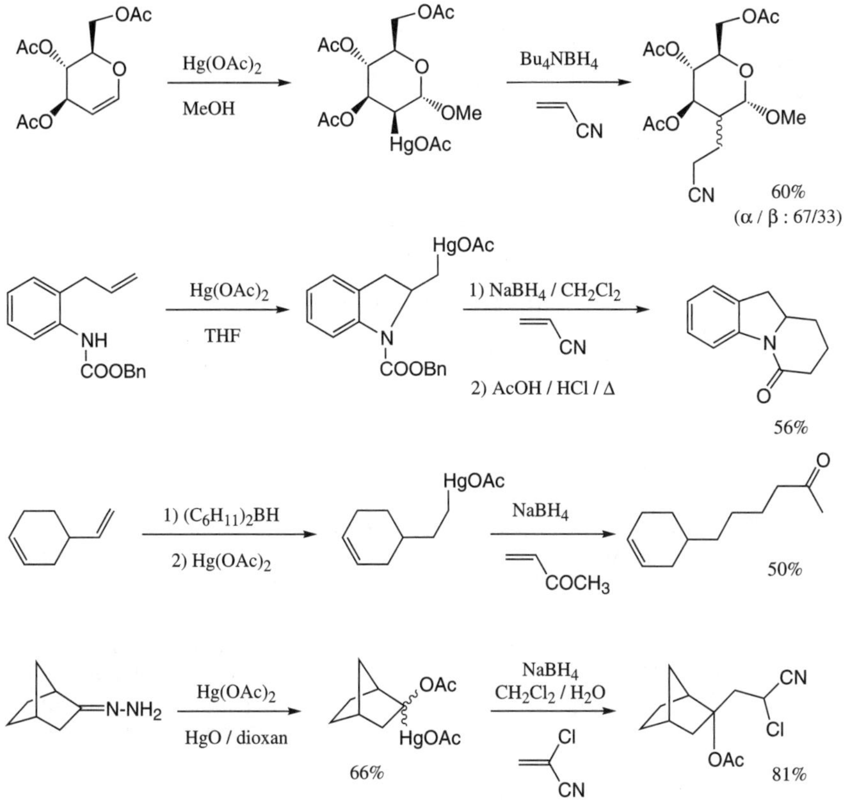

**Scheme 4.18**
Variations on the mercuration–demercuration reaction

(Danishefsky *et al.* 1983). Heating the adduct with acid provides a lactam directly. In the third example (Giese and Kretzschmar 1981), advantage is taken of the selectivity in the hydroboration process to place the mercury on the terminus of the least hindered olefin (direct acetoxymercuration would have involved the more electron-rich *endo*cyclic olefin). It is interesting to note that because the demercuration is performed in a two-phase system, the ketone function survives the presence of borohydride. Finally, acetoxy-mercuration of a hydrazone leads to an intermediate where the mercury and the acetoxy group are on the same carbon (Giese and Erfort 1983). The stereochemical issue is not important here since both epimers will lead to the same radical, which will react from the least hindered *exo*-face. This transformation represents an overall polarity reversal of the normally electrophilic ketone carbon: it now behaves as a nucleophile towards the electrophilic olefin.

To conclude this chapter, we shall illustrate the synthetic potential of the mercuric acetate induced opening of cyclopropane rings with the example pictured in Scheme 4.19 (Giese and Horler 1983). Acetoxymercuration of the trimethylsiloxy cyclopropane in the absence of water gives a silyloxyacetate which undergoes the normal demercurative addition to chloroacrylonitrile. If water is present, however, this compound is rapidly hydrolysed into the corresponding aldehyde and the intermediate radical can now rearrange with an efficiency that depends on the concentration of the olefinic trap: at higher dilution, the initial primary radical has time to

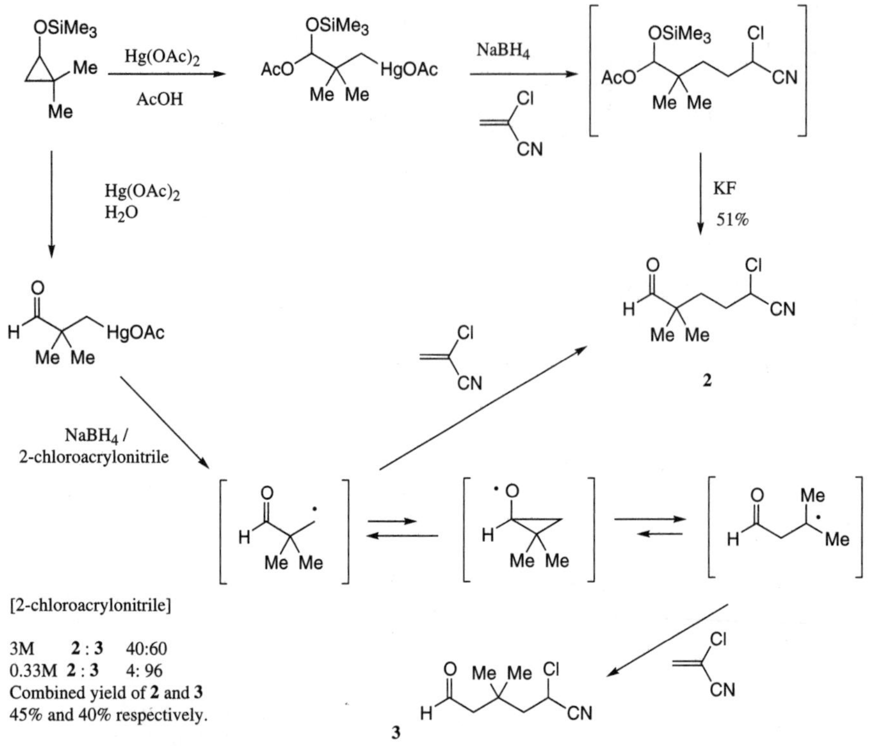

**Scheme 4.19**
Intermolecular additions using the mercuration–demercuration reaction

evolve into the more stable tertiary radical which finally gives adduct **3**. Both isomeric products **2** and **3** can thus be prepared by a simple modification of the experimental conditions.

## References

Alberti, A. and Chatgilialoglu, C. (1990). Addition of tris(trimethylsilyl)silyl radicals to the carbonyl group of ketones and quinones. *Tetrahedron*, **46**, 3963–3972.

Almirante, M. and Cerri, A. (1997). Synthesis of digitoxigenin from 3β[(tert-butyldimethylsilyl)oxy]-17α-iodo-5β-androstan-14β-ol via 17β stereoselective free radical introduction of γ-butyrolactone moiety. *Journal of Organic Chemistry* **62**, 3402–3404.

Ballestri, M., Chatgilialoglu, C., Cardi, N., and Sommazzi, A. (1992). The reaction of tris(trimethylsilyl)silane with acid chlorides. *Tetrahedron Letters*, **33**, 1787–1790.

Barluenga, J. and Yus, M. (1988). Free radical reactions of organomercurials. *Chemical Reviews*, **88**, 487–509.

Barton, D. H. R., Jang, D. O., and Jaszberenyi, J. Cs. (1990). An improved radical chain procedure for the deoxygenation of secondary and primary alcohols using diphenysilane as hydrogen atom donor and triethylborane–air as initiator. *Tetrahedron Letters*, **31**, 4681–4684.

Bordwell, F. G. and Douglass, M. L. (1966). Reduction of alkylmercuric hydroxides by sodium borohydride. *Journal of the American Chemical Society*, **88**, 993–999.

Bürger, H. and Kilian, W. (1969). Spektroskopische untersuchungen an tris (trimethylsilyl) silan und -silan-d1. *Journal of Organometallic Chemistry*, **18**, 299–306.

Chatgilialoglu, C. (1992). Organosilanes as radical-based reducing agents in synthesis. *Accounts of Chemical Research*, **25**, 188–194.

Chatgilialoglu, C. (1995). Structural and chemical properties of silyl radicals. *Chemical Reviews*, **95**, 1229–1251.

Chatgilialoglu, C. and Newcomb, M. (1999). Hydrogen donor abilities of the group 14 hydrides. *Advances in Organometallic Chemistry*, **44**, 67–112.

Chatgilialoglu, C., Ferreri, C., and Lucarini, M. (1993). A comment on the use of triethylsilane as a radical reducing agent. *Journal of Organic Chemistry*, **58**, 249–251.

Chatgilialoglu, C., Griller, D., and Lesage, M. (1988). Tris(trimethylsilyl)silane. A new reducing agent. *Journal of Organic Chemistry*, **53**, 3642–3644.

Chatgilialoglu, C., Griller, D., and Lesage, M. (1989). Rate constants for the reactions of tris(trimethylsil)silyl radicals with organic halides. *Journal of Organic Chemistry*, **54**, 2492–2494.

Chatgilialoglu, C., Guerrini, A., and Seconi, G. (1990). Tris(alkylthio)silanes as new reducing agents via radicals. *Synlett*, **53**, 219–220.

Chatgilialoglu, C., Ingold, K. U., and Scaiano, J. (1983). Absolute rate constants for the addition of triethylsilyl radicals to various unsaturated compounds. *Journal of the American Chemical Society*, **105**, 3292–3296.

Chatgilialoglu, C., Ferreri, C., Ballestri, M., and Curran, D. P. (1996). 2-Functionalized allyl tris(trimethylsilyl)silanes as radical based allylating agents. *Tetrahedron Letters*, **37**, 6387–6390.

Cole, S. J., Kirwan, J. N., Roberts, B. P., and Willis, C. R. (1991). Radical chain reductions of alkyl halides, dialkyl sulphides and *O*-alkyl-*S*-methyl dithiocarbonates to alkanes by trialkylsilanes. *Journal of the Chemical Society, Perkin Transactions 1*, 103–112.

Curran, D. P., Xu, J., and Lazzarini, E. (1995). Controlling radical chain reactions by unimolecular chain transfer. Intramolecular hydrogen transfer of silicon hydrides. *Journal of the American Chemical Society*, **117**, 6603–6604.

Danishefsky, S., Taniyama, E., and Webb II, R. R. (1983). Tetrahydropyridines via intramolecular ureidomercuration. *Tetrahedron Letters*, **24**, 11–14.

Giese, B. and Erfort, U. (1983). Umpolungsreactionenen über radikale: CC bindungsbildung zwischen Ketonen und Alkanen. *Chemische Berichte*, **116**, 1240–1251.

Giese, B. and Gröninger, K. S. (1984). Diastereoselective synthesis of branched 2-deoxysugars via radical CC bond formation reactions. *Tetrahedron Letters*, **25**, 2743–2746.

Giese, B. and Horler, H. (1983). CC Bond formation reactions with umpolung of aldehydes via radicals. *Tetrahedron Letters*, **24**, 3221–3224.

Giese, B. and Kretzschmar, G. (1981). CC Bond formation between 'electron-rich' and 'electron-deficient' alkenes. *Angewandte Chemie International Edition in English*, **20**, 965–966.

Giese, B. and Kretzschmar, G. (1984). Bestimmung absoluter Geschwindigkeitskonstanten bei der radikalischen Addition an Alkene nach der 'Quecksilber-methode'. *Chemische Berichte*, **117**, 3160–3164.

Gimisis, T., Ballestri, M., Ferreri, C., Chatgilialoglu, C., Boukherroub, R., and Manuel, G. (1995). 5,6-Dihydrosilanthrene as a reagent for the Barton–McCombie reaction. *Tetrahedron Letters*, **36**, 3897–3900.

Harris, E. F. P. and Waters, W. A. (1952). Thiol catalysis of the homolytic decomposition of aldehydes. *Nature*, **70**, 212–213.

Hill, C. L. and Whitesides, G. M. (1974). Reactions of alkylmercuric halides with sodium borohydride in the presence of molecular oxygen. *Journal of the American Chemical Society*, **96**, 870–876.

Johnston, L. J., Lusztyk, J., Wayner, D. D. M., Abeywickreyma, A. N., Beckwith, A. L. J., Scaiano, J., and Ingold, K. U. (1985). Absolute rate constants for reaction of phenyl, 2,2-dimethylvinyl, cyclopropyl, and neopentyl radicals with tri-*n*-butylstannane. Comparison of the radical trapping abilities of tri-*n*-butylstannane and -germane. *Journal of the American Chemical Society*, **107**, 4594–4596.

Kanabus-Kaminska, J.-M., Hawari, J. A., Griller, D., and Chatgilialoglu, C. (1987). Reduction of silicon–hydrogen bond strengths. *Journal of the American Chemical Society*, **109**, 5267–5268.

Kopping, B., Chatgilialoglu, C., Zehnden, M., and Giese, B. (1992). (Me₃Si)₃SiH: an efficient hydrosilylating agent. *Journal of Organic Chemistry*, **57**, 3994–4000.

Kosugi, M., Kurata, H., Kawata, K., and Migita, T. (1991). Allyltris(trimethylsilyl)silane as a radical allylating agent for organic halides. *Chemistry Letters*, 1327–1328.

Light, J. P. II, Ridenour, M., Beard, L., and Hershberger, J. W. (1987). Reactivity of allylic and vinylic silanes, germanes, stannanes and plumbanes toward S$_H$2′ and S$_H$2 substitution by carbon or heteroatom-centered radicals. *Journal of Organometallic Chemistry*, **326**, 17–24.

Lusztyk, J., Lusztyk, E., Maillard, B., and Ingold, K. U. (1984). Reaction of organotin hydrides with acid chlorides. Mechanism of aldehyde and ester formation. *Journal of the American Chemical Society*, **106**, 2923–2931.

Matsuki, Y., Mitsuaki, K., and Itô, S. (1979). C–C Bond formation on reduction of organomercurials and its application to biomimetic synthesis of strobane carbon skeleton. *Tetrahedron Letters*, **20**, 4081–4084.

Miura, K., Saito, H., Nagakawa, T., Hondo, T., Tateiwa, J., and Sonoda, M. (1998). Allylsilylation of carbon–carbon and carbon–oxygen unsaturated bonds via a radical process. *Journal of Organic Chemistry*, **63**, 5740–5741.

Musa, O. M., Horner, J., and Newcomb, M. (1999). Laser flash photolysis measurementsof the kinetics of carbon nitrogen bond rotations in α-amide radicals. *Journal of Organic Chemistry*, **64**, 1022–1025.

Newcomb, M. (1993). Competition methods and scales for alkyl radical reaction kinetics. *Tetrahedron*, **49**, 1151–1176.

Newcombe, N. J., Ya, F., Vijn, R. J., Hiemstra, H., and Speckamp, W. N. (1994). The total synthesis of (±)-gelsemine. *Journal of the Chemical Society, Chemical Communications*, 767–768.

Nozaki, K., Ichinose, Y., Wakamatsu, K., Oshima, K., and Utimoto, K. (1990). Et₃B-induced stereoselective radical addition of Ph₃GeH to carbon–carbon multiple bonds and its application to isomerisation of olefins. *Bulletin of the Chemical Society of Japan*, **63**, 2268–2272.

Oba, M. and Nishiyama, K. (1994). 9,10-Dihydro-9,10-disilaanthracene as a new radical-based reducing agent: importance of transannular interaction between silyl radical and silicon atom. *Journal of the Chemical Society, Chemical Communications*, 1703–1704.

Pasto, D. J. and Gontarz, J. (1969). The mechanism of the reduction of organomercurials with sodium borohydride. *Journal of the American Chemical Society*, **91**, 719–721.

Pike, P., Hershberger, S., and Hershberger, J. (1988). Evaluation of tributylgermanium hydride as a reagent for the reductive alkylation of active olefins with alkyl halides. *Tetrahedron*, **44**, 6295–6304.

Stork, G. and Mah, R. (1989). Radical cyclisation of allylic haloacetamides. A route to *cis*-fused 2-pyrrolidones and piperidones. *Heterocycles*, **28**, 723–727.

Quirante, J., Escolano, C., Costejà, L, and Bonjoch, J. (1997). Cyclisation of 1-(carbamoyl) dichloromethyl radicals upon activated alkenes. A new entry to 2-azabicyclo[3.3.1]nonanes. *Tetrahedron Letters*, **38**, 6901–6904.

Quirante, J., Escolano, C., Merino, A., and Bonjoch, J. (1998). First total synthesis of (±)-melinonine and (±)-strychnoxanthine using a radical cyclisation process as the core ring forming step. *Journal of Organic Chemistry*, **63**, 968–976.

Quirk, R. P. (1972). The sodium borohydride reduction of 2,2,2-triphenylethylmercuric chloride. *Journal of Organic Chemistry*, **37**, 3554–3555.

Roberts, B. P. and Winter, J. N. (1999). Polarity-reversal catalysis of hydrogen-atom abstraction reactions: concepts and applications in organic chemistry. *Chemical Society Reviews*, 25–35.

Russell, G. A. (1990). Free radical chain reactions involving saturated and unsaturated organomercurials. In *Advances in free radical chemistry*, Vol. 1 (ed. D. D., Tanner), pp. 1–52. JAI Press Inc. Stamford.

Togo, H., Matsubayashi, S., Yamazaki, O., and Yokoyama, M. (2000). Deoxygenative functionalization of hydroxy groups via xanthates with tetraphenyldisilane. *Journal of Organic Chemistry*, **65**, 2816–2819.

Whitesides, G. M. and San Filippo, J. (1970). The mechanism of reduction of alkylmercuric halides. *Journal of the American Chemical Society*, **92**, 6611–6624.

Yamazaki, O., Togo, H., Matsubayashi, S., and Yokoyama, M. (1999a). Tetraaryldisilane as a novel strategic radical reagent. *Tetrahedron*, **55**, 3735–3747.

Yamazaki, O., Togo, H., and Yokoyama, M. (1999b). Synthetic utility of 1,1,2,2-tetraaryldisilanes: radical reduction of alkyl phenyl chalcogenides. *Journal of the Chemical Society, Perkin Transactions 1*, 2891–2896.

Yamazaki, O., Togo, H., Matsubayashi, S., and Yokoyama, M. (1998). 1,1,2,2-tetraphenyldisilane as a diversified radical reagent. *Tetrahedron Letters*, **39**, 1921–1924.

# 5  The Barton decarboxylation and related reactions

## 5.1  Some early radical decarboxylation processes

The carboxylic acid functional group is ubiquitous in nature, and the possibility of manipulating it using radical chemistry opens tremendous synthetic opportunities. It was known from the work of Kolbe in the middle of the nineteenth century that the electrolysis of carboxylic acid salts gave alkanes which seemed to arise from the joining of two of the alkyl groups of the starting carboxylic acid (Scheme 5.1). This reaction proceeds through an acyloxyl (also called carboxylic) radical which undergoes loss of $CO_2$. This fragmentation is exceedingly fast for aliphatic and alicyclic carboxyl radicals ($k > 10^9 \, s^{-1}$ at 25 °C; Kochi *et al.* 1997) but much slower in the case of radicals derived from aromatic or vinylic carboxylic acids ($k \approx 10^6 \, s^{-1}$ at 25 °C; Chateauneuf *et al.* 1988*a,b*). The Kolbe reaction is not a chain process and has mostly been applied to make symmetrical derivatives as shown by the recent example displayed in Scheme 5.1 (Seebach and Renaud 1985). An alkane may also be obtained if the electrolysis is conducted in the presence of a good hydrogen atom donor such as a thiol.

Another reaction which was discovered later is the Borodin–Hunsdiecker reaction (it is the same Alexander Borodin, the music composer) where a silver (or mercury or thallium) salt of a carboxylic acid is exposed to bromine or (less commonly) iodine to give the corresponding alkyl or aryl bromide or iodide, respectively, as shown in Scheme 5.2 (Johnson and Ingham 1956; Wilson 1957; Sheldon and Kochi 1972).

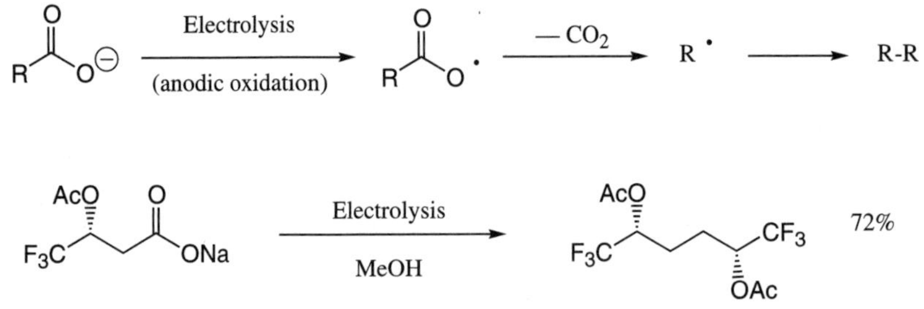

Scheme 5.1  The Kolbe decarboxylation reaction

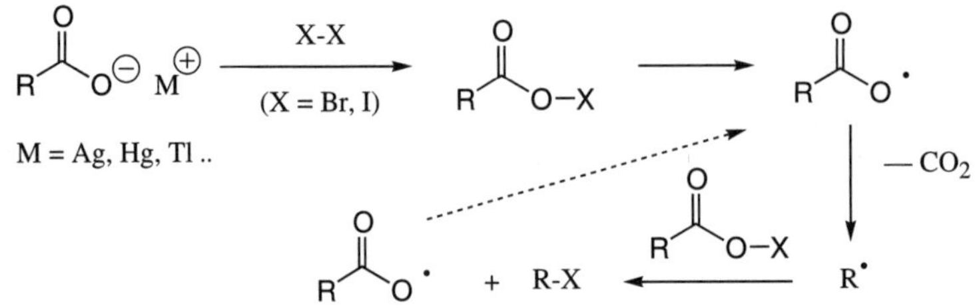

Scheme 5.2  The Borodin–Hunsdiecker reaction

A radical chain appears to be involved but the reaction is often capricious: pure, anhydrous salts are often required for an efficient and reproducible transformation. These practical drawbacks along with cost and toxicity considerations have greatly hindered the utility of this otherwise mechanistically interesting process.

The thermal decomposition of acyl peroxides (RC(O)O–O(O)R) also proceeds through carboxyl radicals and such derivatives are commonly used as initiators (Fujimori 1992; Ryzhkov 1996), but their potential for the radical decarboxylation of carboxylic acids has not yet been fully exploited. One limitation is their oxidizing power and the possibility of a strongly exothermic decomposition that can be induced by various organic substances or by metallic salts.

The problem of decarboxylating carboxylic acids had to wait for nearly a century before a general, elegant solution was discovered by Barton and his group (Barton *et al.* 1983). This powerful method emerged, like many others in this book, through a combination of rational design and serendipity. The study of the Barton decarboxylation will constitute the subject matter of much of this chapter.

## 5.2 The Barton decarboxylation: early experiments

As we have seen, $\beta$-scission of a carbon–oxygen bond is usually slow on the timescale of radical reactions. This permits the generation and manipulation, with impunity, of radicals located on carbohydrate and related structures where carbon–oxygen bonds $\beta$- to the radical centre are almost always present. $\beta$-Scission is sluggish even when the C–O bond is of the benzylic type as in compound **1** which yields only the reduced material **2** and no stilbene when produced by the stannane method (Scheme 5.3). However, further weakening of the C–O bond can reach a point where $\beta$-scission becomes competitive and hence useful as a source of acyloxy radicals. This point is attained with esters of type **3**, where formation of the aromatic phenanthrene provides the extra push in favour of the desired $\beta$-scission. A carboxylic acid can therefore be converted into the corresponding derivative **3** and subjected to

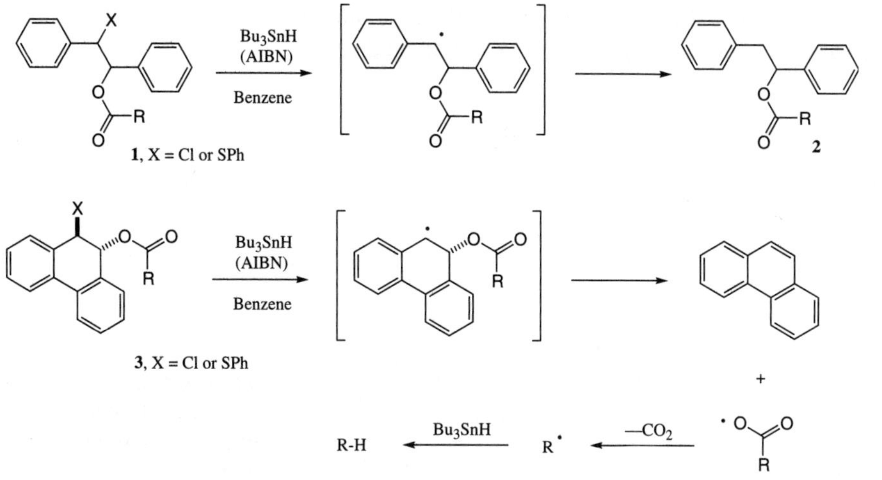

**Scheme 5.3** A phenanthrene-based decarboxylation reaction

the action of tributyltin hydride in the presence of small amounts of AIBN leading to the smooth formation of the corresponding nor-alkane (Barton *et al.* 1980).

In practice, this conceptually appealing method for reductive decarboxylation suffers from the fact that esters **3**, as well as their corresponding alcohol precursors, are fragile substances, tending to rearomatize through ionic elimination at the slightest provocation. Nevertheless, the feasibility of the approach was demonstrated and the way cleared for the development of a far more practical and incredibly versatile reaction based on hydroxamate esters.

## 5.3  The radical chemistry of Barton thiohydroxamate esters

Esters derived from *N*-hydroxy-2-thiopyridone embody all the features required for a successful interaction with stannyl radicals: the high affinity of stannyl radicals for the sulfur of a thiocarbonyl group; the weakness of the N–O bond; and, finally, the formation of an aromatic pyridine ring (Barton *et al.* 1983, 1985*a*; Barton 1993). One example, the reductive decarboxylation of a bile acid, is outlined in Scheme 5.4.

This second method for decarboxylating carboxylic acids, works beautifully for a variety of primary, secondary, or tertiary carboxylic acids. In the early part of the work, the intermediate esters (now often referred to as Barton esters) were not isolated but made and treated *in situ* with tributyltin hydride. When the reaction with the very same bile ester **4** pictured in Scheme 5.5 was repeated in refluxing toluene, the reaction, surprisingly, appeared to proceed much more slowly than in benzene. It turned out that the initial ester **4** had undergone partial transformation into a pyridyl sulfide **6** (with very similar chromatographic mobility on silica gel as the starting ester) and the latter reacted only sluggishly with the stannane to give the same nor-alkane **5** (Barton *et al.* 1983, 1985*a*). Heating or irradiating in the absence of stannane gives cleanly the same pyridyl sulfide **6** (Scheme 5.5). Later studies indicated that pure solutions of Barton esters are stable in refluxing benzene if light or oxygen are excluded but start decomposing in refluxing toluene (Boivin *et al.* 1991).

More generally, this 'decarboxylative rearrangement' of *N*-hydroxy-2-thiopyridone esters **7** occurs by a chain mechanism, as outlined in Scheme 5.6. Carboxylic radicals produced in the initiation step rapidly extrude a molecule of carbon dioxide to give radicals R• which in turn add to the thiocarbonyl group to give intermediate **8**.

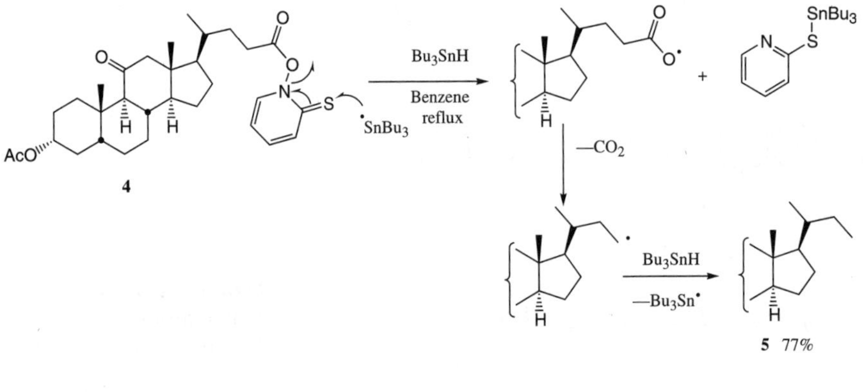

**Scheme 5.4** The Barton decarboxylation via *N*-hydroxy-thiopyridone esters

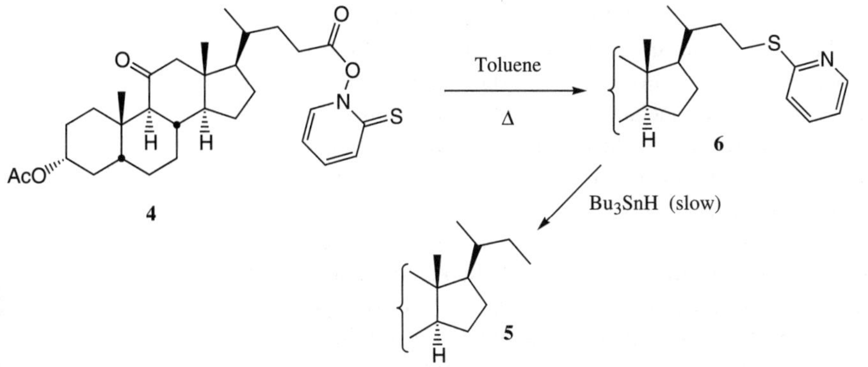

**Scheme 5.5**
Decarboxylative
rearrangement of
*N*-hydroxy-thiopyridone
esters

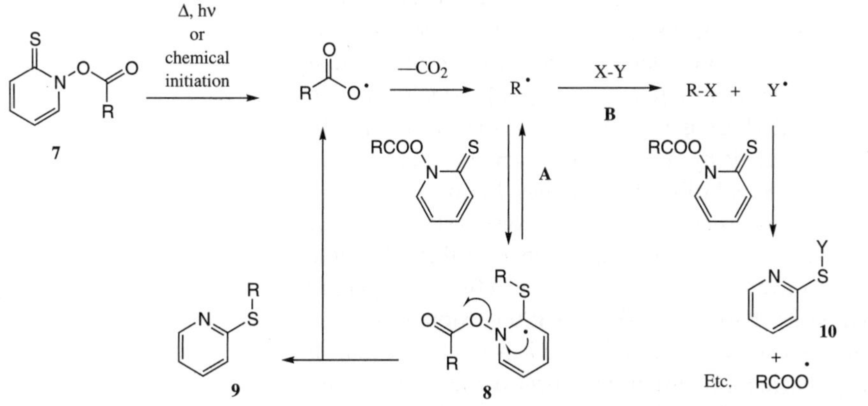

**Scheme 5.6**
Mechanism of the
decarboxylative
rearrangement of
*N*-hydroxy-thiopyridone
esters

This adduct then undergoes $\beta$-scission to sulfide **9** (or R–SPy), with concomitant liberation of a carboxyl radical to propagate the chain (path **A**). The addition of alkyl radicals to the thiocarbonyl group was later shown to be reversible (Barton *et al.* 1987*a*).

The overall process represents a convenient way for generating radicals from carboxylic acids since the intermediate radical R$^\bullet$ can be intercepted by various radicophiles represented by X–Y to give R–X via path **B**, instead of pyridyl sulfide **9**. The Y$^\bullet$ radical produced in step **B** in turn reacts with ester **7** to give PyS–Y **10**, thus also propagating the chain. One example we have just seen is X–Y=HSnBu$_3$ in Scheme 5.4 above. The inevitable competition between paths **A** and **B** can be biased in favour of the latter by exploiting all the theoretical or practical devices discussed in previous chapters (polarity matching between R$^\bullet$ and X–Y, concentration, mode of addition etc.). The fact that the decarboxylative rearrangement involves a fast *but reversible bimolecular* addition of R$^\bullet$ to the thiocarbonyl group of the thiopyridone moiety, followed by an irreversible *unimolecular* $\beta$-scission of the N–O bond with a large positive entropy term provides another powerful lever, and that is temperature effects. Lowering the temperature will slow down considerably the fragmentation step but will not affect very much the addition to the thiocarbonyl which, being reversible, does not consume R$^\bullet$. Nor will the temperature change greatly affect the bimolecular reaction with the X–Y trap. Thus, overall, lowering the temperature will be more detrimental to the decarboxylative rearrangement than to the desired trapping reaction. The selective influence of temperature in the Barton decarboxylation reflects in a way the subtleties inherent in the mechanism, and translate in practice into a simple and easily implemented contrivance for favouring one pathway against the other. But, before proceeding to the many synthetic applications, it is perhaps useful to discuss in more detail some practical aspects.

N-Hydroxy-2-thiopyridone, as its sodium salt, is a speciality chemical, prepared industrially on a multi-ton scale. Its zinc salt ('zinc pyrithione') is often incorporated in shampoos as an anti-dandruff agent, whereas the sodium salt is an anti-fungal and anti-microbial preserving additive in many commercial preparations. The industrial sodium salt is a 40% aqueous solution (sometimes called sodium omadine, available from Olin Corp) from which both the salt and the free N-hydroxy-2-thiopyridone can be easily recovered (Barton *et al.* 1987*a*). The esters used in the Barton decarboxylation reaction are prepared by applying the usual methods for ester synthesis (Scheme 5.7). In addition, N-hydroxy-2-thiopyridone can be activated by reaction with phosgene and related imidoyl chlorides to give salts (e.g. **11**) which combine readily with the carboxylic acid itself (Barton *et al.* 1985*a*; Garner *et al.* 1998). Another convenient route to Barton esters is by reaction of the acid with a combination of 2,2′-dithio-bis(pyridine-N-oxide) and tributyl phosphine (Barton and Samadi 1992). It must be remembered that Barton esters are of the 'activated ester' type, that is their ionic chemistry is more akin to that of anhydrides than to that of simple esters. In other words, they tend to hydrolyse easily and act as efficient acylating agents for amines, alcohols, and thiols (Barton and Ferreira 1996). Moreover, these yellow coloured, often crystalline substances, are sensitive to visible light: the radical chain can be triggered by simply placing a tungsten lamp near the reaction vessel. Their light sensitivity, combined with their hydrolytic liability, make their isolation and storage somewhat tricky: reaction vessels, chromatography columns etc. have to be covered with aluminium foil, and the laboratory lighting must be subdued. In most instances, isolation of the Barton esters is not necessary and many of these complications can be avoided. The decarboxylation process may also be initiated

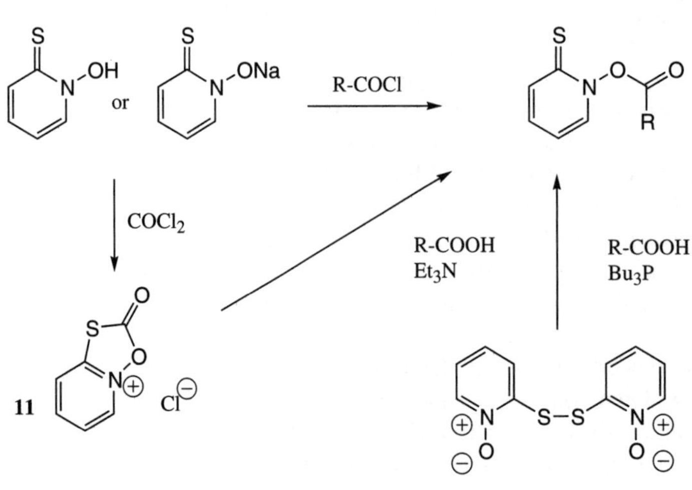

**Scheme 5.7** Synthesis of N-hydroxy-thiopyridone esters

thermally, with or without added chemical initiators. The selection of the initiation method will depend on the substrate, the efficiency of the trapping agent, if any, the scale, the temperature etc. Some specific cases where such 'details' are important will appear in the following sections.

## 5.4   Functional group transformations via Barton esters

In its simplest form, the Barton decarboxylation allows the replacement of a carboxylic function by a pyridyl-sulfide group (RCOOH → RSPy). Primary, secondary, and tertiary aliphatic or alicyclic carboxylic acids are good substrates (Barton *et al.* 1983, 1985*a*). Aromatic or $\alpha,\beta$-unsaturated carboxylic acids are more difficult to transform because the decarboxylation step leads to high-energy aromatic or vinylic radicals and is comparatively slow. The carboxylic radical oxidises the thiocarbonyl group of the starting esters before losing carbon dioxide, causing the formation of products where no decarboxylation has taken place (chiefly the acid anhydride, dipyridyl disulfide, and *N*-oxides thereof).

The pyridyl sulfides (RSPy), obtained through what we shall call the decarboxylative rearrangement for want of a better term (path **A** in Scheme 5.6), are good starting points for a number of synthetically interesting transformations. The corresponding sulfoxides for example, can be converted to olefins by a thermal syn-elimination or to ketones via a Pummerer rearrangement. The former process is illustrated by the sequence in Scheme 5.8, starting from isolongifolic acid derivative **12** (Boivin *et al.* 1990*a*). A 1,5-hydrogen atom migration of rare efficiency between two saturated carbons intervenes before transfer of the pyridyl-sulfide moiety. The hydrogen atom at C-3 is very close in space to the norbornyl radical at C-7, which is slightly bent and possesses therefore a somewhat greater $\sigma$-character. Oxidation to the sulfoxide and heating in pyridine gives a quantitative yield of the olefin.

We have seen above (Scheme 5.5) that capture by tributyltin hydride leads to nor-alkanes. The same transformation can be accomplished with thiols. Hindered tertiary thiols are best because simpler thiols (R′SH) react with Barton esters to form

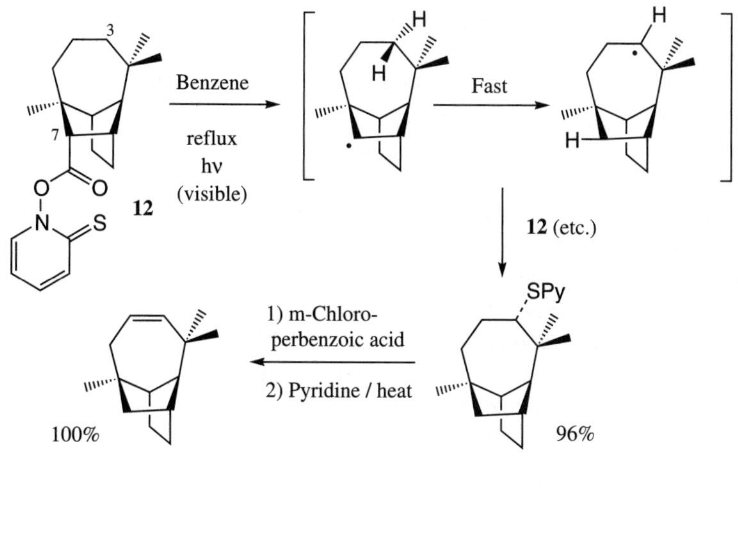

**Scheme 5.8** Example of decarboxylative rearrangement and radical translocation

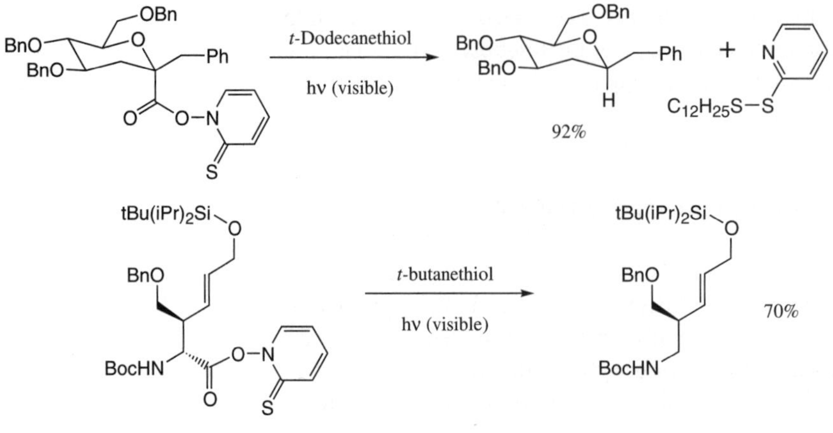

**Scheme 5.9** Reductive decarboxylation with thiols

thioesters [RC(=O)SR′] by an ionic mechanism. Initial experiments involved *t*-butylthiol (Barton *et al.* 1983, 1985a) but this evil smelling substance can be advantageously replaced by *t*-dodecanethiol (Crich and Ritchie 1988), a commercially available mercaptan with a strong but tolerable odour of grapefruits. The reductive decarboxylation of a saccharide (Crich and Lim 1990) and of an amino-acid derivative (Kazmaier and Schneider 1998) are two recent applications recorded in Scheme 5.9. The anomeric radical in the first example reacts selectively from the α-side and the co-product, *t*-dodecyl pyridyl disulfide, corresponding to Y–SPy in Scheme 5.6, is non-polar and easily separated.

Another, very useful modification of the Barton decarboxylation consists in performing the reaction in the presence of carbon tetrachloride or bromo-trichloromethane, as solvent or co-solvent (Barton *et al.* 1985a). This leads cleanly to the formation of the corresponding chloride or bromide, in what is a vastly more powerful equivalent of the Borodin–Hunsdiecker reaction. Iodotrichloromethane, in contrast, is an unstable substance, but analogous decarboxylative iodinations may be performed using iodoform, ethyl iodoacetate, or 2,2,2-trifluoroethyl iodide as the iodine transfer agent. Three typical examples are set out in Scheme 5.10. The first, by Drost and Cava (1991), concerns the synthesis of an analogue of CC-1065, a powerful anti-cancer agent. The second is a key step in an approach to fortamine, the 1,4-diaminocyclitol subunit of fortimycin A (Kobayashi *et al.* 1990); it is interesting to note that none of the previously known methods could accomplish this transformation. Finally, a double decarboxylative iodination allowed Tsanaktsidis and Eaton (1989) to gain an expedient entry into diiodocubane. The co-product in the first two reactions is trichloromethyl pyridyl sulfide and trifluoroethylpyridyl sulfide in the third (these again corresponding to Y–SPy in Scheme 5.6).

Of all the numerous transformations feasible through the Barton technology, decarboxylative halogenation is so far the only one that can be applied with reasonable efficiency to aromatic or α,β-unsaturated carboxylic acids (Scheme 5.11). One reason for success is that carbon tetrachloride, bromotrichloromethane, and iodoform are impervious to the reactive carboxylic radical intermediate which, it was noted above, decarboxylates comparatively slowly to give aryl or vinyl radicals.

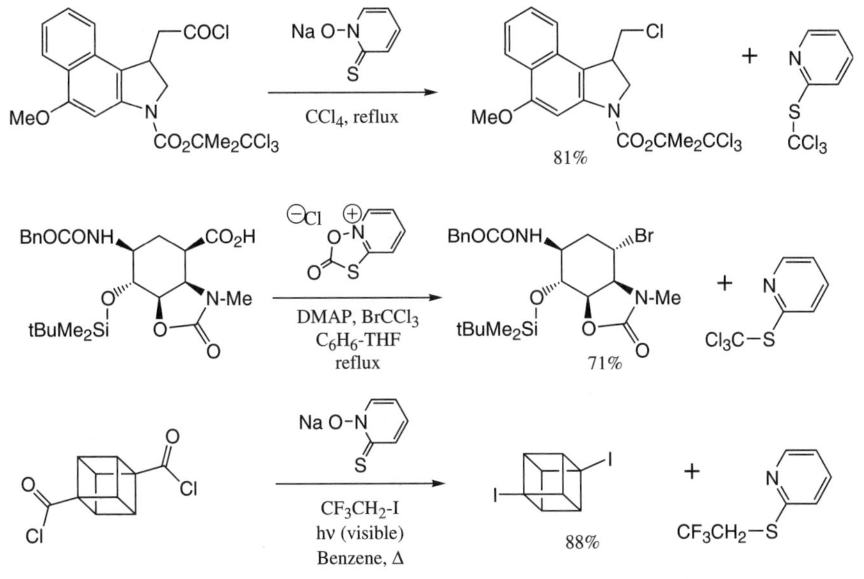

**Scheme 5.10**
Decarboxylative
halogenation

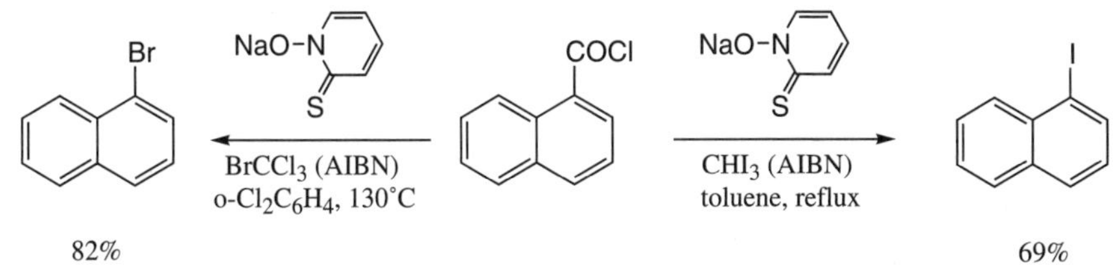

**Scheme 5.11**
Decarboxylative
halogenation of aromatic
carboxylic acids

The unimolecular expulsion of carbon dioxide is speeded up by increasing the temperature and, to avoid premature attack on the thiocarbonyl group, the build-up of the thiohydroxamate ester concentration is curtailed by slow addition of the acid chloride to a refluxing suspension of the sodium salt of *N*-hydroxythiopyridone (Barton *et al.* 1985*b*, 1987*b*). Incorporation of some AIBN to the acid chloride further improves the modestly efficient thermal initiation process.

Yet another variation consists in intercepting the intermediate carbon radical with a disulfide, a diselenide or even a ditulleride (i.e. X–Y in Scheme 5.6 is R'S–SR', R'Se–SeR', or R'Te–TeR'). A good yield of the various chalcogenides can be secured by irradiating with visible light a solution of the Barton ester with the dichalcogenide at room temperature or below. It is worthy of note, as an illustration of the temperature effect discussed above, that under simple thermal conditions, it is mostly the decarboxylative rearrangement product (R–SPy) that predominates, especially with the less reactive disulfides (Barton *et al.* 1984).

The decarboxylative chalcogenation has found some interesting applications. The first transformation in Scheme 5.12 represents the introduction of a methyl sulfide group and is the key step in the preparation of a new class of anti-asthma steroid derivatives (Ashton *et al.* 1996). The second example concerns the synthesis of optically

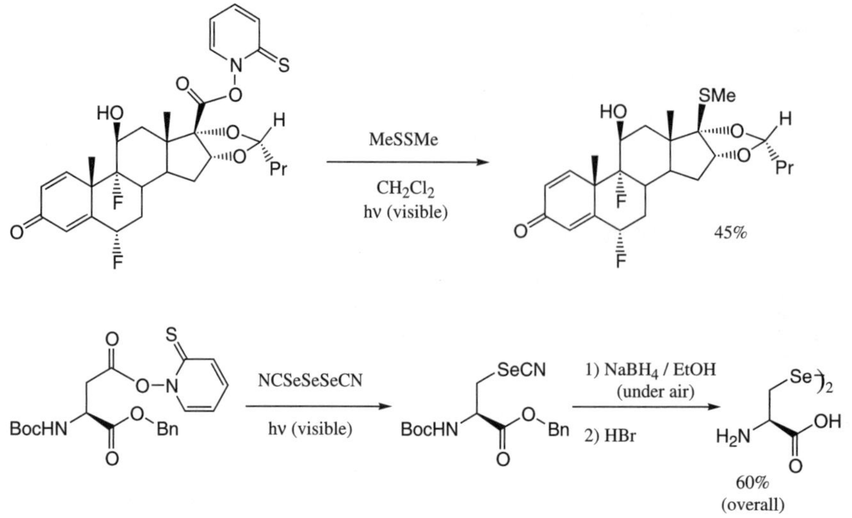

Scheme 5.12
Decarboxylative
chalcogenation

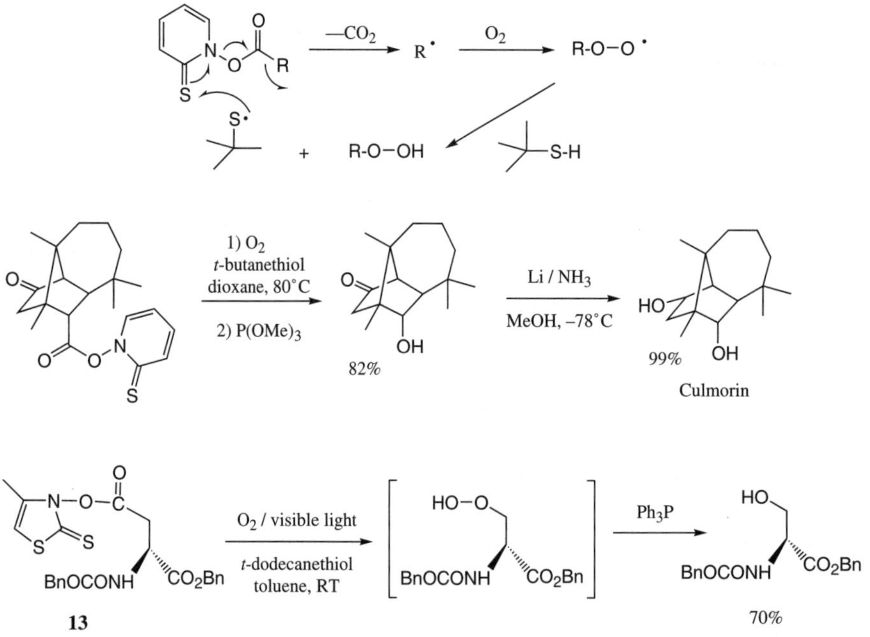

Scheme 5.13
Decarboxylative
oxygenation

pure selenocystine by decarboxylative selenation using the exotic triselenide trap (Barton *et al.* 1986*a*). The selenol initially formed upon reduction of the selenocyanate with sodium borohydride is rapidly oxidized by air to the diselenide.

Replacing the carboxylic acid group with a C–O bond may be accomplished by performing the decarboxylation in the presence of oxygen and a thiol (Barton *et al.* 1985*a*). The peroxy radical (R–O–O•) is quenched by the thiol, and the ensuing thiyl radical propagates the chain (Scheme 5.13). The hydroperoxide may be transformed into a variety of products, but the most interesting is perhaps its reduction to the corresponding alcohol, either by leaving it in prolonged contact with the excess thiol in the medium or, more expediently, by adding a phosphine or a phosphite. Several reactions are in competition in this decarboxylative hydroxylation, and the efficacy is highly dependent on the experimental conditions. One recent application of this transformation is in the total synthesis of (±)-culmorin by Takasu *et al.* (1999). The key step is displayed in Scheme 5.13; the use of dioxane as solvent was found to be important in this case. Incidentally, it is Barton who first determined the structure of culmorin in 1968! (Barton and Westiuk 1968).

*N*-hydroxy-thiazolinethione analogues are sometimes more convenient than the original *N*-hydroxythiopyridone esters in the decarboxylative hydroxylation (Barton *et al.* 1998). They are less sensitive to light and to hydrolysis, and can be easily prepared and stored. Mere stirring with *t*-dodecanethiol in various inert solvents under air or oxygen, with either exposure to laboratory lighting or deliberate irradiation with a tungsten lamp, is sufficient to bring about the formation of the hydroperoxide, which is finally reduced to the alcohol with triphenylphosphine. The second example in Scheme 5.13 shows the conversion of aspartic acid derivative **13** into a protected serine (Barton *et al.* 1998).

An alternative route to nor-alcohols is via organoantimony derivatives (Barton *et al.* 1985c, 1989a). For example, heating the thiazolinethione ester derived from palmitic acid with tris(triphenythio)antimony gave an organoantimony species, of presumed structure **14**, which underwent autoxidation upon work up to give pentadecanol (Scheme 5.14). The existence of an organoantimony intermediate was demonstrated by protonolysis with HCl and iodination with iodine. This method for

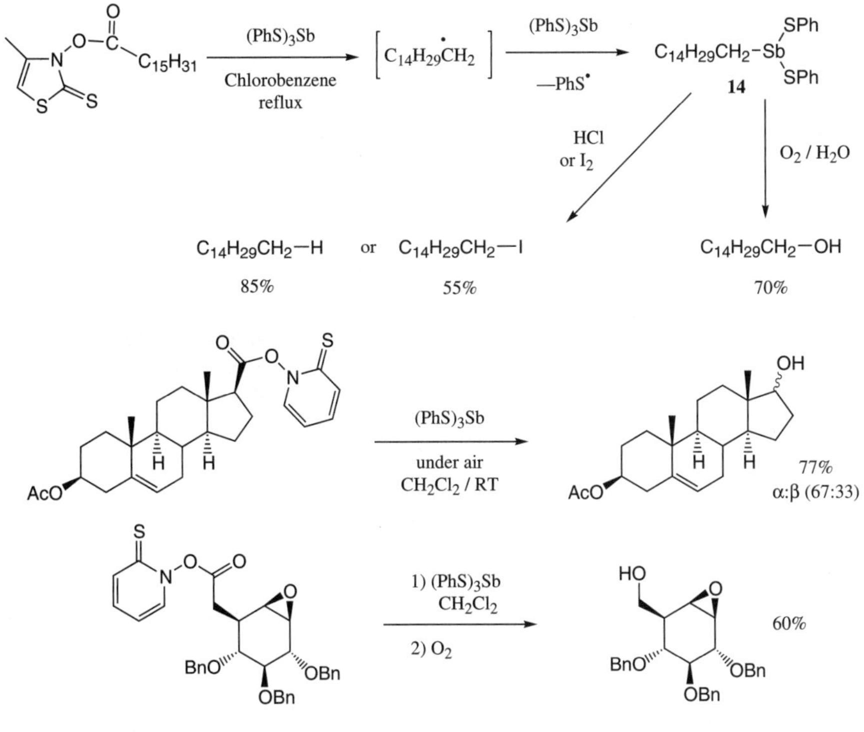

**Scheme 5.14**
Decarboxylative oxygenation using tris (phenythio) antimony

decarboxylative hydroxylation works well for various aliphatic and alicyclic carboxylic acids, as illustrated by the two further examples in the same scheme, the last being one step in a synthesis of (+)-cyclophellitol by Ziegler and Wang (1998). In practice, this method suffers from the need to work with a pure reagent: tris(triphenythio)antimony is water sensitive and its hydrolysis liberates thiophenol which is detrimental to the radical reaction. The use of TEMPO to capture the carbon radical (cf. Scheme 2.48) in a non-chain process would be another way to introduce a carbon–oxygen bond. Barton esters can be decomposed stoicheometrically with light and the implementation of a chain is therefore not strictly necessary.

Organophosphorus derivatives may be obtained by forming the C–P bond by way of an $S_H2$ reaction on tris(phenylthio)phosphine, much in the same way as for the organoantimony derivative **14** (Barton *et al.* 1986*b*). The corresponding trivalent phosphorus intermediate initially produced is oxidized *in situ* by the combined action of the disulfide co-product and water (Scheme 5.15). Yields are moderate and the reaction apparently fails with tertiary carboxylic acids. A much more efficient synthesis of phosphonic acids involves the use of highly reactive white phosphorus (caution!) and oxidation of the ill-defined organophosphorus intermediate with hydrogen peroxide (Barton and Zhu 1993).

If the thiohydroxamate ester is irradiated in the presence of sulfur dioxide as shown in Scheme 5.16, the intermediate carbon radical is (reversibly) captured to give a sulfonyl radical which ultimately evolves into a thiosulfonate (Barton *et al.* 1988). These substances are convenient precursors for sulfonic acids, sulfonyl chlorides, sulfonamides, etc. by application of conventional ionic reactions.

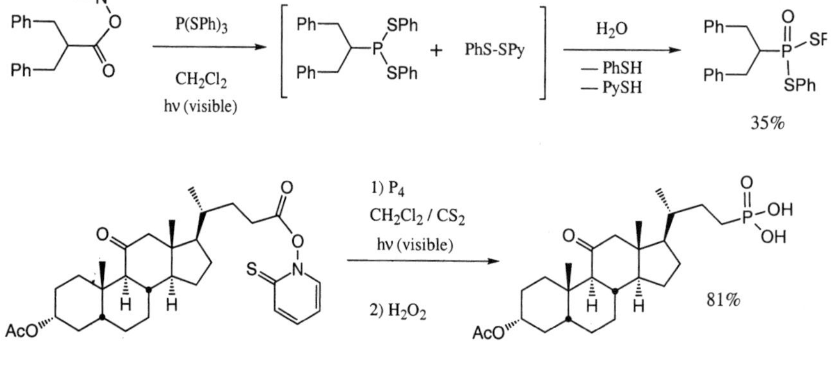

**Scheme 5.15** Decarboxylative phosphorylation

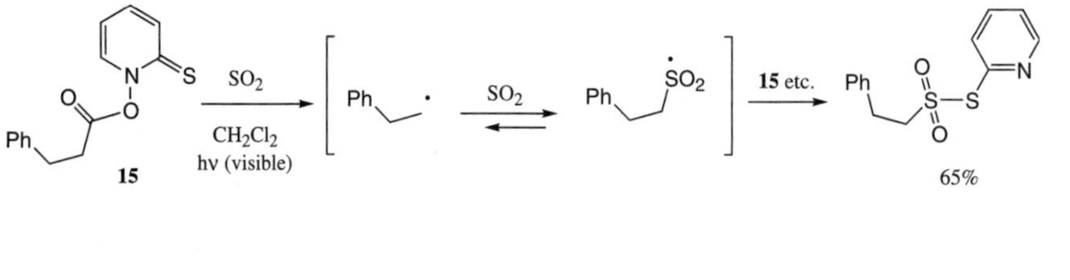

**Scheme 5.16** Synthesis of thiosulfonates

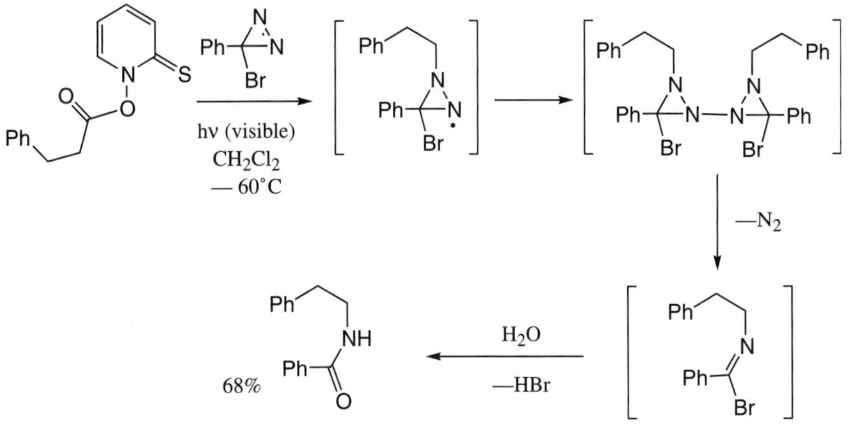

**Scheme 5.17**
Decarboxylative
amination

Other functional group interconversions starting with Barton esters of carboxylic acids include the formation of thiols and nitriles, using elemental sulfur or tosyl cyanide as radical traps respectively (Barton *et al.* 1991a, 1994), and the introduction of C–N bonds. Exchange of the carboxylic unit with a nitrogen functional group can in principle be accomplished by irradiation of Barton esters in the presence of nitric oxide ($R^\bullet + {}^\bullet NO \rightarrow R–NO$) or a thionitrite, as was shown recently (Girard *et al.* 1995). Another interesting and ingenious solution involves the use of diazirines (Barton *et al.* 1993a). One example showing the conversion of hydrocinnamic acid into the nor-amide is displayed in Scheme 5.17. Irradiation is performed at very low temperature and the process in this case is not a chain. Overall, this transformation may be considered as the radical equivalent of the Curtius or Hoffmann rearrangements.

## 5.5 Carbon–carbon bond formation

Interception, by an internal or external olefin, of the carbon radical formed upon decarboxylation represents a mild yet powerful means of creating new C–C bonds. The general reaction manifold is outlined in Scheme 5.18. The olefinic trap has to be quite reactive in order to compete with the background decarboxylative rearrangement. One important feature in comparison with the tin-, silicon-, or mercury-based systems, is that a greater variety of olefins can be used and more highly functionalized adducts (**15**) are generally obtained. The incorporation of the pyridylsulfide unit geminal to the activating group **E** profoundly alters and tremendously enriches the chemistry of both functionalities.

As would be expected, intramolecular C–C bond formation is the easiest to accomplish. The example displayed in Scheme 5.19 illustrates the generation and capture of an oxiranyl radical, a species quite difficult to obtain by other means (Ziegler and Wang 1996).

The intermolecular variant is exemplified in Scheme 5.20 by the efficient decarboxylative addition to methyl acrylate starting from a protected tartaric acid

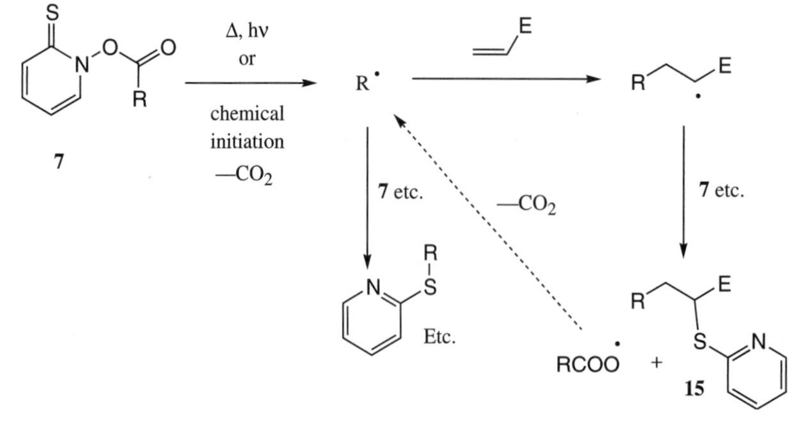

**Scheme 5.18**
Mechanism of
decarboxylative addition
to an olefin

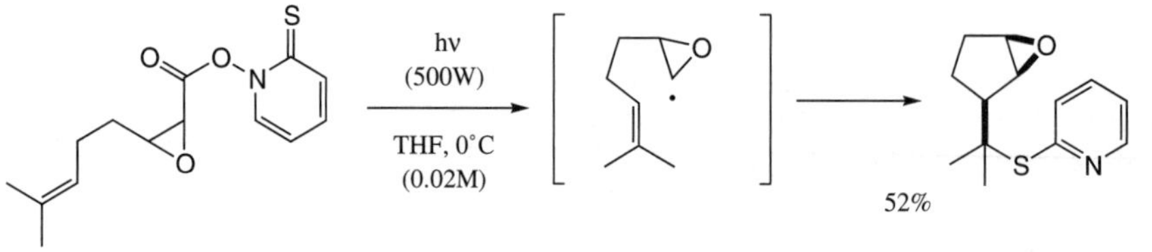

**Scheme 5.19**    Cyclization
of an $\alpha$-oxiranyl radical

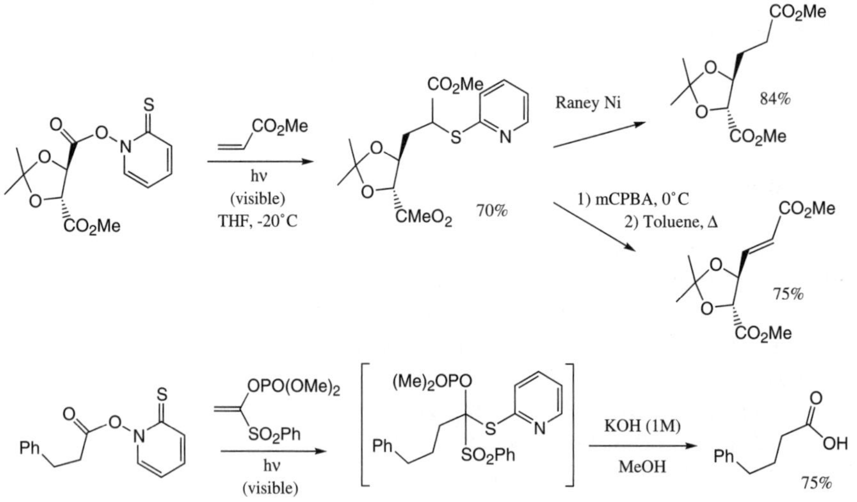

**Scheme 5.20**
Intermolecular addition to
electron-poor olefins

(Barton *et al.* 1993*b*). It represents a two-carbon elongation with retention of configuration in this case because the addition takes place from the least hindered side of
the dioxolane ring, opposite to the carboxymethyl group. The pyridyl sulfide can be
reduced away or oxidatively eliminated to leave behind an olefin. A one carbon
homologation, equivalent to the traditional Arndt-Eistert reaction, is displayed in the
same scheme; it is accomplished by using 1-phosphonoxy-1-phenylvinyl sulfone as

the radical trap (Barton *et al.* 1991*b*). The terminal carbon in the adduct is now at the oxidation level of a carboxylic acid and mere exposure to dilute KOH suffices for this conversion. Various procedures for the one-, two-, and three-carbon homologation of carboxylic acids have been devised (Barton *et al.* 1992, 1993*c*).

Adducts derived from unsubstituted phenylvinyl sulfone itself can undergo an amazing number of subsequent transformations (Barton *et al.* 1985*d*, 1991*c*; Boivin *et al.* 1995), due in a large measure to the geminal disposition of the sulfone and sulfide fragments. The chemoselective replacement of only one of the two phenyl sulfones in bicyclic compound **17** by a methyl group, as shown in Scheme 5.21, is one illustration of the special reactivity imparted on the sulfone group by the geminal pyridyl sulfide. Other groups besides a methyl (e.g. allyl or hydrogen) can be selectively introduced by simply replacing trimethylaluminum with other reagents (Barton *et al.* 1991).

Two sequential radical additions to phenylvinyl sulfone are involved in the construction of bicyclic derivative **17**. Another, more complex and ingenious sequence, outlined in Scheme 5.22, was devised by Saicic and Cekovic (1992, 1994). It uses acrylonitrile as the external trap and allows the construction of a linear triquinane from an open chain precursor. The phenylthiyl radical produced in the last step propagates the chain by reacting with the starting thiohydroxamate ester. Both

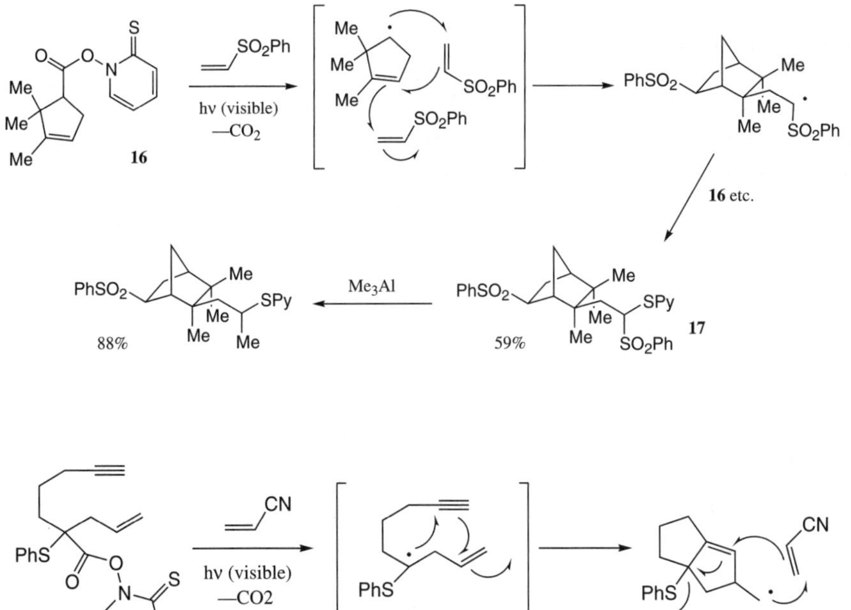

**Scheme 5.21**
Successive additions to phenylvinyl sulfone

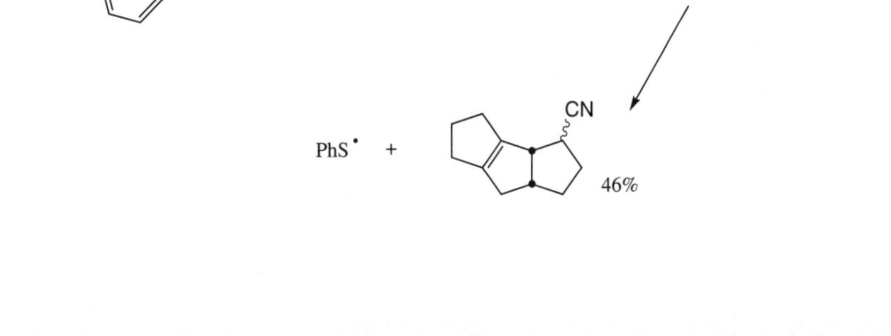

**Scheme 5.22**
Decarboxylative radical cascade

of these radical cascades reflect the extraordinary synthetic potential of the Barton decarboxylation process.

Barton esters tolerate in general the presence of anhydrous acid, and this opens various synthetic variations not available with the hydride-based radical methods. Simple nitroolefins for example are highly sensitive to base- and nucleophile-induced side reactions and cannot usually be employed as radical traps with stannane chemistry. Yet, if anhydrous camphorsulfonic acid is incorporated in the medium to neutralize any adventitious basic impurity, decarboxylative addition on such olefins can be readily accomplished. One interesting application is outlined in Scheme 5.23 whereby a bile acid side chain is elaborated into that of vitamin $D_3$ (Barton *et al.* 1985*e*). The intermediate radical adduct was not isolated but treated immediately with aqueous titanium trichloride. Addition of excess methylmagnesium iodide to the resulting ketone completed the sequence (the acetate group in ring A is also cleaved by this treatment).

The compatibility with anhydrous acid allows the capture of the intermediate carbon radical by pyridinium and other related salts, extending thus the pioneering work of Minisci (Minisci *et al.* 1983). Such radical additions represent a powerful means of functionalizing heteroaromatic rings which are often recalcitrant partners in the classical Friedel–Crafts reaction. The addition of various radicals to lepidinium camphorsulfonate is pictured in Scheme 5.24 (Barton *et al.* 1986*c*; X⁻ represents the camphorsulfonate anion). Pyridinium salts are electrophilic radical traps and often react more efficiently with tertiary than with secondary or primary carbon radicals. The last example shows that quite complicated structures can be assembled rapidly by this technique (Ikegami *et al.* 1994). There is no problem of regiochemistry with the lepidine system; with unsubstituted pyridines and quinolines, it is the 2- and 4-positions which are activated, and mixtures of regioisomers are usually observed. Camphorsulfonate salts are nicely crystalline compounds: they are easy to dry and are quite soluble in dichloromethane, the most common solvent for these reactions.

Because of their oxidizing power, quinones are also not normally compatible with metal hydrides. In contrast, useful yields of addition products can be obtained using

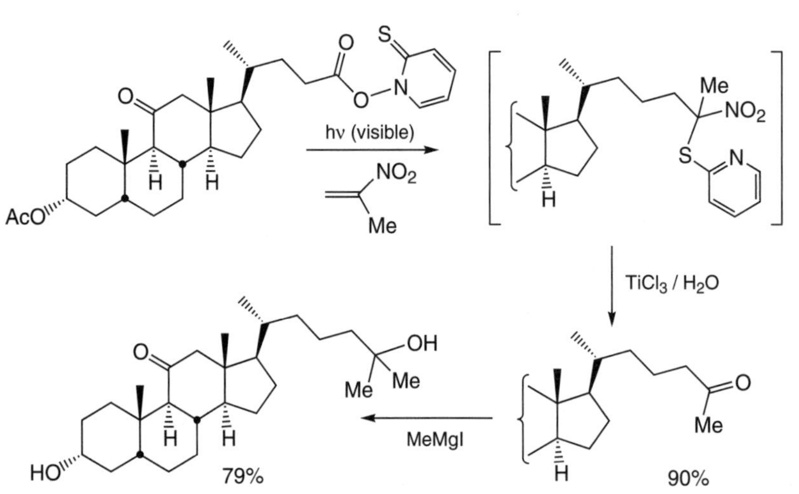

**Scheme 5.23**
Decarboxylative radical addition to nitropropene

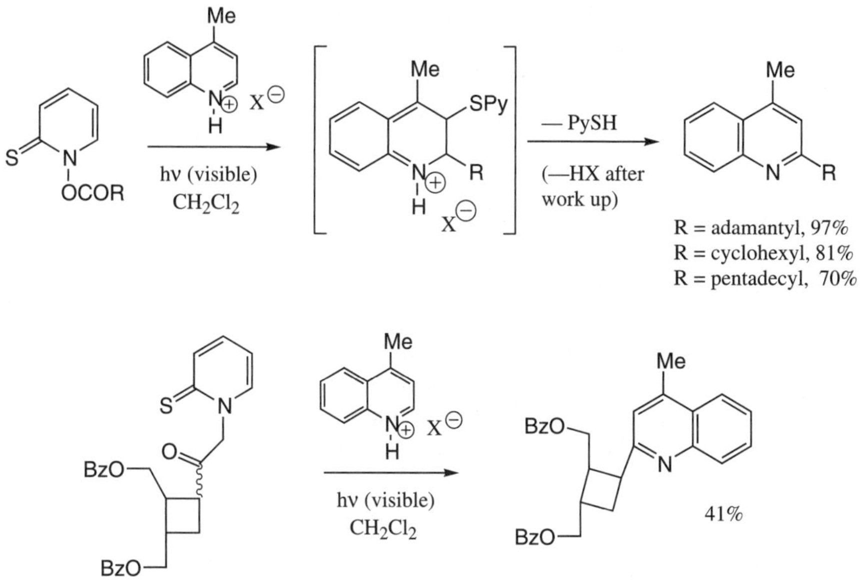

**Scheme 5.24**
Decarboxylative radical addition to protonated heteroaromatics

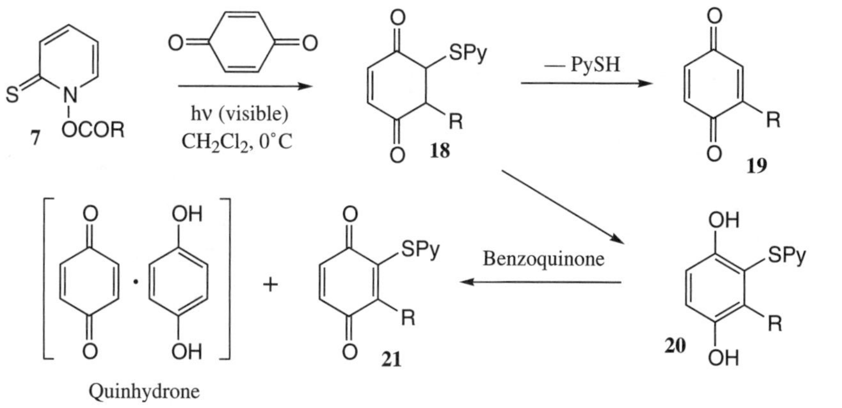

**Scheme 5.25**
Decarboxylative radical addition to quinones

thiohydroxamate esters, as long as the decarboxylation is performed at low temperature. The reaction manifold, delineated in Scheme 5.25 for benzoquinone itself, is fairly complex and the outcome of an experiment depends on the structure of the starting acid, the nature and amount of quinone used, and the exact conditions (Barton *et al.* 1987*c*).

In the case of a primary acid such as palmitic acid (R=$n$-C$_{15}$H$_{31}$- in Scheme 5.25) and using 5 eq. of benzoquinone, the major product is **21** (77%) along with a small amount of **19** (10%). Hydroquinone **20**, arising by aromatization of the 'normal' adduct **18**, is oxidized by the excess benzoquinone (the co-product is hydroquinone which immediately forms quinhydrone, a dark-green, charge-transfer complex with benzoquinone). This can be easily shown by performing the reaction with only a slight excess of benzoquinone. Hydroquinone **20** then becomes the predominant product; it can be isolated and separately oxidized into **21** with benzoquinone. With a tertiary acid derivative (e.g. R=1-methylcyclohexyl), the reaction stops at the hydroquinone **20** stage. The oxidation step in this case is presumably slowed down by the steric bulk of the R⁻ group. Fairly hindered, substituted quinones have been prepared by this route (Barton and Sas 1990).

## 5.6   Generation of carbon radicals from alcohols

The Barton decarboxylation can be modified in such a way as to allow radical generation from alcohols. The conception underlining this alternative way of manipulating thiohydroxamate esters hinges on the decarboxylation of oxalate derivatives (Scheme 5.26). The intermediate alkoxycarbonyl radical can either be captured by a suitably located internal olefin to give a lactone or, in turn, expel another molecule of carbon dioxide to give an alkyl radical, which can be intercepted by an internal or

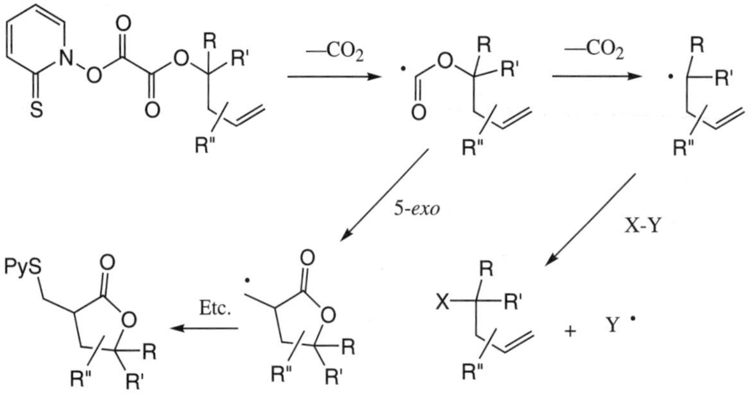

**Scheme 5.26**
Deoxygenation of alcohols via their oxalate derivatives

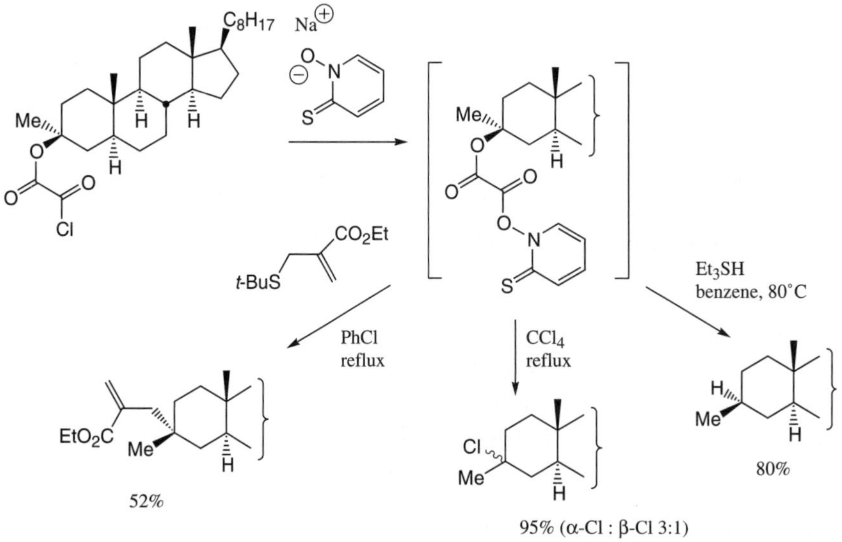

**Scheme 5.27**
Modification of alcohols through decarboxylation of their oxalate derivatives

external trap (Barton and Crich 1986; Togo and Yokoyama 1990). The rate of the second loss of carbon dioxide is not very high and depends strongly on the nature of R and R′, as these determine the stability of the ensuing alkyl radical (Simakov *et al.* 1998). Extrusion of carbon monoxide from an alkoxycarbonyl to produce a highly reactive alkoxy radical is energetically unfavourable and does not normally occur.

Three examples are displayed in Scheme 5.27 where the tertiary 3-methyl-cholestan-3-yl is converted into an alkane, a chloride, or captured by an activated olefin through an addition–elimination process (Barton and Crich 1986; Crich and Fortt 1987). These transformations are not easily transposable to derivatives of simple primary and secondary alcohols where extrusion of carbon dioxide from the corresponding alkoxycarbonyl radical is not sufficiently rapid (cf. Scheme 2.10).

## 5.7   Generation of nitrogen- and oxygen-centred radicals

Barton esters can be modified to provide access to a variety of nitrogen and oxygen-centred radicals. We have seen that neutral aminyl radicals are not very reactive, and that their additions to olefins are often reversible (cf. Scheme 2.49). The reasonable compatibility of thiohydroxamate derivatives with an anhydrous acidic medium allows the generation of the much more reactive and synthetically useful aminium radicals. This is accomplished by irradiating carbamate derivatives of *N*-hydroxy-2-thiopyridone (e.g. **22**) in the presence of trifluoroacetic acid (Newcomb and Deeb 1987; Esker and Newcomb 1993*b*). In the example shown in Scheme 5.28, the aza[3.2.1]bicyclic system of tropanes is constructed quite efficiently (Newcomb *et al.* 1990)

The precursors are readily prepared and various nitrogen containing structures can be readily assembled. A similar approach may be applied to generate amidyl radicals but the corresponding substrates are more delicate to handle. Iminyl radicals, which have a reactivity that is intermediate between amidyls and neutral aminyls (Le Tadic-Biadatti *et al.* 1997) can be produced by a different route which hinges on the

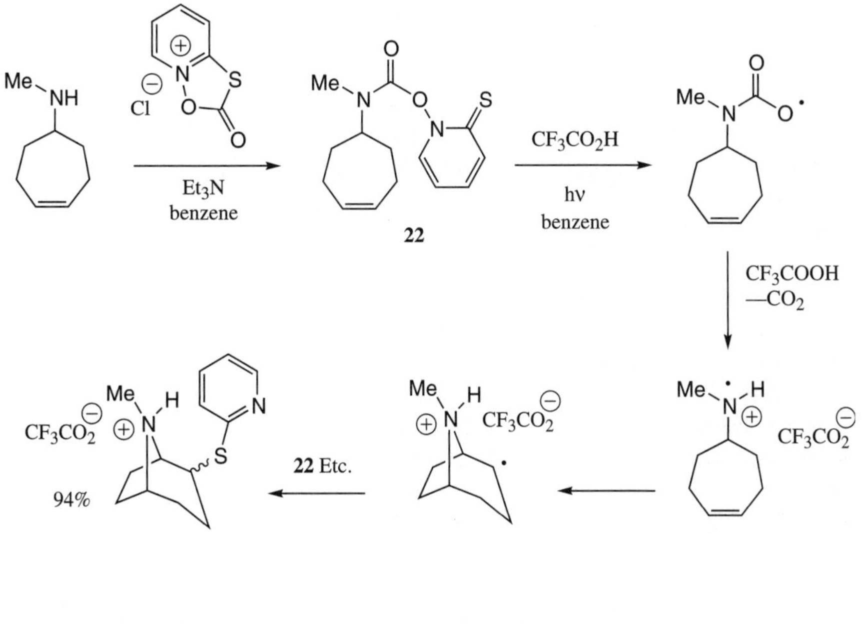

**Scheme 5.28**
Generation and capture of aminyl radicals

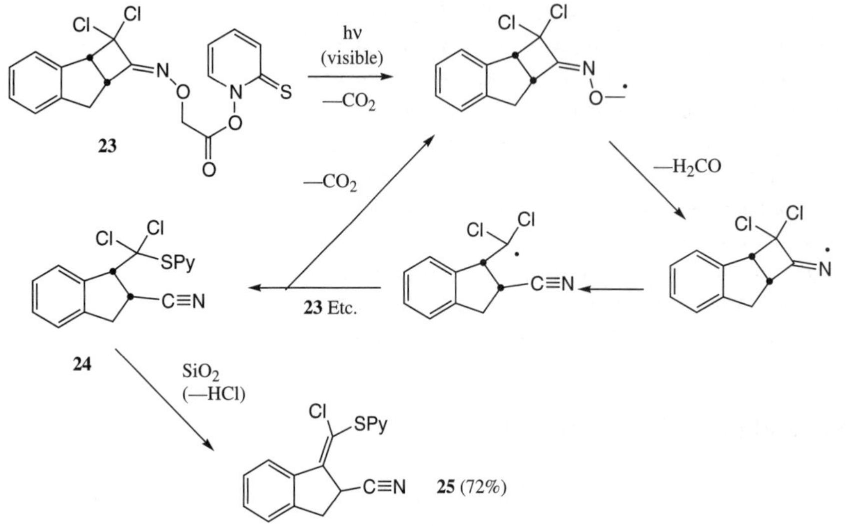

**Scheme 5.29**
Generation and fragmentation of iminyl radicals

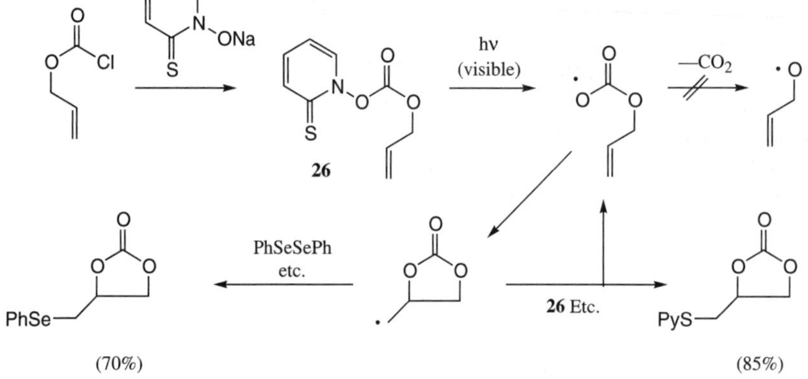

**Scheme 5.30**
Generation and capture of oxygen-centred radicals

extrusion of formaldehyde from an oxime derivative such as **23**, as illustrated by the example in Scheme 5.29 (Boivin *et al.* 1994*b*; Zard 1996). The initial adduct **24** is unstable to silica and undergoes elimination of HCl to give **25**, as only one isomer, presumably the one drawn. It is interesting to note that in the absence of the geminal chlorines, the regiochemistry of the opening of the intermediate cyclobutyliminyl is different.

Oxygen-centred radicals can be obtained by irradiation or thermal decomposition of the appropriate precursor. Hydroxyl radicals can thus be produced from *N*-hydroxy-2-thiopyridone itself (Boivin *et al.* 1990*b*). These species are too reactive and undiscriminating to be generally useful synthetically but seem to be implicated in a number of important biological processes such as ageing and cancerogenesis. DNA strands can indeed be cleaved by irradiation in the presence of *N*-hydroxy-2-thiopyridone and some of its esters (Adam *et al.* 1995; Theodorakis *et al.* 1997). Alkoxy radicals can be made in a similar way by photolytic cleavage of *N*-alkoxypyridinethiones (Beckwith and Hay 1988). From a synthetic standpoint, the most interesting species appear to be (alkoxycarbonyl)oxyl radicals, $RO(C{=}O)O^\bullet$ which, unlike carboxylic $R(C{=}O)O^\bullet$ or carbamyloxy $RR'N(C{=}O)O^\bullet$ radicals, do not readily extrude carbon dioxide (Chateauneuf *et al.* 1988*b*). Their intramolecular capture constitutes a route to vicinal diols protected as the cyclic carbonate, as shown by example in Scheme 5.30 (Beckwith and Davison 1991; Newcomb *et al.* 1991).

## 5.8 Some related processes

The principle of exploiting the weakness of the N–O bond to generate the highly energetic carboxylic radicals can be extended to esters of various hydroxylamine derivatives such as *N*-hydroxy-phthalimide, selenohydroximates, or benzophenone oxime. Thus, ethylbenzene can be obtained in good yield by treatment of *N*-(3-phenylpropanoyloxy)-phthalimide with tributylstannane (Scheme 5.31; Barton *et al.* 1989*b*). The tributylstannane in this reaction can be replaced by of a combination of a tertiary thiol, *N*-benzyl-1,4-dihydronicotinamide, a catalytic amount of a ruthenium bipyridyl complex [Ru(bpy)$_3$Cl$_2$] and irradiation with visible light (Okada *et al.* 1988, 1992). The process appears to involve a redox-initiated chain mechanism.

The use of selenohydroximates in conjunction with tributylstannane is illustrated by the second example in the same Scheme (Kim and Lee 1997). Finally, decarboxylation of carboxylic acids by irradiation of the corresponding benzophenone oxime esters has more scope since either a transfer of a hydrogen (Hasebe and Tsuchiya 1987) or a halogen atom (Hasebe and Tsuchiya 1988) can be incorporated into the non-chain sequence. One example of the latter case which is applicable to the formation of aromatic chlorides is shown in Scheme 5.31.

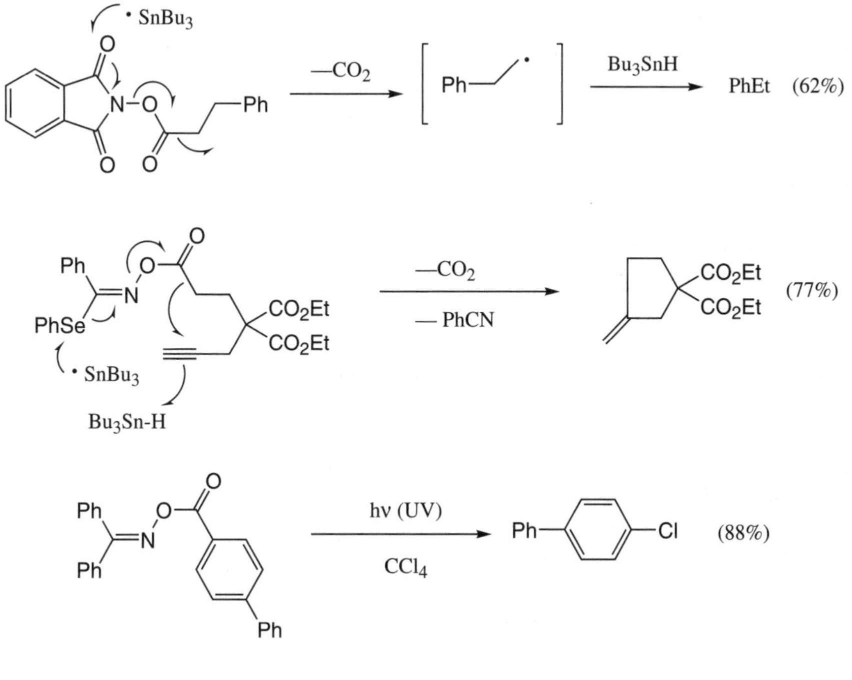

**Scheme 5.31** Variations on the Barton decarboxylation reaction

These variations nevertheless pale next to the tremendously powerful and versatile Barton decarboxylation via thiohydroxamate esters. This elegant reaction will certainly stand out as the most important radical process discovered in recent times.

## References

Ashton, M. J., Laurence, C., Karlsson, J.-A., Stuttle, K. A. J., Newton, C. G., Vacher, B. Y. J., Webber, S., and Whitnall, M. J. (1996). Anti-inflammatory 17$\beta$-thioalkyl-16$\alpha$, 17$\alpha$-ketal and -acetal androstanes: a new class of airway selective steroid for the treatment of asthma. *Journal of Medicinal Chemistry*, **39**, 4888–4896.

Adam, W., Ballmaier, D., Epe, B., Grimm, G. N., and Saha-Möller, C. R. (1995). *N*-Hydroxypyridinethiones as photochemical hydroxyl radical sources for oxidative DNA damage. *Angewandte Chemie International Edition in English*, **34**, 2156–2158.

Barton, D. H. R. (1993). *Half a century of free radical chemistry*, pp. 91–147. Cambridge University Press, Cambridge.

Barton, D. H. R. and Crich, D. (1986). The invention of new radical reactions. Part 11. A new method for the generation of tertiary radicals from alcohols. *Journal of the Chemical Society, Perkin Transactions,* **1**, 1603–1619.

Barton, D. H. R. and Ferreira, J. A. (1996). *N*-Hydroxypyridine-2(1*H*)-thione derivatives of carboxylic acids as activated esters. Part I. The synthesis of carboxamides. *Tetrahedron*, **52**, 9347–9366.

Barton, D. H. R. and Samadi, M. (1992). The invention of radical reactions. Part XXV. A convenient method for the synthesis of the acyl derivatives of *N*-hydroxy-pyridine-2-thione. *Tetrahedron*, **48**, 7083–7090.

Barton, D. H. R. and Sas, W. (1990). The invention of radical reactions. Part XIX. The synthesis of very hindered quinones. *Tetrahedron*, **46**, 3419–3430.

Barton, D. H. R. and Westiuk, N. H. (1968). Sesquiterpenoids. Part XIV. The constitution and stereochemistry of culmorin. *Journal of the Chemical Society (C)*, 148–155.

Barton, D. H. R. and Zhu, J. (1993). Elemental white phosphorus as a radical trap: a new and general route to phosphonic acids. *Journal of the American Chemical Society*, **115**, 2071–2072.

Barton, D. H. R., Bridon, D., and Zard, S. Z. (1989*a*). The invention of radical reactions. Part XVIII. Decarboxylative radical addition to arsenic, antimony, and bismuth phenylsulphides: a novel synthesis of noralcohols from carboxylic acids. *Tetrahedron*, **45**, 2615–2626.

Barton, D. H. R., Blundell, P., and Jaszberenyi, J. Cs. (1989*b*). Acyl derivatives of hydroxamic acids as a source of carbon radicals. *Tetrahedron Letters*, **30**, 2341–2344.

Barton, D. H. R., Bridon, D., and Zard, S. Z. (1984). New decarboxylative chalcogenation of aliphatic and alicyclic carboxylic acids. *Tetrahedron Letters*, **25**, 5777–5780.

Barton, D. H. R., Bridon, D., and Zard, S. Z. (1985*c*). A convenient high yielding synthesis of noralcohols from carboxylic acids. *Journal of the Chemical Society, Chemical Communications*, 1066–1068.

Barton, D. H. R., Bridon, D., and Zard, S. Z. (1986*b*). Radical decarboxylative phosphorylation of carboxylic acids. *Tetrahedron Letters*, **27**, 4309–4312.

Barton, D. H. R., Bridon, D., and Zard, S. Z. (1987*c*). The invention of radical reactions. Part XIV. A decarboxylative radical addition to quinones. *Tetrahedron*, **43**, 5307–5314.

Barton, D. H. R., Castagnino, E., and Jaszberenyi, J., Cs. (1994). The reaction of carbon radicals with sulfur. A convenient synthesis of thiols from carboxylic acids. *Tetrahedron Letters*, **35**, 6057–6060.

Barton, D. H. R., Chern, C.-Y., and Jaszberenyi, J. Cs. (1991*b*). Homologation of acids via carbon radicals generated from the acyl derivatives of *N*-hydroxy-2-thiopyridone. (The two carbon problem.) *Tetrahedron Letters*, **32**, 3309–3312.

Barton, D. H. R., Chern, C.-Y., and Jaszberenyi, J. Cs. (1992). Homologation of carboxylic acids by improved methods based on radical chain chemistry of acyl derivatives of *N*-hydroxy-2-thiopyridone. *Tetrahedron Letters*, **33**, 5013–5016.

Barton, D. H. R., Crich, D., and Motherwell, W. B. (1983). New and improved methods for the radical decarboxylation of acids. *Journal of the Chemical Society, Chemical Communications*, 939–941.

Barton, D. H. R., Crich, D., and Motherwell, W. B. (1985*a*). The invention of new radical chain reactions. Part VIII. Radical chemistry of thiohydroxamic esters. A new method for the generation of carbon radicals from carboxylic acids. *Tetrahedron*, **41**, 3901–3924.

Barton, D. H. R., Lacher, B., and Zard, S. Z. (1985*b*). Radical decarboxylative bromination of carboxylic acids. *Tetrahedron Letters*, **26**, 5939–5942.

Barton, D. H. R., Lacher, B., and Zard, S. Z. (1987*b*). The invention of radical reactions. Part XVI. Radical decarboxylative bromination and iodination of aromatic acids. *Tetrahedron*, **43**, 4321–4328.

Barton, D. H. R., Jaszberenyi, J. Cs., and Theodorakis, E. A. (1991*a*). Radical nitrile transfer with methanesulfonyl cyanide or *p*-toluenesulfonyl cyanide to carbon radicals generated from the acyl derivatives of *N*-hydroxy-2-thiopyridone. *Tetrahedron Letters*, **32**, 3321–3324.

Barton, D. H. R., Togo, H., and Zard, S. Z. (1985*d*). Radical addition to vinyl sulphones and vinyl phosphonium Salts. *Tetrahedron Letters*, **26**, 6349–6352.

Barton, D. H. R., Togo, H., and Zard, S. Z. (1985*e*). The invention of new radical reactions. Part X. High yield radical addition reactions of $\alpha,\beta$-unsaturated nitroolefins. An expedient construction of the 25-hydoxy-vitamin $D_3$ side chain from bile acids. *Tetrahedron*, **41**, 5507–5516.

Barton, D. H. R., Bridon, D., Fernandez-Picot, I., and Zard, S. Z. (1987*a*). The Invention of radical reactions. Part XV. Some mechanistic aspects of the decarboxylative rearrangement of thiohydroxamic esters. *Tetrahedron*, **43**, 2733–2740.

Barton, D. H. R., Chern, C.-Y., Jaszberenyi, J., Cs., and Shinada, T. (1993*c*). Chain elongation and degradation of carboxylic acids by Barton ester based radical chemistry. *Tetrahedron Letters*, **34**, 6505–6508.

Barton, D. H. R., Dowlatshahi, H. A., Motherwell, W. B., and Villemin, D. (1980). A new radical decarboxylation for the conversion of carboxylic acids into hydrocarbons. *Journal of the Chemical Society, Chemical Communications*, 732–733.

Barton, D. H. R., Garcia, B., Togo, H., and Zard, S. Z. (1986*c*). Radical decarboxylative addition onto protonated heteroaromatics (and related) compounds. *Tetrahedron Letters*, **27**, 1327–1330.

Barton, D. H. R., Jaszberenyi, J. Cs., Theodorakis, E. A., and Reibenspies, J. H. (1993*a*). The invention of radical reactions. 30. Diazirines as carbon radical traps. Mechanistic aspects and synthetic applications of a novel and efficient amination process. *Journal of the American Chemical Society*, **115**, 8050–8059.

Barton, D. H. R., Lacher, B., Misterkiewicz, B., and Zard, S. Z. (1988). The invention of Radical reactions. Part XVII. A decarboxylative sulphonylation of carboxylic acids. *Tetrahedron*, **44**, 1153–1158.

Barton, D. H. R., Boivin, J., Sarma, J., da Silva, E., and Zard, S. Z. (1991*c*). Decarboxylative radical additions to vinylsulphones and vinylphosphonium bromide: some further novel transformations of geminal (pyridine-2-thiyl) phenylsulphones. *Tetrahedron*, **47**, 7091–7108.

Barton, D. H. R., Géro, S. D., Holliday, P., Quiclet-Sire, B., and Zard, S. Z. (1998). A practical decarboxylative hydroxylation of carboxylic Acids. *Tetrahedron*, **54**, 6751–6756.

Barton, D. H. R., Bridon, D., Hervé, Y., Potier, P., Thierry, J., and Zard, S. Z. (1986a). Concise syntheses of *L*-selenomethionine and *L*-selenocystine using radical chain reactions. *Tetrahedron*, **42**, 4983–4990.

Barton, D. H. R., Gateau-Olesker, A., Géro, S. D., Lacher, B., Tachdjian, C., and Zard, S. Z. (1993b). Radical decarboxylative alkylation of tartaric acid. *Tetrahedron*, **49**, 4589–4602.

Beckwith, A. L. J. and Davison, I. G. E. (1991). Diastereoselective formation of cyclic carbonates by cyclisation of alkenoxycarbonyloxy radicals. *Tetrahedron Letters*, **32**, 49–52.

Beckwith, A. L. J. and Hay, B. P. (1988). Generation of alkoxy radicals from *N*-alkoxypyridinethiones. *Journal of the American Chemical Society*, **110**, 4415–4416.

Boivin, J., Crépon, E., and Zard, S. Z. (1990b). *N*-Hydroxy-2-pyridinethione: a mild and convenient source of hydroxyl radicals. *Tetrahedron Letters*, **31**, 6869–6872.

Boivin, J., Crépon, E., and Zard, S. Z. (1991). Further observations on the thermal stability of *N*-hydroxy-2-thiopyridone esters: improved decarboxylative radical additions to olefins. *Tetrahedron Letters*, **32**, 199–202.

Boivin, J., Fouquet, E., and Zard, S. Z. (1994b). Iminyl radicals: Part III. Further synthetically useful sources of iminyl radicals. *Tetrahedron*, **50**, 1769–1776.

Boivin, J., Schmitt, A., Lallemand, J.-Y., and Zard, S. Z. (1995). Reduction of activated thiopyridine compounds by zinc metal. *Tetrahedron Letters*, **36**, 7243–7246.

Boivin, J., da Silva, E., Ourisson, G., and Zard, S. Z. (1990a). Proof of a radical transannular hydrogen migration in the longifolene series. *Tetrahedron Letters*, **31**, 2501–2504.

Chateauneuf, J., Lusztyk, J., and Ingold, K. U. (1988a). Spectroscopic and kinetic characteristics of aroyloxyl radicals. 2. Benzoyloxyl and ring-substituted aroyloxyl radicals. *Journal of the American Chemical Society*, **110**, 2886–2893.

Chateauneuf, J., Lusztyk, J., Maillard, B., and Ingold, K. U. (1988b). First spectroscopic and absolute kinetic studies on (alkoxylcarbonyl)oxyl radicals and an unsuccessful attempt to observe carbamoyloxyl radicals. *Journal of the American Chemical Society*, **110**, 6727–6731.

Crich, D. and Fortt, S. M. (1987). A new method for the synthesis of tert-alkyl chlorides from tert-alcohols. *Synthesis*, 35–37.

Crich, D. and Lim, L. B. L. (1990). Synthesis of 2-deoxy-β-*C*-pyranosides by diastereoselective hydrogen atom transfer. *Tetrahedron Letters*, **31**, 1897–1900.

Crich, D. and Ritchie, T. J. (1988). Stereoselective free radical reactions in the preparation of 2-deoxy-β-*D*-glucosides. *Journal of the Chemical Society, Chemical Communications*, 1461–1463.

Drost, K. J. and Cava, M. P. (1991). A photochemically based synthesis of the benzannelated analogue of the CC-1065 unit. *Journal of Organic Chemistry*, **56**, 2240–2244.

Esker, J. L. and Newcomb, M. (1993). The generation of nitrogen radicals and their cyclisations for the construction of the pyrrolidine nucleus. *Advances in Heterocyclic Chemistry*, **58**, 1–45.

Fujimori, K. (1992). Diacyl peroxides. In *Organic peroxides* (ed. W. Ando), pp. 319–385. John Wiley & Sons, Chichester.

Garner, P., Anderson, J. T., Dey, S., Youngs, W. J., and Galat, K. (1998). *S*-(1-oxido-2-pyridinyl)-1,1,33-tetramethythiouronium hexafluorophosphate. A new reagent for preparing hindered Barton esters. *Journal of Organic Chemistry*, **63**, 5732–5733.

Girard, P., Guillot, N., Motherwell, W. B., and Potier, P. (1995). Observations on the reactions of of *O*-acyl thiohydroxamates with thionitrite esters: a novel free radical chain reaction for decarboxylative amination. *Journal of the Chemical Society, Chemical Communications*, 2385–2386.

Hasebe, M. and Tsuchiya, T. (1987). Photoreductive decarboxylation of carboxylic acids via their benzophenone oxime esters. *Tetrahedron Letters*, **28**, 6207–6210.

Hasebe, M. and Tsuchiya, T. (1988). Photodecarboxylative chlorination of carboxylic acids via their benzophenone oxime esters. *Tetrahedron Letters*, **29**, 6287–6290.

Ikegami, S., Togo, H., and Yokoyama, M. (1994). Reactivity of carbocyclic four-membered radicals for the preparation of carbocyclic analogues of oxetanocins. *Journal of the Chemical Society, Perkin Transactions,* **1**, 2407–2412.

Johnson, R. G. and Ingham, R. K. (1956). The degradation of carboxylic acid salts by means of halogen. *Chemical Reviews*, **56**, 219–269.

Kazmaier, U. and Schneider, C. (1998). Application of the asymmetric chelate-enolate Claisen rearrangement to the synthesis of 5-*epi*-isofagomine. *Tetrahedron Letters*, **39**, 817–818.

Kim, S. and Lee, T. A. (1997). Facile generation of alkyl, aminyl, and alkoxy radicals from Se–phenyl benzoselenohydroximate derivatives. *Synlett*, 950–952.

Kobayashi, S., Kamiyama, K., and Ohno, M. (1990). The first enantioselective synthesis of fortamine, the 1,4-diaminocyclitol moiety of fortimycin A, by chemico-enzymatic approach. *Journal of Organic Chemistry*, **55**, 1169–1177.

Kochi, J. K., Bockman, T. M., and Hubig, S. M. (1997). Direct observation of ultrafast decarboxylation of acyloxy radicals via photoinduced electron transfer in carboxylate ion pairs. *Journal of Organic Chemistry*, **62**, 2210–2221.

Le Tadic-Biadatti, M.-H., Callier-Dublanchet, A.-C., Horner, J. H., Quiclet-Sire, B., Zard, S. Z., and Newcomb, M. (1997). Absolute rate constants for iminyl radical reactions. *Journal of Organic Chemistry*, **62**, 559–577.

Minisci, F., Citterio, A., and Giordano, C. (1983). Electron-transfer processes: peroxydisulfate, a useful and versatile reagent in organic chemistry. *Accounts of Chemical Research*, **16**, 27–32.

Newcomb, M. and Deeb, T. M. (1987). *N*-Hydroxypyridine-2-thione carbamates as aminyl and aminium radical precursors. Cyclizations for synthesis of the pyrrolidine nucleus. *Journal of the American Chemical Society*, **109**, 3163–3165.

Newcomb, M., Marquardt, D. J., and Deeb, T. M. (1990). *N*-hydroxypyridine-2-thione carbamate.V. Synthesis of alkaloid skeletons by aminium cation cyclizations. *Tetrahedron*, **46**, 2329–2344.

Newcomb, M., Udaya Kumar, M., Boivin, J., Crépon, E., and Zard, S. Z. (1991). Cyclisations and intermolecular additions of alkoycarbonyloxy radicals from *N*-hydroxypyridine-2-thione carbonates. *Tetrahedron Letters*, **32**, 45–48.

Okada, K., Okamoto, K., and Oda, M. (1988). A new and practical method of decarboxylation: photosensitized decarboxylation of *N*-acyloxy phthalimides via electron transfer mechanism. *Journal of the American Chemical Society*, **110**, 8736–8738.

Okada, K., Okubo, K., Morita, N., and Oda, M. (1992). Reductive decarboxylation of *N*-acyloxy phthalimides via redox-initiated radical chain mechanism. *Tetrahedron Letters*, **33**, 7377–7380.

Ryzhkov, L. R. (1996). Radical nature of pathways to alkene and ester from thermal decomposition of primary alkyl diacyl peroxide. *Journal of Organic Chemistry*, **61**, 2801–2808.

Saicic, R. N. and Cekovic, Z. (1992). Sequential radical addition/cyclisation/β-elimination reactions. 3-*Exo* and 5-*exo* cycloaddition reactions of 5-phenythio-3-pentenyl and 5-phenylthio-3-pentynyl radicals. *Tetrahedron*, **48**, 8975–8992.

Saicic, R. N. and Cekovic, Z. (1994). Sequential free radical synthesis of a linear triquinane skeleton from an acyclic synthon. *Tetrahedron Letters*, **35**, 7845–7848.

Seebach, D. and Renaud, P. (1985). Chirale Synthesebausteine durch *Kolbe*-Elektrolyse enantiomerenreiner β-Hydroxy-Carbonsaüre derivate (R) und (S)-Methyl- sowie (R)-trifluoromethyl-γ-Butyrolacton und -δ-Valerolacton. *Helvetica Chimica Acta*, **68**, 2342–2349.

Sheldon, R. A. and Kochi, J. K. (1972). Oxidative decarboxylation of acids by lead tetraacetate. *Organic Reactions*, **19**, 279–421.

Simakov, P. A., Martinez, F. N., Horner, J. H., and Newcomb, M. (1998). Absolute rate constants for alkoxycarbonyl radical reactions. *Journal of Organic Chemistry*, **63**, 1226–1232.

Takasu, K., Mizutani, S., Nogushi, M., Makita, K., and Ihara, M. (1999). Stereocontrolled synthesis of (±)-culmorin via the intramolecular double Michael addition. *Organic Letters*, **1**, 391–393.

Theodorakis, E. A., Xiang, X., and Blom, P. (1997). Photochemical cleavage of the duplex DNA by N-benzoyloxy-2-thiopyridone linked to 9-aminoacridone. *Journal of the Chemical Society, Chemical Communications*, 1463–1464.

Togo, H. and Yokoyama, M. (1990). One-pot preparation of $\gamma$-butyrolactone derivatives from olefinic alcohols via intramolecular radical cyclisation reaction. *Heterocycles*, **31**, 437–441.

Tsanaktsidis, J. and Eaton, P. E. (1989). Synthesis of iodocubanes by decarboxylative iodination. *Tetrahedron Letters*, **30**, 6967–6968.

Wilson, C. V. (1957). The reaction of halogens with silver salts of carboxylic acids. *Organic Reactions*, **9**, 332–387.

Zard, S. Z. (1996). Iminyl radicals: a fresh look at a forgotten species (and some of its relatives). *Synlett*, 1148–1155.

Ziegler, F. E. and Wang, Y. (1996). Carbon–carbon bond forming reactions with oxiranyl radicals. *Tetrahedron Letters*, **37**, 6299–6302.

Ziegler, F. E. and Wang, Y. (1998). Free radical studies and solutions to the synthesis of (+)-cyclophellitol. *Journal of Organic Chemistry*, **63**, 7920–7930.

# 6 Atom and group transfer reactions

## 6.1 Kharasch-type processes: an overview

The accidental discovery by Kharasch and his students of the dramatic effect of oxygen or peroxides on the course of the addition of hydrogen bromide to olefins (cf. Scheme 1.4) is an important watershed in the history of radical chemistry, not only because it established conclusively the role of free radicals as reactive intermediates in solution, but also because this early example of a chain process is the prototype for a large number of exceedingly useful reactions (reviews: Stacey and Harris 1963; Walling and Huyser 1963; Sosnovsky 1964). In fact, many of the radical reactions of organo-tin, -germanium, and -silicon hydrides that we have seen in the earlier chapters may be viewed as particular cases of the 'Kharasch-type' atom or, more generally, group transfer process. Understanding the features that govern such transformations is therefore of special importance.

A Kharasch reaction may be written in a more general form as depicted in Scheme 6.1. Following an initiation step, a small amount of radicals $X^\bullet$ is produced from a reagent $X–Y$. These can add, possibly reversibly, to an olefinic trap to give adduct radical 1 (eqn 1, Scheme 6.1) which, in turn, can abstract $Y$ from $Y–X$ (also reversible in principle) to furnish the addition product (eqn 2, Scheme 6.1) and regenerate at the same time radical $X^\bullet$ to propagate the chain. Overall, we have added the elements of $X–Y$ across the olefin. If we use hydrogen bromide, as in the original Kharasch reaction, one has to simply replace $X–Y$ by $Br–H$. Like any radical chain process, it is the combined efficiency of the individual propagation steps that will determine the overall performance. Let us thus consider each of the two elementary steps in Scheme 6.1, assuming that the solvent does not interfere and that the steady-state concentration of the radicals remains low so that we can neglect, for the time being, the various possible radical–radical interactions (termination steps).

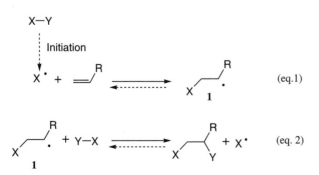

**Scheme 6.1**
Generalized Kharasch-type mechanism

## 6.2   Degeneracy and the lifetime of the intermediate radicals

The first step in Scheme 6.1 is simply the addition of a radical to an olefin, with the crucial difference now that there is no major competing pathway to consume radical $X^•$: its reaction with its precursor $X–Y$ is not only reversible but *redundant* (or *degenerate*, i.e. $X^• + Y–X <\!\!=\!\!=\!\!> X–Y + X^•$). This may sound obvious or even trivial, but it has far reaching consequences because it means that $X^•$ will acquire a comparatively long effective lifetime in the medium, allowing it to react for example with unactivated olefins. Apart from possible radical interactions, the main limiting factors are now interaction with the solvent and hydrogen abstraction, either from the solvent or from the substrate or product; but these are not generally fast processes. This is in contrast to the rapid and irreversible quench of the radical species by stannanes, organomercury hydrides, or Barton esters we have seen so far. In the present case, favourable polar and other factors that will speed up the addition step represented by eqn 1 (Scheme 6.1) will of course be beneficial. In other words, suppression of the main competing pathway endows the key intermediate radicals with a larger window of reactivity and expands considerably the synthetic scope of the process. If hydro-stannylations, -germylations, or -silylations are successful with unactivated alkenes or alkynes, it is precisely because these are strict Kharasch processes (simply replace $X–Y$ in Scheme 6.1 with $R_3Sn–H$, $R_3Ge–H$, or $R_3Si–H$): the reaction of stannyl (respectively germyl or silyl) radicals with their organostannane (respectively organogermane or organosilane) precursor is a degenerate process. These radicals acquire therefore sufficient 'persistence' in the medium to allow them to interact with unactivated olefins or other relatively unreactive substrates such as chlorides or sulfides (in this case, an $S_H2$ reaction of the stannyl, germyl, or silyl radical with the halide or sulfide replaces the addition to olefin in eqn 1). Thus, simple reductions with triorgano-tin, -germanium, or -silicon hydrides in a sense also belong to the Kharasch family. Allylations with allyl stannanes and related reactions are feasible for the same reason, namely the overall degeneracy of the addition–fragmentation of the stannyl radical to the allylstannane.

One important practical consequence is that the question of the concentration of the various components is tremendously simplified: there is usually no need to ensure that one of the reagents is added slowly (the perennial syringe pump!); in fact Kharasch reactions will often work better under concentrated conditions, and mere mixing of all the ingredients under heating with a chemical initiator or irradiation will suffice in most cases. This is not a negligible advantage when large-scale work is being contemplated.

As for the second propagation step in Scheme 6.1 (eqn 2), its rate will of course depend on the reactivity of the adduct radical, the weakness of the $X–Y$ bond, and also on the polarity matching between the two. The exchange of group $Y$ is usually reversible; in such cases, the system must be designed so as to drive the equilibrium as much forward as possible (i.e. radical **1** has to be less stable than $X^•$). An unfavourable equilibrium in the exchange represented by eqn 2 will result in a build-up in the concentration of adduct radical **1** with deleterious consequences on the efficiency and the yield, caused by possible telomerization of the olefin and increased chances of unwanted radical–radical interactions. If such a situation is unavoidable, it may still be possible to prevail by draining the unfavourable equilibrium in the desired direction through the incorporation of a fast, preferably irreversible, subsequent step (e.g. a cyclization or opening of a strained ring).

X and Y can represent a variety of atoms or functional groups, and more than two propagation steps can be involved. This chapter will be structured according to the nature of Y: we shall thus examine the cases where Y is a hydrogen, a halogen, a selenide, or a dithiocarbonyl group. In the last section, the principle of degeneracy as a selectivity controlling element will be extended to allylations and vinylations using sulfones. We shall also include reactions involving boranes and a few non-chain processes where the degeneracy of one or more steps still allows a synthetically useful overall control of the selectivity.

## 6.3   Transfer of hydrogen atoms

The addition of silicon-, germanium-, and tin-hydrides to alkenes and alkynes were discussed for convenience in the previous chapters, even though, mechanistically, they really belong under this heading. The various examples that were presented must now be viewed in the light of the above discussion. From a synthetic perspective, it would be desirable to include in this list of group IV hydrides, the 'hydrides' of carbon, the lightest but most important member. Unfortunately, alkanes do not generally participate in efficient radical chain additions to olefins, mainly because the second step in the Kharasch system (eqn 2 in Scheme 6.1) fails: the C–H bond is too strong and the exchange, now essentially thermoneutral, is much too slow. However, in some instances, the C–H bond is sufficiently weak to permit an additive hydrogen atom transfer. This is case for example with 2-propanol where the tertiary C–H bond is weakened by the partial filling of its antibonding $\sigma^*$ orbital with electrons flowing from one of the lone pairs on oxygen (an anomeric type effect which causes a lengthening and therefore weakening of the C–H bond). This interaction enhances the 'hydride' character of the hydrogen atom; indeed, 2-propanol can be used to reduce ketones by a hydride transfer catalysed by aluminium or titanium alkoxides—the well-known Meerwein–Ponndorf–Verley reduction. The rate of hydrogen abstraction by an electrophilic radical is consequently increased because of a correct polarity matching between the reacting partners. One early example of application is the expedient, highly efficient one-step synthesis of terebic acid, devised by Schenck *et al.* (1957). Merely irradiating (at 366 nm) a solution of maleic acid in 2-propanol in the presence of benzophenone gives a 96% yield of terebic acid (Scheme 6.2). Excited triplet benzophenone abstracts a hydrogen from 2-propanol to give the nucleophilic dimethylcarbinyl radical, which readily adds to the electron poor olefin. Hydrogen atom transfer from 2-propanol to the electrophilic adduct radical is the second propagation step. The alcohol has to be used as the solvent because the rate of hydrogen abstraction, despite the favourable factors mentioned above, remains comparatively low. Incidentally, the diphenylcarbinyl radical formed in the initiation step can also act as the hydrogen atom donor, by a disproportionation reaction that leads back to benzophenone.

Two further examples are displayed in Scheme 6.3 illustrating additions of alcohols to carbohydrates, to give highly functionalized, complex structures. The first concerns a reaction with methanol, a less reactive alcohol in comparison to

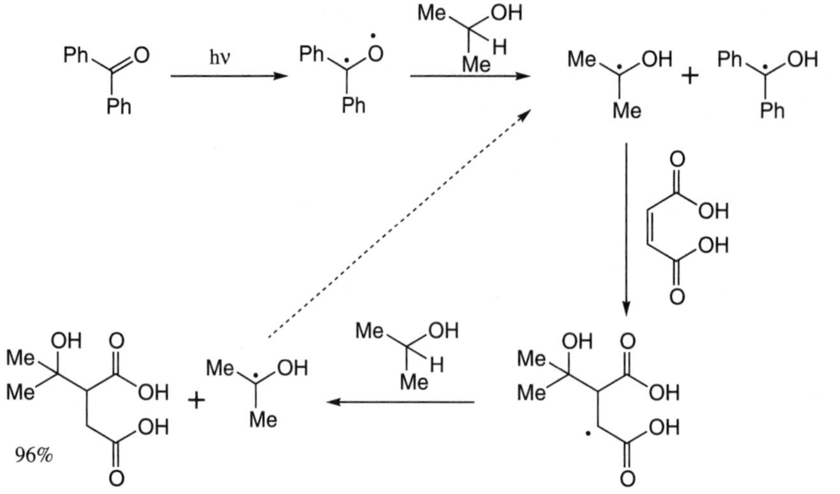

**Scheme 6.2** Synthesis of terebic acid by radical addition of 2-propanol to maleic acid

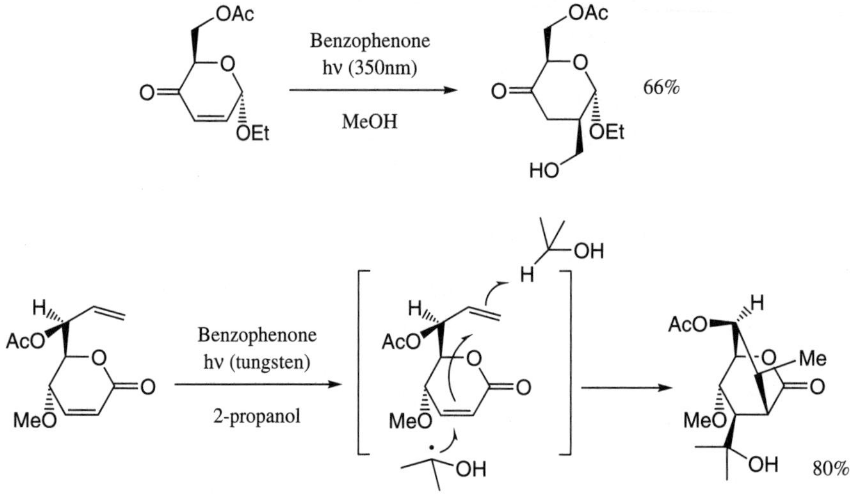

**Scheme 6.3** Radical addition of alcohols to carbohydrate enones

2-propanol (Fraser-Reid *et al.* 1977; Benko *et al.* 1988); as expected, the addition occurs from the more accessible β-face of the sugar. In the second transformation, the ketyl radical from 2-propanol adds first to the more electrophilic olefin, followed by a 6-*exo*-cyclization onto the terminal alkene (Gomez *et al.* 1997).

The alcohol in these transformations may be replaced by other derivatives containing a weakened C–H bond, as long as they are cheap enough to be employed in large excess, and sufficiently volatile to be removable by distillation. Ethers, acetals, ortho-formates, and even aldehydes and amines can be suitable candidates (Walling and Huyser 1963). The first example in Scheme 6.4 shows the remarkable chain addition of a dioxolane to diketene, an early step in a short synthesis of (−)-tetrahydrolipstatin

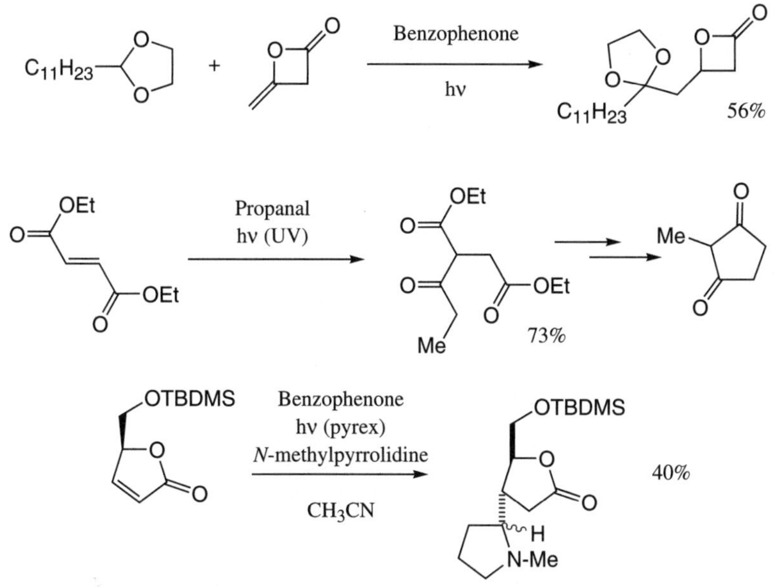

**Scheme 6.4** Radical addition of acetals, aldehydes, and amines to electrophilic olefins

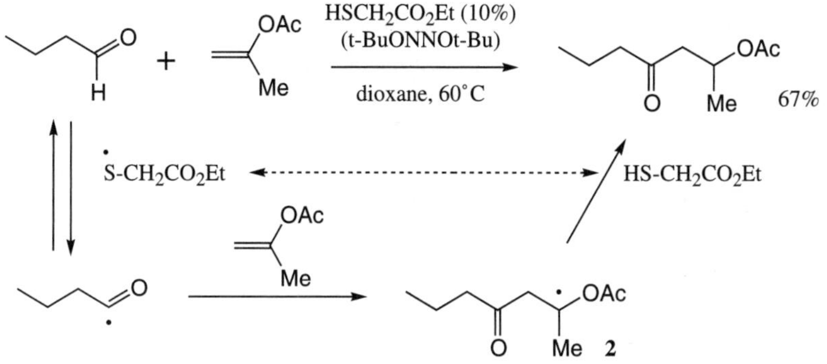

**Scheme 6.5** Polarity reversal catalysis by a thiol of the radical addition of an aldehyde to an olefin

(Parsons and Cowell 2000). The second transformation illustrates the case of an aldehyde as the radical source and corresponds to the first step in the synthesis of 2-methyl-1,3-cyclopentanedione, a key substance in many industrial syntheses of steroids (Patrick 1952; for rates of hydrogen abstraction from aldehydes, see Chatgilialoglu *et al.* 1984). Finally, addition of *N*-methyl pyrrolidine to a butenolide occurs with high facial selectivity as far as the lactone ring is concerned but, as would be expected, with little selectivity at the level of the pyrrolidine nucleus (de Alvarenga and Mann 1993; for α-CH bond dissociation energies, see Pombrowski *et al.* 1999). Additions of amines, even to unactivated olefins, were accomplished earlier by Urry and Juveland (1958).

It is possible to apply to aldehydes the polarity reversal concept involving thiols, which was used in the case of silane reactions (cf. Section 4.5). This is illustrated by the mercaptoacetate-catalysed addition of butanal to isopropenyl acetate outlined in Scheme 6.5 (Dang and Roberts 1996; Roberts and Winter 1999). The mildly electrophilic thiyl radicals, generated following thermal decomposition of the initiator, di-*t*-butyl hyponitrite, are capable of reversibly removing the aldehyde hydrogen to give the corresponding acyl radical. Addition to 2-acetoxypropene furnishes tertiary radical **2**, which is too stabilized and nucleophilic in character to efficiently abstract a hydrogen atom from the aldehyde; it is however rapidly quenched by the thiol, regenerating the thiyl radical to propagate the chain. The thiol thus acts as a relay, improving considerably an otherwise sluggish transformation. It must be pointed out that the addition of the thiyl radical to the olefin is not a complicating factor because this addition is highly reversible (*vide infra*).

Cleavage of the C–H bond is the limiting step, as demonstrated clearly by the experiment outlined in Scheme 6.6 (Julia 1971; Julia and Maumy 1976). Heating malonate **3** in cyclohexane in the presence of dibenzoyl peroxide affords the cyclohexane diester **8**, instead of the expected methylcyclopentyl isomer **6**. The thermodynamically favoured 6-*endo-trig* mode of cyclization prevails over the kinetically preferred 5-*exo* mode because of the reversibility of the latter under the reaction conditions. Radical **4** is stabilized by the two ester groups, thus facilitating the reverse, ring-opening process, while hydrogen abstraction from the starting material **3** (or even from the solvent, cyclohexane) is too slow to prevent the equilibration between **4** and **5** from setting in. Ultimately, it is the essentially irreversible 6-*endo* ring closure to give **7** and thence **8** which carries the day. We shall see in the following section that it is possible to obtain the kinetic product by replacing the slow hydrogen atom transfer by the much faster iodine exchange.

Hydrogen abstraction from thiols is especially fast (of the order of $10^8$ $M^{-1}s^{-1}$ for thiophenol; Newcomb 1993). Unfortunately, the overall efficiency of the addition of thiols to olefins is diminished considerably by the strong reversibility of the first addition step, corresponding to eqn 1 in Scheme 6.1 (for relative rates of $\beta$-cleavages of halides and sulfur-based functional groups, see Wagner *et al.* 1978). It is therefore often advantageous to lower the reaction temperature in order to slow down the reverse step (this step is a unimolecular fragmentation with a large entropy term and is therefore more sensitive to temperature), and to use unhindered olefinic traps with a stabilizing group, so as to push the equilibrium in the first step as much forward as possible. Styrenes and dienes are thus especially interesting substrates because the

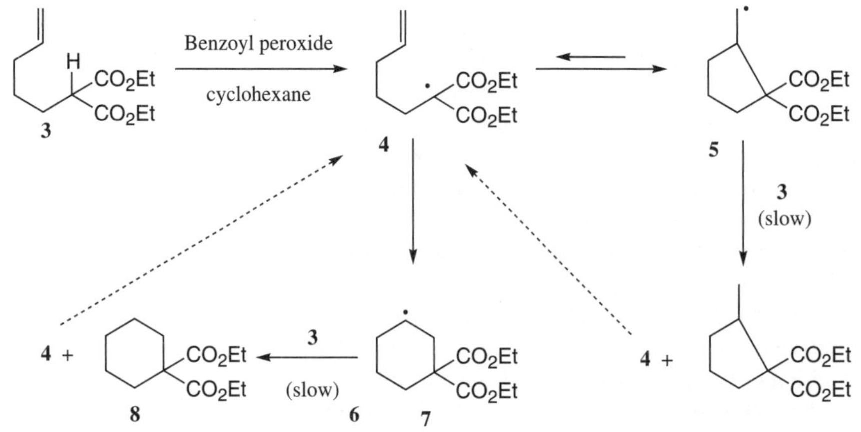

**Scheme 6.6** Radical cyclization under thermodynamic control

adduct radical is benzylic or allylic. The addition of thiophenol to indene (Oswald 1960) and to 2,3-dimethyl-butadiene (Oswald *et al.* 1962) is pictured in Scheme 6.7. Addition to the diene gives only '1,4' addition, with transfer of hydrogen taking place at the terminal carbon. The '1,2' isomer is observed as a very minor product when butadiene itself is used. Initiation can be achieved either by irradiation or by exploiting the slow oxidation of thiols with molecular oxygen or with a peroxide. Strained olefins (cyclobutenes, cyclopropenes, etc.) and terminal acetylenes are also good substrates because the back-fragmentation is slowed down by the higher energetic cost incurred in regenerating a strained alkene or an alkyne. In the last example in Scheme 6.7, a double addition of ethyl mercaptan to propargyl alcohol took place after two weeks of irradiation (Blomquist and Wolinsky 1958).

Finally, in order to drain the equilibrium in the desired direction, it is possible to couple the *reversible* addition of the thiyl radical with a fast *irreversible* process such as a cyclization or an opening of a small ring. The first reaction in Scheme 6.8 is a key step in a recent synthesis of kainic acid (Miyata *et al.* 1997), whereas the second is a more complex formal [3 + 2] annulation process (Miura *et al.* 1988). In both examples, advantage is taken of the β-elimination of the thiyl radical to contrive a cycle that is catalytic in thiol. The reversibility of the addition of a thiyl radical to alkenes may be exploited to isomerise *cis-* to more stable *trans*-olefins (e.g. see Thalman *et al.* 1984 or Smith *et al.* 1990).

Another approach is to combine the reversible thiyl addition with a fast capture of the adduct radical by molecular oxygen. An exceedingly elegant application is pictured in Scheme 6.9, whereby a highly complex *endo*-peroxide, structurally related to the anti-malarial natural product yingzhaosu A, was assembled in one operation from limonene (Bachi and Khorshin 1998). In the process, oxygen is used

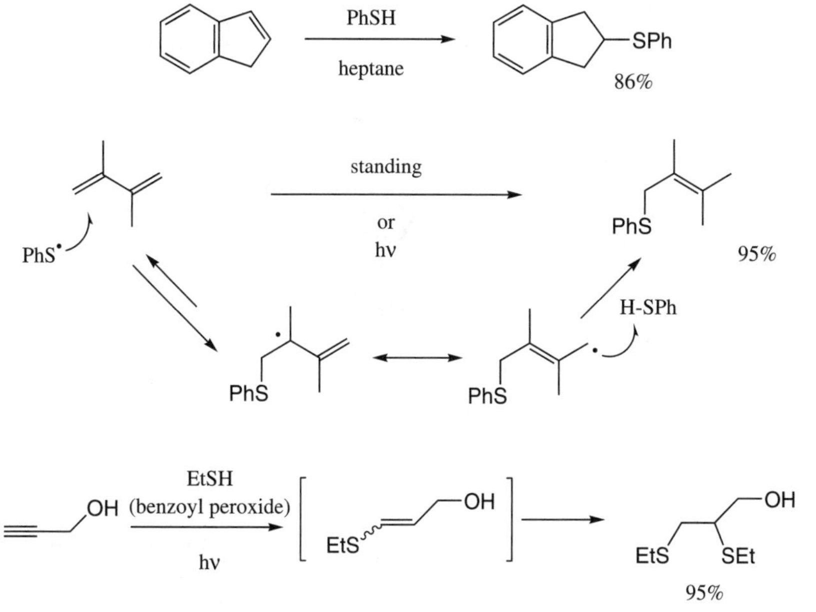

**Scheme 6.7** Radical addition of thiols to alkenes and alkynes

twice, in a manner reminiscent of the postulated pathway in the biosynthesis of prostacyclins from arachidonic acid.

Sulfonyl radicals suffer less from the reversibility of the first addition step and various sequences have been implemented using these species (Bertrand 1994). The efficient and very practical synthesis of a sulfonate salt by addition of sodium bisulfite to diallylamine is shown in Scheme 6.10 (Schmitt 1995). The reaction is complete in less than an hour at 0 °C and air is sufficient to initiate the process. The synthesis of sulfonic acids is of great importance to the detergent and surfactant industry. The second transformation is a more elaborate cascade involving isomerization,

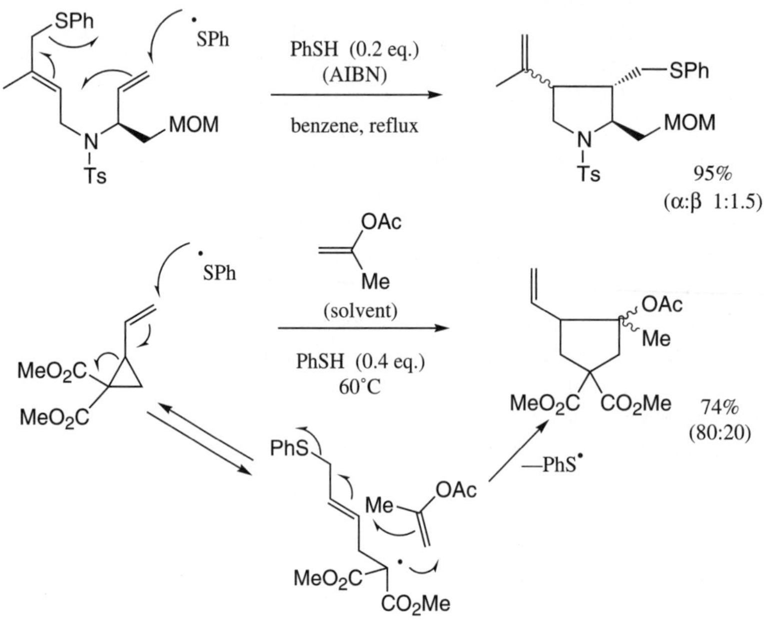

**Scheme 6.8** Sequential radical reactions involving thiols

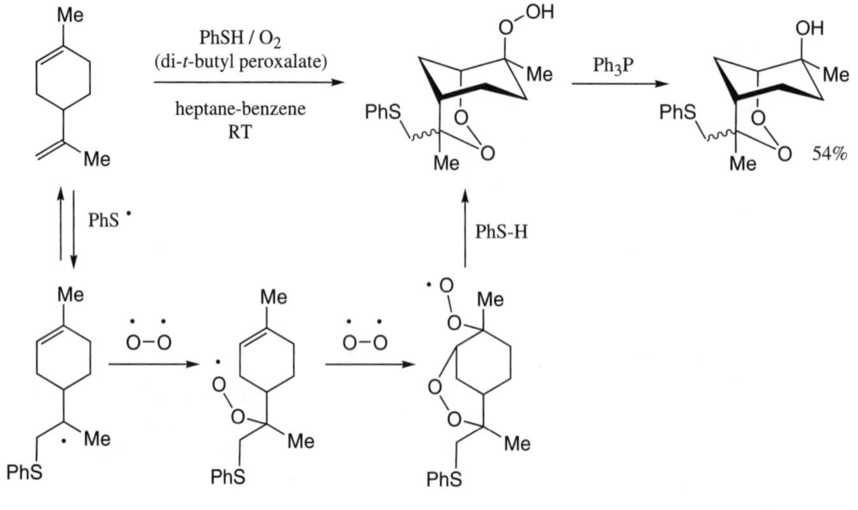

**Scheme 6.9** Sequential radical reactions involving thiols and triplet oxygen

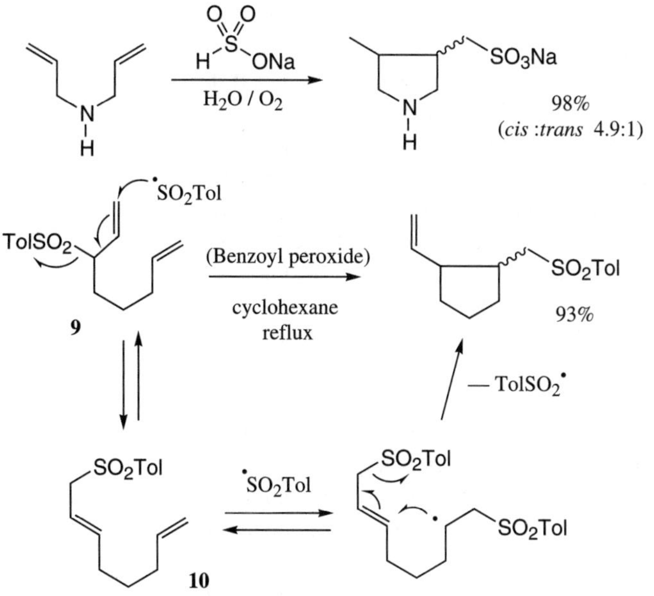

**Scheme 6.10** Addition of sulfite and sulfonyl radicals

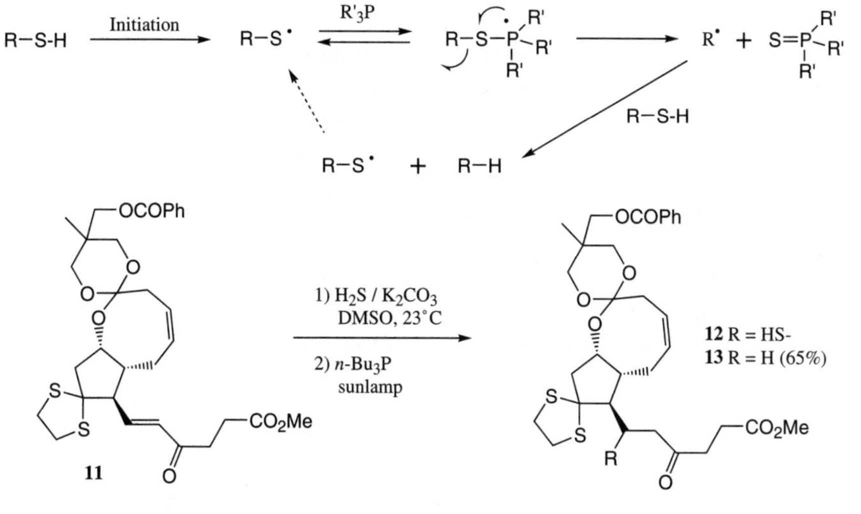

**Scheme 6.11** Reductive desulfurylation of thiols with phosphites

cyclization, and β-scission (Smith and Whitham 1989). The addition of the toluene-sulfonyl radical can and does occur reversibly on both of the terminal olefins of the starting material **9**, but only the one shown is not a dead end and leads to isomer **10** of the initial sulfone. Addition to the remaining terminal olefin now proceeds in a constructive manner to give the observed product in high yield.

Thiyl radicals also add reversibly to trivalent phosphorus derivatives. This 'α-addition' produces a phosphoranyl radical, which may undergo β-scission of the C–S bond to give an alkyl radical (Scheme 6.11). In the presence of the parent thiol, fast hydrogen abstraction takes place to give the corresponding alkane and a thiyl radical that propagates the chain. Overall, the sequence represents a mild reductive desulfurylation process or, more generally, a way for generating carbon radicals from aliphatic thiols (Hoffman *et al.* 1956). This neat reaction has not attracted the attention it deserves from the synthetic community. The example in Scheme 6.11 highlights one spectacular application, where the powerful nucleophilic properties of the thiolate anion were exploited to selectively reduce compound **11**, a key intermediate in the synthesis of a urinary metabolite of prostaglandin $D_2$ (Corey and Shimoji 1983). In this case, reduction of the hindered, conjugated enone by more common methods was foiled by the presence of a number of fragile functional groups in the molecule; however, base-induced Michael addition of hydrogen sulfide to give thiol **12** and irradiation in the presence of tributyl phosphine provided the reduced material **13** in good overall yield.

Kharasch-type processes involving the addition of phosphorus-centred radicals across activated or unactivated olefins followed by hydrogen atom transfer from a P–H bond can be applied to the preparation of a wide variety of organophosphorus compounds. The first example in Scheme 6.12 represents yet another addition to diketene (Dingwall and Tuck 1986) whereas the second is an addition–fragmentation involving $\beta$-pinene (Kenney and Fisher 1974).

The possibility of quenching an alkyl radical by hydrogen abstraction from a P–H bond, the radicophilicity of the thiocarbonyl group, and the strength of a P–S linkage can be exploited to deoxygenate secondary alcohols by the Barton–McCombie protocol, but without using a stannane reagent, as demonstrated by the first example in Scheme 6.13. The hydrogen atom donor is the cheap, water soluble hypophosphorus acid or its amine salts. Reductive debrominations and deiodinations, and deaminations may be readily accomplished (Barton *et al.* 1992, 1993; Jang 1996) and the intermediate radical may be captured by an internal olefin as illustrated by the second example (Graham *et al.* 1999*a*,*b*; Gonzalez Martin *et al.* 2000). Even if the chains are not very long and substantial amounts of initiator may be required to drive the reaction to completion, this modification is quite attractive for large-scale work.

It is appropriate to end this section by a brief look at reactions involving hydrogen halides. It is the capricious behaviour of hydrogen bromide towards alkenes that led to the discovery of the Kharasch reactions in the first place. But

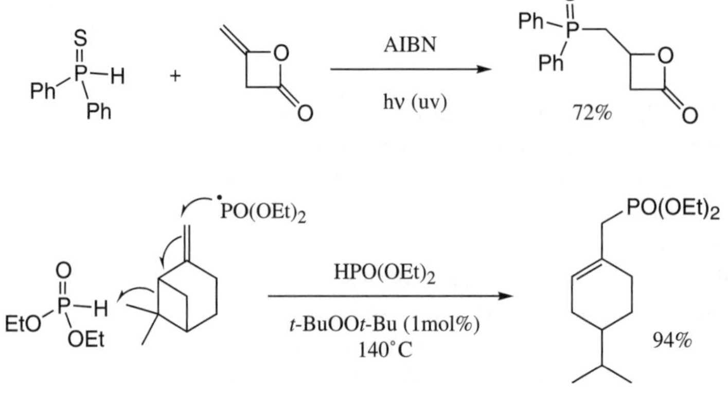

**Scheme 6.12** Addition of phosphorus-centred radicals

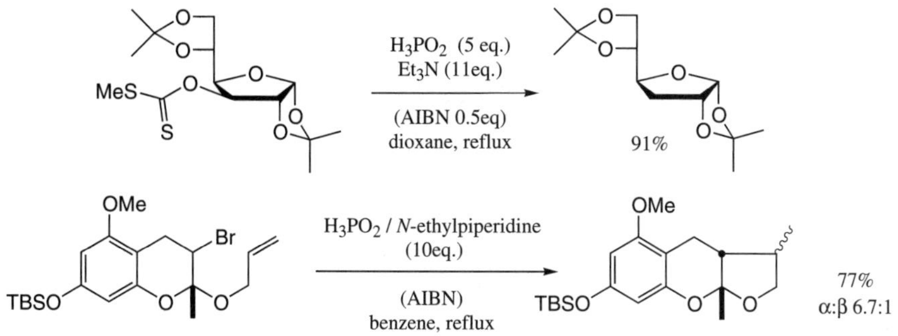

**Scheme 6.13** Reductions with hypophosphites

despite its historical importance, the anti-Markovnikov radical chain addition of hydrogen bromide is seldom used in synthesis. The reagent is much too acidic to be compatible with many of the functional groups encountered in modern synthesis, and unwanted ionic side reactions can be difficult to suppress, especially in the presence of moisture. Most of the applications concern therefore additions to simple, robust substrates. This is illustrated by the addition of hydrogen bromide to methylcycloheptene (Scheme 6.14) to give mostly *cis*-1-bromo-2-methyl-cycloheptane; the hydrogen transfer in the second propagation step occurs from the least hindered side, opposite to the bromine atom (Abell and Bohm 1961).

Reactions with HF and HCl fail because the halogen–hydrogen bond is too strong. With hydrogen iodide, the problem now is the high reversibility of the addition of iodine atoms to olefins (Mayo and Walling 1944). Hydrogen bromide appears to be a good compromise in having a reasonably weak Br–H bond and a decent efficiency in the (reversible) addition step.

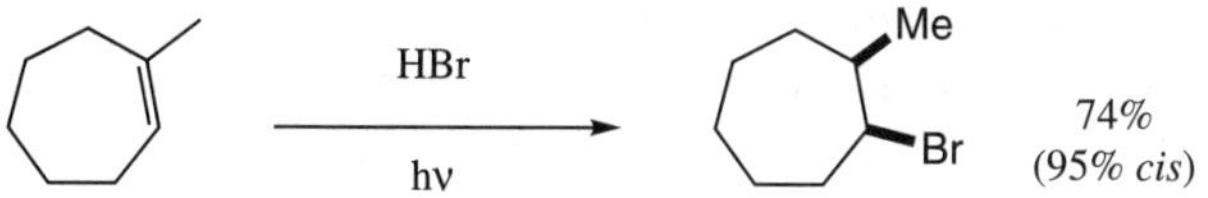

**Scheme 6.14** Addition of hydrogen bromide to 1-methylcycloheptene

## 6.4 Halogen atom transfer

Transfer of halogen atoms ($Y = Cl$, Br, I in Scheme 6.1), also discovered by Kharasch and his students (Kharasch *et al.* 1945, 1948*a*,*b*; Schiesser and Wild 1996), is of greater synthetic scope as compared with hydrogen transfer, especially as concerns the intermolecular formation of C–C bonds. Both of the two propagation steps can be made efficient, and the introduction of a halide in the product opens a variety of synthetic possibilities. As would be expected from bond strengths and polarizabilities of the halogen atoms, the reactivity increases in the order $Cl < Br < I$ (Curran *et al.* 1989; Schiesser and Wild 1996). With the exception of reactions involving elemental fluorine itself or exceedingly reactive derivatives thereof such as

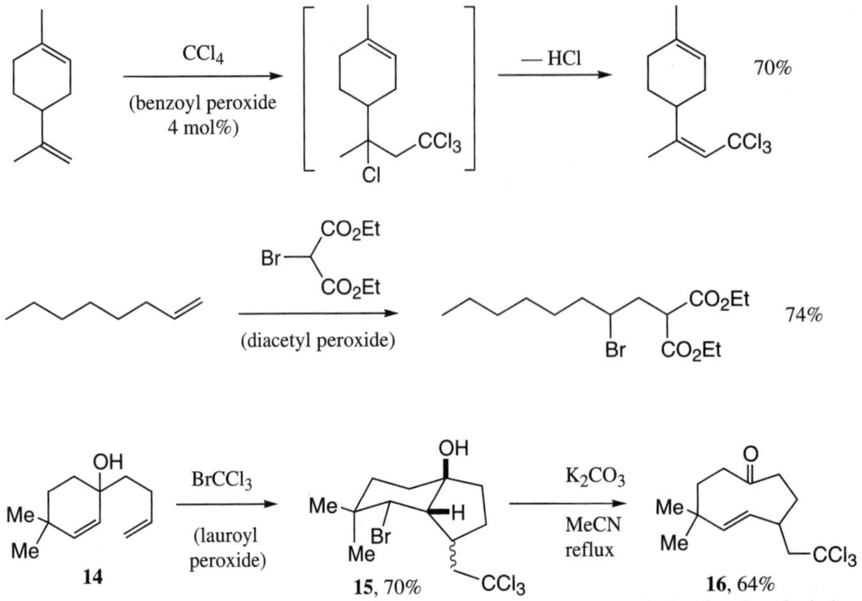

**Scheme 6.15**  Radical addition of halides to alkenes

xenon difluoride, no fluorine atom transfers appear to have been recorded in the literature. Indeed, the creation of C–F bonds under mild conditions using radical chemistry remains an unsolved problem and a formidable challenge.

Many examples of simple additions to olefins of chloro- or bromo-derivatives have been reported (Stacey and Harris 1963); most involve carbon tetrachloride, bromotrichloromethane, or other polyhalo-derivatives, but bromoesters and related compounds have also been employed. A few examples are displayed in Scheme 6.15. The first illustrates the selectivity of addition of $CCl_4$ to the least hindered olefin in limonene with concomitant elimination of HCl from the adduct, which is apparently unstable under the reaction conditions (Israelashvili and Diamant 1952). The trichloromethyl group in the product may be converted into a carboxylic acid by treatment with hot alkali. The addition of diethyl bromomalonate to 1-octene is an example of a non-polyhalogenated component in the Kharasch halogen transfer process (Kharasch *et al.* 1948*b*). In this early work, diacetyl peroxide was used as initiator but this compound is dangerous to handle; safer initiators such as benzoyl or lauroyl peroxide would probably have been as efficient. Indeed, lauroyl peroxide was used to trigger the addition of bromotrichloromethane to 1,6-diene **14** to give bicyclic bromide **15** by a succession of inter- and intra- molecular radical additions. Exposure of the adduct to potassium carbonate in refluxing acetonitrile induced a Grob fragmentation to furnish *trans*-cyclononenone **16** (Boivin *et al.* 1999*a*).

The important difference in the rate of transfer of a chlorine and a bromine atom, and its consequence on the outcome of a reaction, is nicely illuminated by examining the addition of carbon tetrachloride and bromotrichloromethane to optically pure sabinene (Batey *et al.* 1992). As shown in Scheme 6.16, two completely different products are observed. In both transformations, the first step is an irreversible

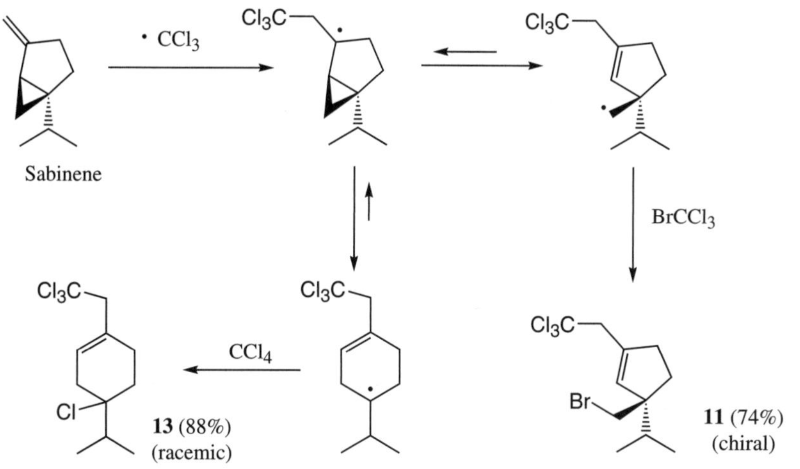

**Scheme 6.16** Kinetic and thermodynamic control in the radical addition of polyhalides to sabinene

addition of a trichloromethyl radical to the exocyclic double bond of sabinene, followed by a rapid *but reversible* opening of the cyclopropane ring. Because of better orbital overlap, it is the formation of the least stable primary radical that is kinetically favoured. This radical is promptly captured by bromotrichloromethane to give the *optically active* cyclopentyl derivative. In contrast, chlorine transfer from carbon tetrachloride is slow enough to allow the other mode of ring opening to occur, leading to the thermodynamically more stable tertiary radical. Chlorine transfer finally furnishes the *racemic* cyclohexene derivative. This is yet another instance where either the thermodynamic or the kinetic product can be made to prevail by simply modifying the reagent.

Iodine atom transfers have also been known for a long time (Walling and Huyser 1963). Indeed, radical additions starting with perfluoroiodoalkanes are used commercially for the production of a host of fluorinated speciality chemicals and building blocks for the pharmaceutical and agrochemical industries (Hudlicky 1992). Kharasch-type transformations involving iodides can be achieved from a greater variety of substrates as compared with bromides and chlorides, and the process is finding increasing synthetic applications (Curran and Chang 1989). The reaction can be initiated with light, peroxides (Ollivier *et al.* 2000), or even using triethylborane/oxygen (Ichinose *et al.* 1989; Baciocchi and Muraglia 1994; Yorimitsu *et al.* 1998). Sometimes, however, complications are encountered because of the formation of small amounts of molecular iodine, hydrogen iodide, etc., which inhibit the chain reaction. These difficulties can often be overcome by addition of hexabutylditin, a scavenger for both iodine and hydrogen iodide (Curran and Chang 1989). Addition of bicarbonate to remove hydrogen iodide may also be useful sometimes (Ollivier *et al.* 2000).

The first two examples in Scheme 6.17 illustrate the synthesis of a fluorinated molecule by UV-induced addition of difluoroiodomethyl phenyl ketone to ethyl acrylate (Qiu and Burton 1994) and a ring closure leading to a lactam (Ollivier *et al.* 2000). The third transformation starting with the iodomalonate **17** involves the same intermediates **4** and **5** as in Scheme 6.6 above and serves to further clarify the importance of kinetic versus thermodynamic control. Because the iodine transfer is so rapid

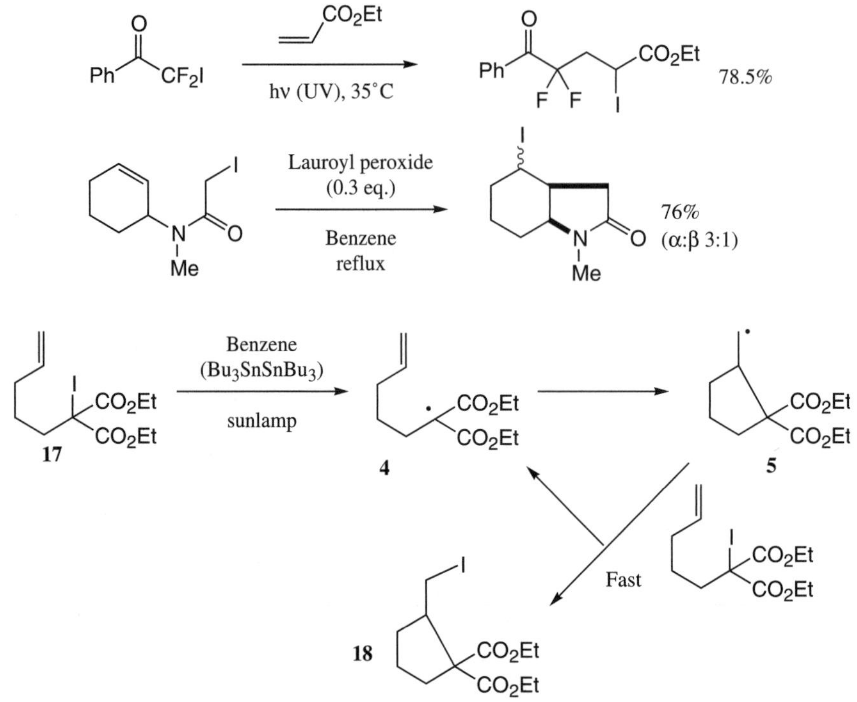

**Scheme 6.17**
Examples of iodine atom transfer reactions

$(10^8–10^9$ M$^{-1}$s$^{-1}$ in this case; Curran *et al.* 1989), equilibration between the cyclopentylmethyl radical and its open chain precursor does not have time to take place; the system is therefore under kinetic control and only the cyclopentyl derivative **18** is obtained. This in sharp contrast to the cyclization of the non-iodinated malonate **3** discussed above (Scheme 6.6) which, being under thermodynamic control, leads to the cyclohexane diester **8**.

The relative long lifetime of radicals generated under Kharasch-type systems manifests itself in the key step of an elegant approach to silphinene shown in Scheme 6.18 (Crimmins and Mascarella 1987). Attempts to introduce a methyl group with a defined stereochemistry by ring opening of the cyclobutane ring (cf. Scheme 3.4) using tributyltin hydride were not satisfactory: they foundered because of premature reduction to give unwanted compound **20**. This difficulty was overcome by application of iodine atom transfer to force the opening of the reluctant cyclobutane ring, followed by a classical reduction of iodide **23** in good overall yield. Radical **22** has only a fleeting existence in the medium: it is quenched by a fast and essentially irreversible iodine transfer from the starting material **19**. Radical **21**, in contrast, is fairly long-lived: it is unable to abstract iodine from the product **23** (this would give the thermodynamically less stable primary radical **22**), and exchange of iodine with the starting material is degenerate. Its extended lifetime allows it therefore to undergo the slow ring opening, as it is now the main option left.

Iodine transfer cyclization to alkynes is especially interesting because the intermediate vinyl radical is highly reactive: it rapidly abstracts iodine from the starting material, pushing the reaction forward. This is a convenient way to produce

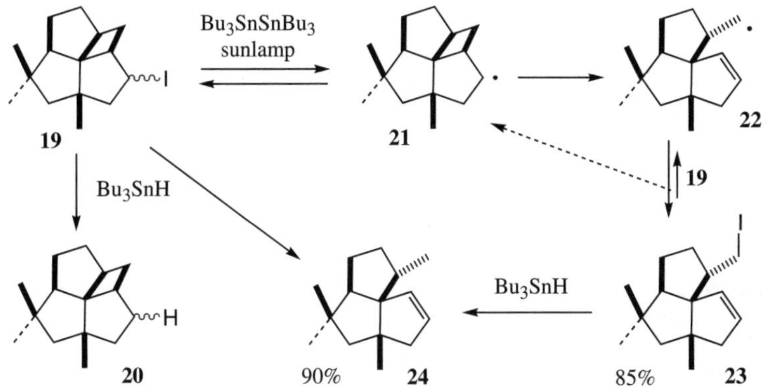

**Scheme 6.18** Iodine atom transfer triggered opening of a cyclobutane ring

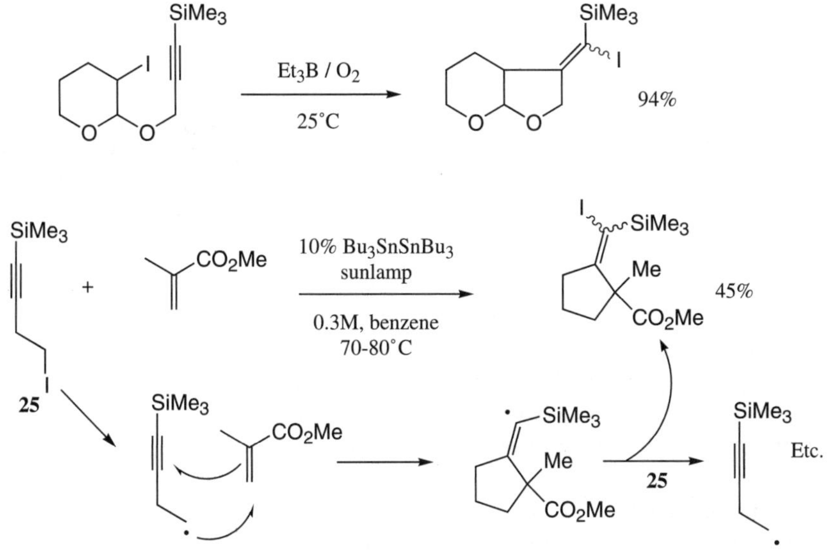

**Scheme 6.19** Iodine atom transfer additions to alkynes

iodoalkenes which lie at the heart of many transition-metal-mediated reactions. The first example in Scheme 6.19 illustrates a simple ring closure (Ishinose *et al.* 1989), whereas the second combines an intermolecular addition to methyl methacrylate with a cyclization step (Curran and Chen 1987). If the alkyne group is not substituted then, in addition to the expected cyclopentane derivative, a small amount of the cyclohexenyl iodide isomer (arising from a 6-*endo-dig* cyclization) is observed. Telomer formation is not a problem because the essentially irreversible iodine transfer from the primary iodide **25** to the vinyl radical in the last propagation step is much faster than addition of the vinyl radical to methyl acrylate.

The conversion of aliphatic iodides into esters can be accomplished by using carbon monoxide as the external radical trap and visible light to initiate the chain

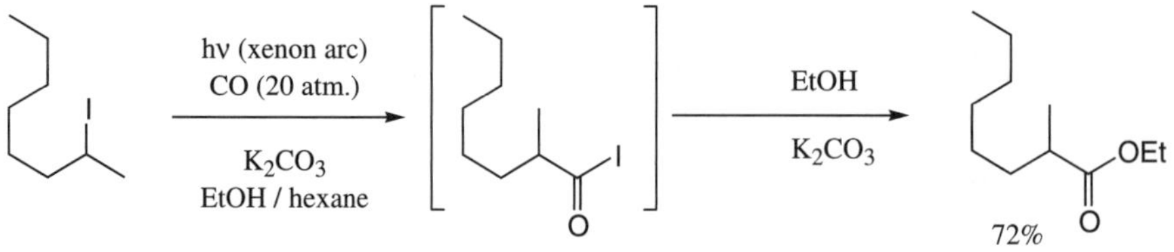

**Scheme 6.20**  Synthesis of esters from iodides by radical addition to carbon monoxide

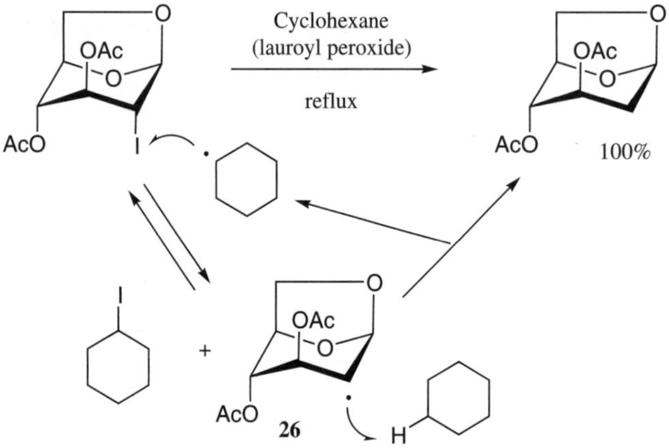

**Scheme 6.21**  Reductive deiodination by hydrogen atom transfer from cyclohexane

(Scheme 6.20). The reaction leads first to a highly reactive acyl iodide, which is immediately quenched by ethanol; the hydrogen iodide formed in the ionic step is neutralized by added potassium carbonate (Nagahara *et al*. 1997). The presence of base is essential for success, for otherwise the free hydrogen iodide would inhibit the chain process (for a discussion of the beneficial effect of base in bromine atom transfer reactions, see Curran and Ko 1998).

The fast and reversible iodine exchange may be exploited to replace the iodine atom with hydrogen using cyclohexane as the hydrogen atom source. Cyclohexane is a rather poor donor of hydrogen atoms but if the radical derived from the iodide is especially electrophilic in character *yet not stabilized by resonance*, then the hydrogen abstraction does occur at a useful rate. This conception is illustrated by the reductive deiodination of the glucose derivative in Scheme 6.21 (Boivin *et al*. 1997). Thus, reversible iodine abstraction by the cyclohexyl radical gives radical **26** and iodocyclohexane. Intermediate **26** is an unstabilized secondary radical with an electrophilic character imparted by the combined inductive effect of the neighbouring electron-withdrawing C–O bonds (acetoxy and acetal groups). The polarity matching improves the efficiency of the hydrogen abstraction from the cyclohexane solvent leading to the reduced product and a cyclohexyl radical to propagate the chain. This economical deiodination process is especially adapted to carbohydrates, since radicals derived therefrom are usually flanked by electron-attracting oxygen containing functions.

Halogen atom transfer can be performed from sulfur or phosphorus halides (Stacey and Harris 1963). Sulfonyl halides are by far the most commonly employed, and their chain addition to alkenes represents a convenient entry to vinyl sulfones, since the halide in the adduct is easily removed by base induced β-elimination (Liu *et al.* 1980). One such example, involving toluenesulfonyl iodide (prepared immediately before use or generated *in situ*), is presented in Scheme 6.22 (Najéra *et al.* 1988). An interesting synthetic variation, shown in the same scheme, is to combine the radical addition with a vinylogous Ramberg–Bäcklund reaction, resulting in an overall entry into 1,3-dienes (Block *et al.* 1986). Sulfenyl and sulfinyl halides are not generally suitable: they tend to react rapidly with olefins by a classical ionic electrophilic addition.

It is perhaps the mediation of ring closures of diene systems that constitutes the greatest asset of sulfonyl halides. This illustrated by the reaction of 1,5-cyclooctadiene with toluenesulfonyl chloride and bromide displayed respectively in Scheme 6.23 (De Riggi *et al.* 1988). With the latter, the transfer of the bromine atom is faster than the comparatively slow intracyclic ring closure and only the simple addition product is observed. This example parallels the radical addition to sabinene discussed above in that a bromine and a chlorine atom transfer sequence can lead to two different products because of the large difference in the relative rates of one of the steps.

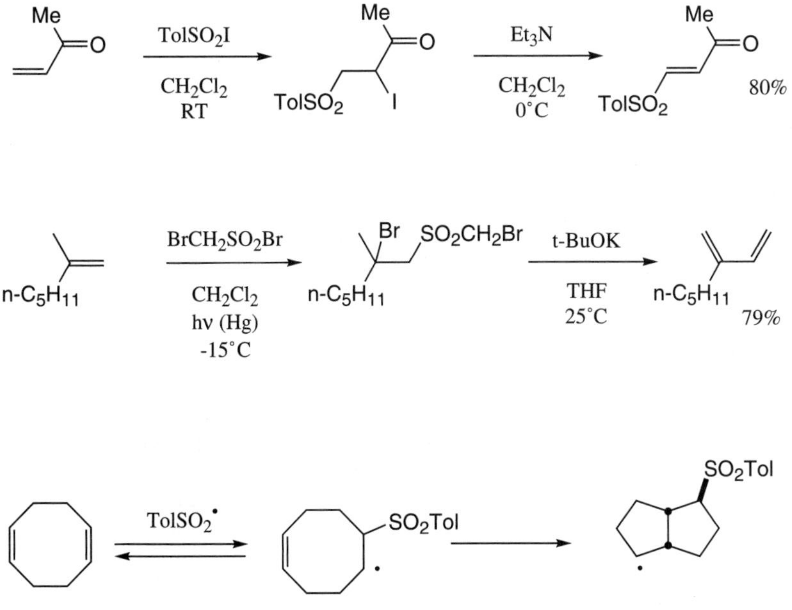

**Scheme 6.22** Halogen atom transfer from sulfonyl halides

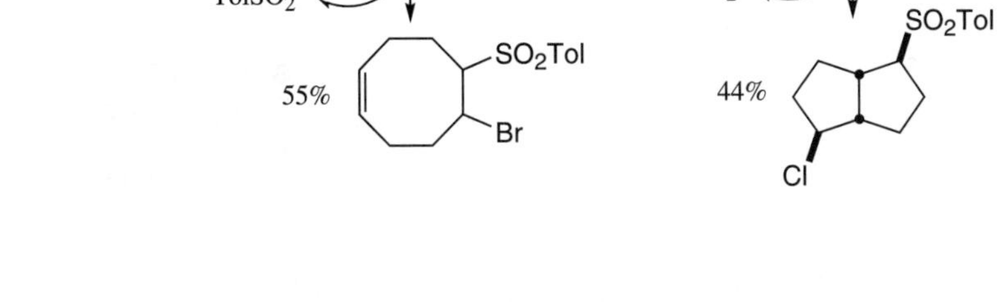

**Scheme 6.23** Influence of rate of halogen transfer on the nature of product

## 6.5    Transfer of sulfides, selenides, tellurides, and nitriles

Sulfides are not sufficiently reactive to be transferred in a general manner, whereas the reactivity of selenides may be compared to that of bromides. Transfer of selenides is feasible if the radical produced is especially stabilized (Byers and Lane 1993). Initiation is often best accomplished by irradiation with a sun lamp. The nice radical cascade in Scheme 6.24 was devised as an entry into prostaglandins via the Corey lactone (Renaud and Vionnet 1993).

Selenide transfer from sulfonyl selenides is especially effective because of the weakness of the S–Se bond (Back and Collins 1981; Gancarz and Kice 1981). The sulfonyl radical intermediate can be made to add either to an alkene (Back and Collins 1981) or to an alkyne (Back *et al.* 1983), as shown by the two examples in Scheme 6.25. Sulfonyl selenides hold much synthetic potential: they are readily available, for example by mixing a sulfonyl hydrazide with phenyl selenenic anhydride (Back and Collins 1981), and the adducts are highly functionalized.

Tellurides are, not unexpectedly, more reactive than selenides. But they are also more difficult to handle because of their light sensitivity, and their chemistry remains to be more fully explored. One example of a telluride transfer involving an acyl radical intermediate is outlined in Scheme 6.26 (Crich *et al.* 1994). Curiously, this reaction

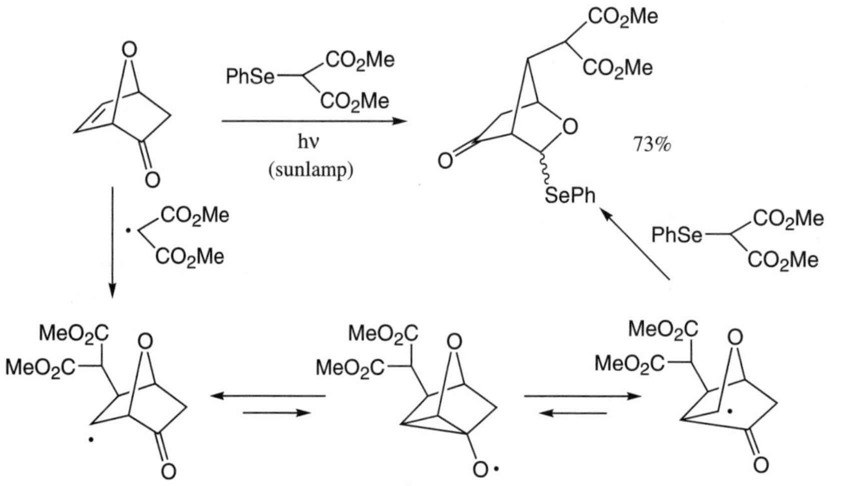

**Scheme 6.24**  Radical transfer of a phenylseleno group

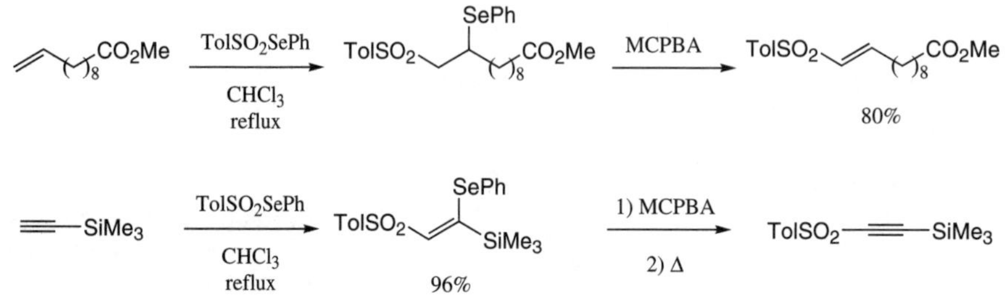

**Scheme 6.25** Phenylselenide transfer from phenylselenosulfonates

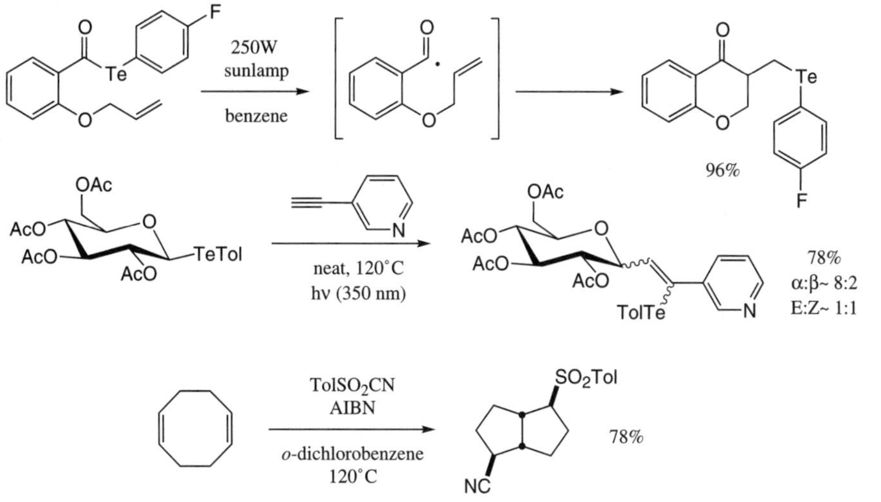

**Scheme 6.26** Transfer of telluride and cyanide groups

appears to work only for aromatic or $\alpha,\beta$-unsaturated acyl tellurides; saturated acyl tellurides failed to undergo the anticipated transformation. In contrast, the aliphatic series seems to be well-behaved, as shown by the second transformation involving a glycosyl telluride (Yamago *et al.* 1999a,b).

Finally, the third example displayed in Scheme 6.26 illustrates a cyanide group transfer (Fang and Chen 1987). The obtention of a bicyclic [3.3.0] structure from cyclooctadiene, in the same manner as in the addition of the corresponding sulfonyl chloride mentioned above, suggests that cyanide and chloride transfers occur at comparable rates.

## 6.6  Transfer of xanthates and related dithiocarbonyl derivatives

It is possible to accomplish an overall transfer of a xanthate group by designing the sequence so as to induce cleavage of the C–S bond rather than the C–O bond, as was the case in the Barton–McCombie deoxygenation discussed earlier in Chapter 2 (Zard 1997; Quiclet-Sire and Zard 1999). The general reaction manifold is outlined in Scheme 6.27. Radicals R$^\bullet$, produced by a chemical (peroxides or Et$_3$B/O$_2$) or photo-chemical initiation step, rapidly add to the starting xanthate **27**, because of the high radicophilicity of the thiocarbonyl group. This *reversible* addition can be made degenerate by placing a group on the oxygen atom that corresponds to a radical of much higher energy than R$^\bullet$. This prevents the adduct radical **28** from undergoing a C–O bond scission leading to *S-,S*-dithiocarbonate isomer **29**. An ethyl group is shown for clarity because most of the examples involve *O*-ethyl xanthates, but other primary or aryl substituents may be used. We now have a situation obeying the general Kharasch mechanism: the reaction of radical R$^\bullet$ with its xanthate precursor is redundant and therefore not in competition with the desired addition to the olefin. The xanthate group simply corresponds to Y in Scheme 6.1, the only, but nonetheless important difference being that its transfer involves a two-step addition–fragmentation sequence rather than a direct S$_H$2 process, as with hydrogen, halide, or chalcogenide abstraction.

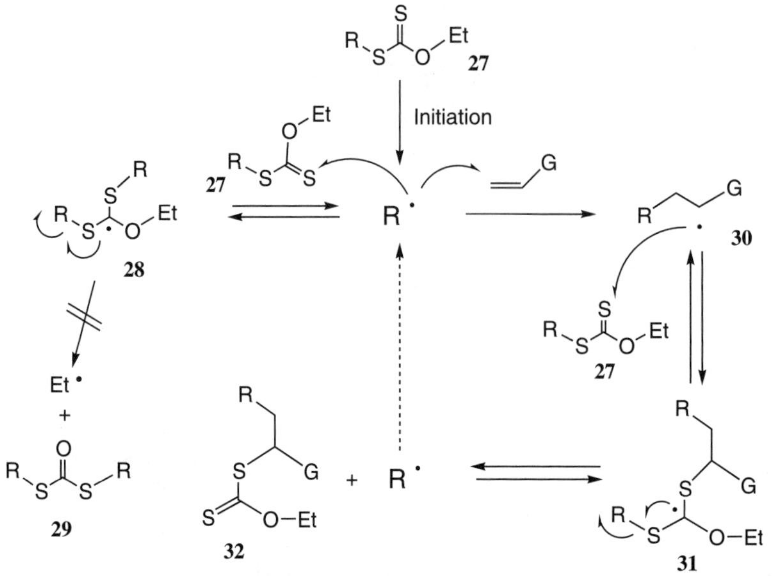

**Scheme 6.27**  Xanthate transfer manifold

The same sequence can be applied to other dithiocarbonyl derivatives such as trithiocarbonates, dithiocarbamates, and dithiocarboxylates, as long as the substituents are chosen so as to maximize the transfer rate and minimize potential side reactions. Most of the work in this area has concerned xanthates because of their ready availability: potassium *O*-ethyl xanthate is commercially available and inexpensive; it is a strong nucleophile, capable of displacing halides, tosylates, etc., and the synthesis of the precursors is generally straightforward. All the benefits of Kharasch processes are found in radical reactions involving xanthates: comparatively long lifetime for the intermediate radicals, the possibility (and even desirability) in many cases of operating under high concentration, the possibility of using unactivated olefins in *intermolecular* additions, the incorporation of a functionality in the end product, which allows an entry into the rich radical and non-radical chemistry of sulfur and, finally, experimental simplicity. There are some additional advantages. The transfer ability of the xanthate group appears to be comparable to that of iodine but xanthates are generally more stable than the corresponding iodides with respect to ionic side reactions: *S*-acyl xanthates or xanthates in the anomeric position of carbohydrates for instance are readily prepared whereas the corresponding iodides are exceedingly fragile. As a consequence, it is not usually necessary to incorporate tin derivatives to prevent inhibition of the chain process. Furthermore, the reversible addition to the thiocarbonyl group means that reactive radicals are in fact 'stored' as the corresponding stabilized adducts **28** or **31**; this lowers their concentration in the medium and limits unwanted side reactions. Such a possibility is not open to iodides and other derivatives where exchange occurs by a one-step homolytic substitution.

Since the transfer of the xanthate group is reversible, the sequence must be designed so as to push the equilibrium forward. This is best done by ensuring that initial radical R• is at least as, and preferably more stable than, intermediate radical **30** arising from addition to the olefin. This ensures that as long as the starting

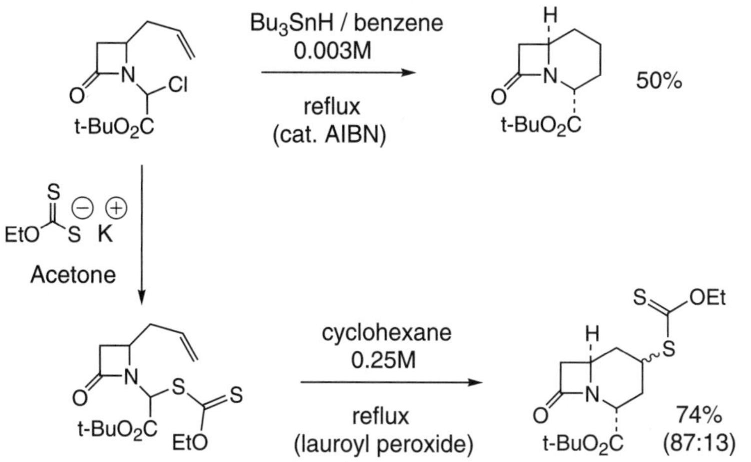

**Scheme 6.28**
Comparative cyclization reactions using tributyltin hydride and xanthate transfer

xanthate **27** is present in the medium, the formation of R⁺ is preferred over adduct radical **30**. This point is of key importance, especially in *intermolecular* additions, and worth emphasizing again. This constraint may at first appear as a limitation, but in practice it turns out to be a controlling element, allowing the obtention of a mono-adduct in preference to oligomerization.

One comparative experiment, illustrating some of the properties of the xanthate transfer technique, is pictured in Scheme 6.28. It concerns an arduous tributyltin-hydride-mediated 6-*endo* ring closure in the β-lactam series first described by Bachi *et al.* (1987). This is a difficult cyclization and high dilution (0.003 M) conditions were required to avoid extensive premature reduction of the intermediate uncyclized radical. If the chloride in the starting material is displaced by potassium *O*-ethyl xan-thate, then the 6-*endo* cyclization can be accomplished at a much higher concentra-tion, merely by heating the xanthate in cyclohexane in the presence of lauroyl peroxide as the initiator (Boiteau *et al.* 1998). In addition to improving the yield and not employing any organotin reagent, the reaction leads to a more highly function-alized product.

The xanthate transfer approach was also successful in the synthesis of oxocanes by a direct 8-*endo* ring closure where other methods apparently failed (Udding *et al.* 1994*a*). In the example in Scheme 6.29 taken from this study, the cyclization is par-tially followed by an intracyclic 1,5-hydrogen shift and fragmentation to give ulti-mately a linear aldehyde. The second transformation in the same scheme is a model for the construction of the skeleton of pleuromutilin (Bacqué *et al.* 2003). Other dif-ficult cyclizations, such as those leading to 4-,5,6-, and even 7-membered lactams, can also be accomplished (Axon *et al.* 1994; Boiteau *et al.* 1998*a*).

Perhaps the greatest asset of the xanthate transfer technique is the wide range of *intermolecular* additions to non-activated olefins that can be achieved. A few repres-entative examples are shown in Scheme 6.30. The first is an addition of a protected glycine radical to decene and constitutes an interesting approach to unnatural α-amino acids (Udding *et al.* 1994*b*). The second illustrates the addition of a tetrazolylmethyl radical to an allyl boronate (Biadatti *et al.* 1998; Lopez-Ruiz and Zard 2001).

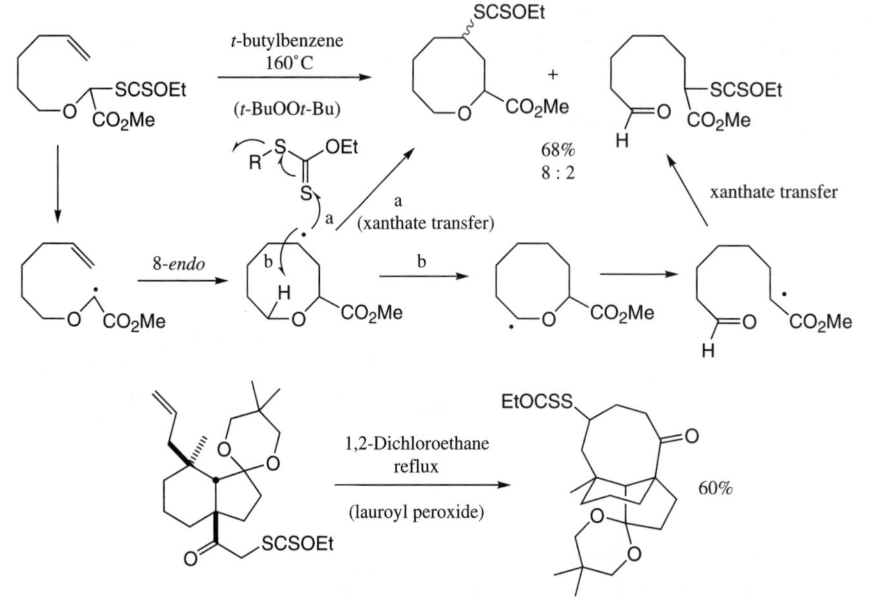

**Scheme 6.29**
8-membered ring formation

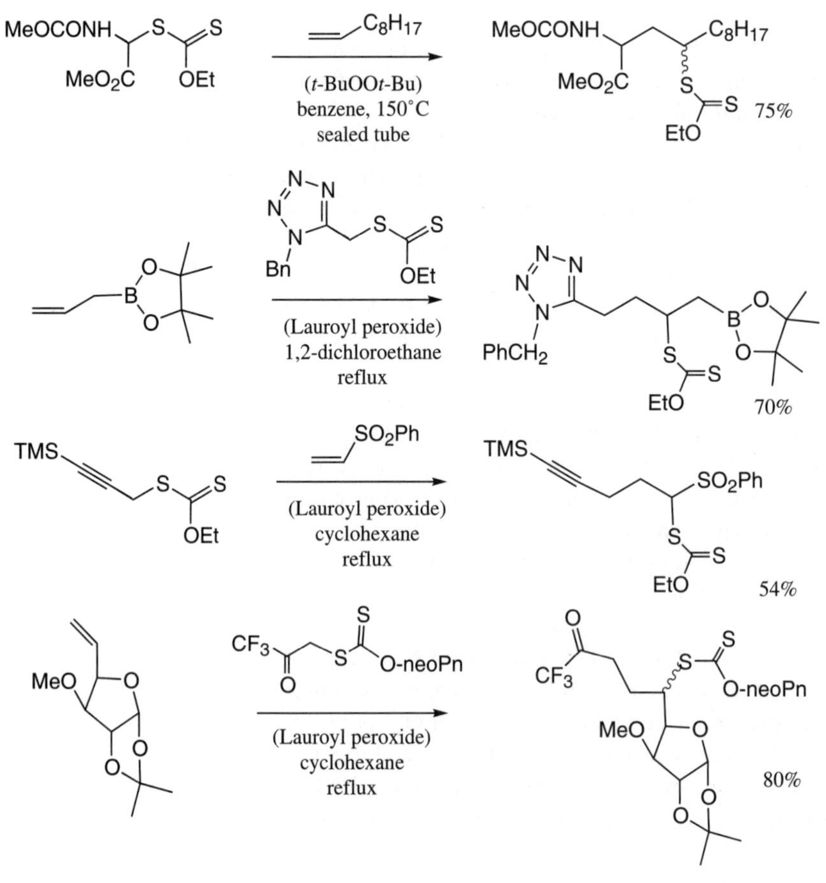

**Scheme 6.30** Examples of intermolecular xanthate additions to alkenes

The third transformation represents an unusual way for introducing an alkyne through the generation and capture of a propargyl radical. Propargyl radicals are relatively unreactive but can be trapped intermolecularly by electrophilic olefins (Denieul *et al.* 1996*a*). The last example concerns the addition of the trifluoroacetonyl radical to a sugar olefin derived from glucose (Denieul *et al.* 1996*b*). The use of an *O*-neopentyl xanthate in this case rather than the ubiquitous *O*-ethyl xanthate was dictated by the tendency of the trifluoromethyl ketone to form a hydrate: the hydrated radical cannot avail itself of the stabilization provided by the carbonyl group. *O*-Neopentyl xanthates are more hydrophobic and therefore alleviate the tendency of the highly electrophilic ketone to add water. They are also less prone to ionic decomposition or to the Chugaev fragmentation reaction and can be used to advantage in situations where such complications may arise.

Intermolecular additions bring together various functional groups that can then be made to react under appropriate conditions, allowing a rapid and convergent assembly of complex architectures. This aspect is highlighted by the three transformations displayed in Scheme 6.31. The first concerns a straightforward synthesis of a *cis*-piperidine by addition of a cyclopropylacetonyl radical to a protected allylamine (Boivin *et al.* 1999*b*). Azepines can similarly be made by using a protected homoallylamine as the radical trap. The xanthate group may be kept as an extra functionality in the product, as shown by the second sequence, representing an interesting strategy for the construction of the CD ring system of steroids (Cholleton *et al.* 2000). Unmasking of the second ketone followed by treatment with base gives the bicyclic product as one diastereoisomer. The intramolecular Wittig–Horner reaction occurs preferentially with the ketone that leads to the product where the xanthate group in the least congested equatorial orientation. In the third transformation, where interestingly a quaternary centre is created in the radical addition, the adduct contains a masked aldehyde in the form of a geminal acetoxy xanthate. Heating with

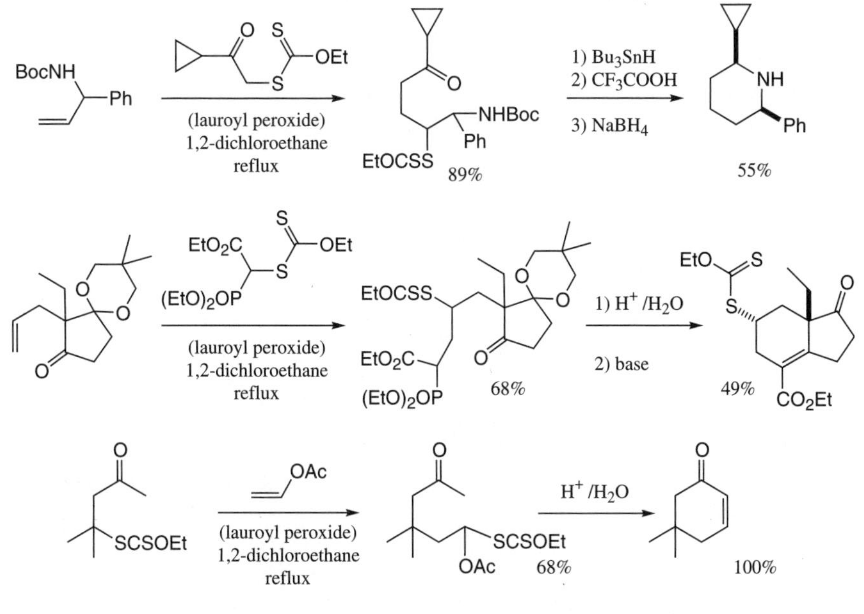

**Scheme 6.31**
Synthesis of piperidines and carbocyclic structures

aqueous acid results in a Robinson-type ring closure to give a cyclohexenone (Binot *et al.* 2003).

Acyl, alkoxycarbonyl, and alkoxythiocarbonyl radicals have been generated using the corresponding xanthate precursors. All these derivatives are yellow in colour and the reactions can be conveniently performed by irradiation with visible light (peroxides can of course also be used) and the progress of the reaction can often be monitored by the fading of the yellow colour. From a synthetic standpoint, acyl and alkoxycarbonyl radicals are the most interesting since their capture leads to ketones and esters or lactones, respectively. The addition of an *S*-acyl xanthate derived from benzoic acid to allyl acetate (Delduc *et al.* 1988) as well as the key step in the synthesis of cinnamolide (Forbes *et al.* 1999) are two examples pictured in Scheme 6.32. The presence of the xanthate in the $\beta$-position to the carbonyl group allows the introduction of the unsaturation present in cinnamolide. *S*-carbonyl xanthates behave, in terms of ionic chemistry, like anhydrides and are relatively easily decomposed by nucleophiles (an interesting ionic chain mechanism is involved; Zard 1997). A bulky hydrophobic group (e.g. neopentyl) on the xanthate moiety improves their resistance against moisture and adventitious nucleophiles.

In the absence of a trap, irradiation of *S*-acyl or *S*-alkoxycarbonyl xanthates can result in the loss of carbon monoxide or carbon dioxide, resulting in overall decarbonylation or decarboxylation, respectively (Barton *et al.* 1962; Delduc *et al.* 1988; Forbes *et al.* 1999). The ease of the extrusion reflects the stability of the ensuing alkyl radical. One synthetic application of this process, outlined in Scheme 6.33, represents an alternative construction of the side chain present in the new family of anti-asthmatic steroids described in the previous chapter (Scheme 5.12). Irradiation of *S*-acyl xanthate **33**, made by reacting the corresponding acid chloride with sodium *O*-neopentyl xanthate, gives rise to nor-xanthate **35**, through loss of carbon monoxide from the intermediate radical **34**. Saponification and alkylation with methyl iodide completes the synthesis of **36** (Quiclet-Sire and Zard 1998*a*).

In the same way, alkoxycarbonyl radicals (ROC$^{\bullet}$=O) generated from the corresponding xanthate can be forced to undergo decarboxylation. This property may be exploited for replacing the C–O bond in an alcohol by a C–S bond (Forbes *et al.* 1999). Because of the relatively long life of the intermediate radicals, unstabilized tertiary, secondary, and even primary alcohols become useful substrates. The ease with which these transformations are accomplished must be contrasted with the

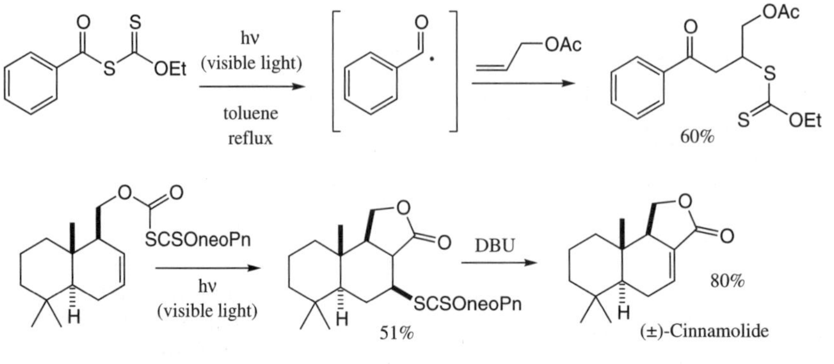

**Scheme 6.32** Generation and capture of acyl radicals.

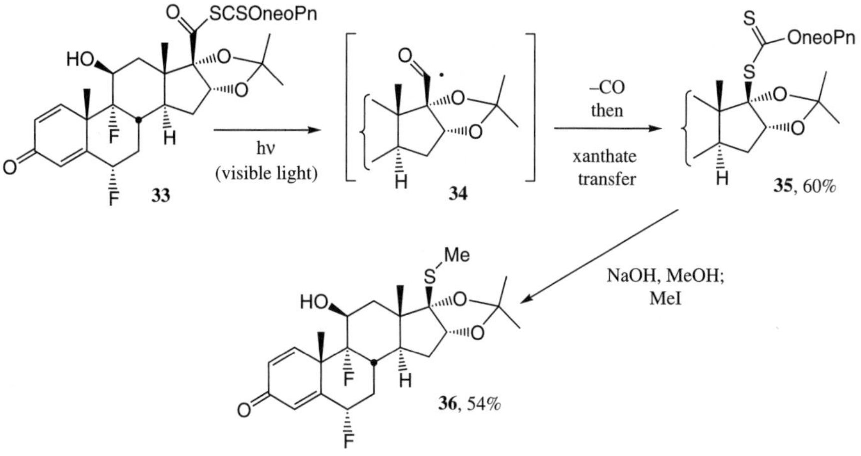

**Scheme 6.33**
Decarbonylation of an *S*-acyl xanthate

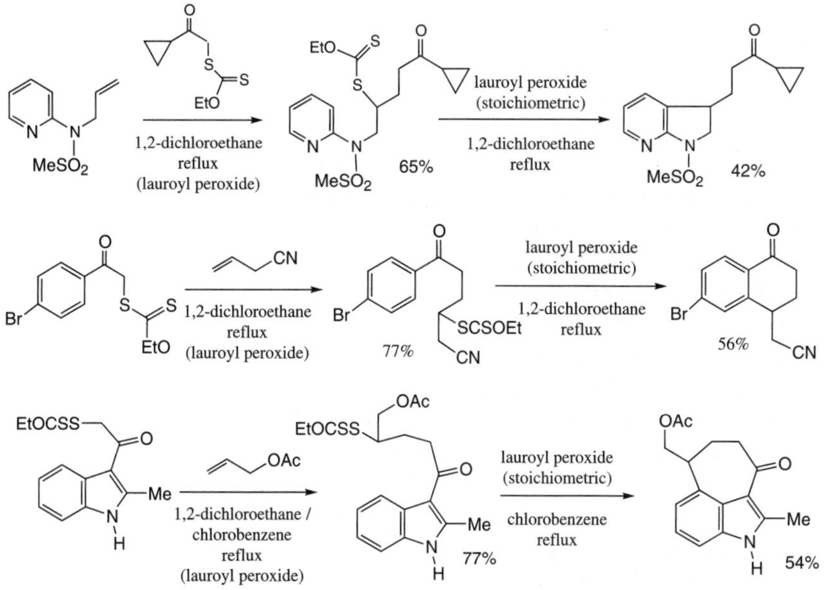

**Scheme 6.34**
Sequential intermolecular addition and cyclization to an aromatic ring

difficulties encountered in the deoxygenation of primary (and even secondary) alcohols using alkoxycarbonyl selenides and tributyltin hydride as the reducing agent, a reaction discussed at length in Section 2.3 (Scheme 2.10).

The fact that the adduct from a xanthate transfer reaction is itself a xanthate offers the opportunity of performing yet another radical sequence. One important application is to combine the intermolecular addition with a ring closure onto an aromatic or heteroaromatic ring. This provides a practical access to a variety of aromatic structures as illustrated by the examples displayed in Scheme 6.34. Oxindoles (Axon *et al.* 1994), indanes and indolines (Ly *et al.* 1999), tetralones (Liard *et al.* 1997), tetrahydroisoquinolones (Cholleton and Zard 1998), and even in some cases 7-membered ring ketones as in the last example (Kaoudi *et al.* 2000). In contrast to the first step which is a true chain, requiring only a small amount of peroxide to go

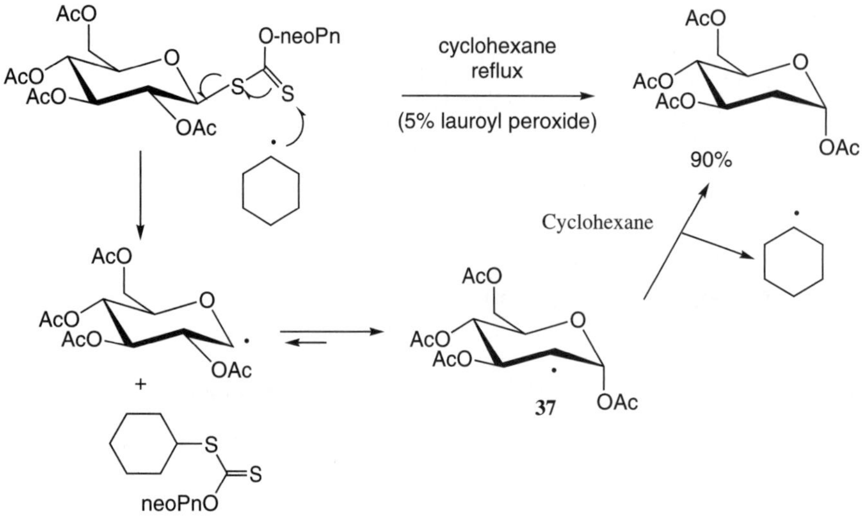

**Scheme 6.35**
Synthesis of
2-deoxysugars

to completion, the second requires stoichiometric quantities of the peroxide in order to oxidize the intermediate cyclohexadienyl radical. This oxidation step involves an electron transfer to the peroxide leading, on one hand, to the cyclohexadienyl cation which aromatizes by loss of a proton and, on the other to the radical anion of the peroxide, which decomposes to a carboxylate anion and a carboxylic radical. The latter rapidly looses carbon dioxide to give an undecyl radical that propagates a chain by reacting with the starting xanthate (cf. the mechanism displayed in Scheme 6.37 below). The peroxide must nonetheless be added portion-wise in order to avoid build up of the radical concentration in the medium, since the electron-transfer step is probably not very efficient. In other words, the peroxide is now both an initiator and a reagent. Lauroyl peroxide is cheap and safe, and the co-products are usually relatively easy to separate from the product. Because both electron-donating and electron-withdrawing groups in the aromatic ring are tolerated, this simple and convergent approach complements nicely the classical ionic (e.g. Friedel–Crafts) and transition-metal-based routes to aromatic and heteroaromatic structures.

As was the case with iodides (cf. Scheme 6.21), xanthates that lead to electrophilic but otherwise unstabilized secondary radicals may be reduced using cyclohexane as the hydrogen atom source and a small amount of the peroxide initiator. The transformation displayed in Scheme 6.35 constitutes an inexpensive, efficient, and metal-free synthesis of 2-deoxy-glucose tetraacetate, through the intermediate migration of the 2-acetoxy group producing electrophilic radical **37** (cf. Scheme 3.19). If the xanthate group is located at a non-anomeric secondary position in the carbohydrate, then a simple reduction is observed with no migration of ester groups (Quiclet-Sire and Zard 1996*a*).

Some evidence for the importance of the electrophilic nature of the radical can be adduced by examining the competition experiment also delineated in Scheme 6.36. When an equimolar mixture of xanthates **38** and **39**, obtained in good yield by radical addition of *S*-cyanomethyl *O*-ethyl xanthate to 1-decene and heptadecafluoro-1-decene, respectively, was heated with lauroyl peroxide in refluxing cyclohexane,

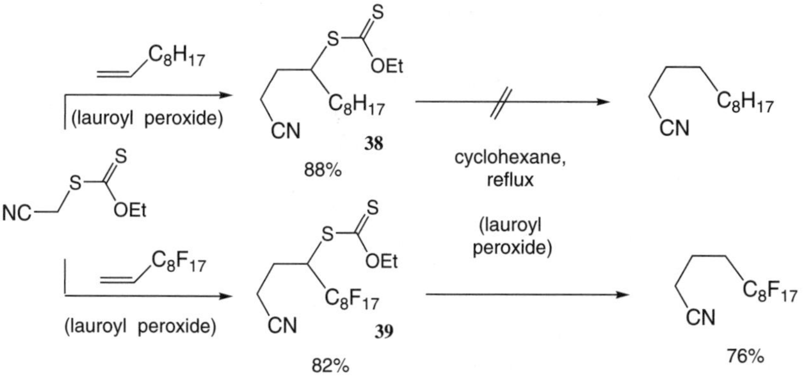

**Scheme 6.36** Effect of fluorine atoms on the ability for hydrogen abstraction

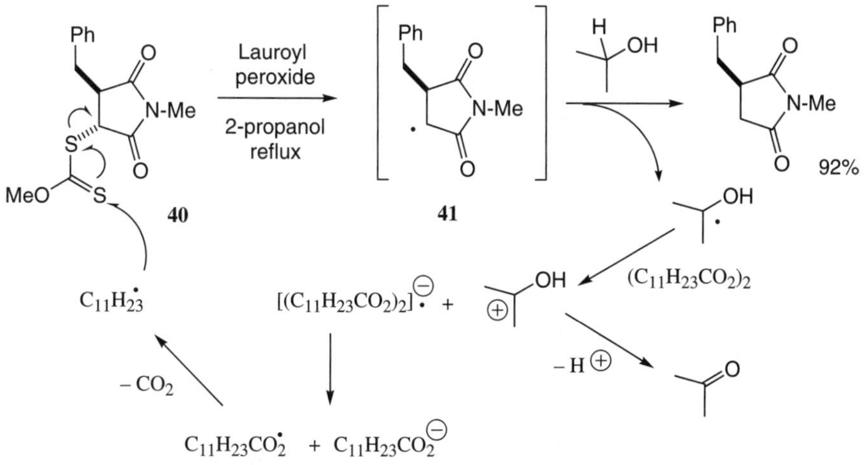

**Scheme 6.37** Reductive dexanthylation with 2-propanol

only the latter underwent reduction. The unfluorinated xanthate **38** was recovered largely unchanged. Thus, even though both carbon radicals are produced in the medium, only the radical derived from **39**, made electrophilic by the powerful inductive pull of the fluorine atoms, is capable of rapidly abstracting hydrogen from the cyclohexane.

In order to reductively remove a xanthate group attached to an 'ordinary' secondary carbon, then a more effective hydrogen atom transfer agent is required. Tributylstannane will do the job swiftly and cleanly. Another way is to use as solvent 2-propanol, a better hydrogen atom donor than cyclohexane (Liard *et al.* 1996). This is illustrated by the efficient reduction of xanthate **40** depicted in Scheme 6.37. The parent radical **41**, produced by the action of the peroxide, can and does react rapidly with xanthate **40**, but the resulting addition–fragmentation is degenerate. Although much slower, abstraction of the tertiary hydrogen from 2-propanol is the only alternative left. The ensuing tertiary radical is too stabilized to propagate the chain and is most likely oxidized to acetone by electron transfer to the peroxide. As in the cyclization to aromatic rings, the peroxide is now also a reagent and must be used in stoichiometric amounts.

The relative sluggishness of the hydrogen abstraction from isopropanol as compared to the radical addition to a thiocarbonyl group may be exploited to accomplish a tin-free Barton–McCombie type deoxygenation. The reaction manifold is displayed in Scheme 6.38. A primary undecyl radical, derived from the thermal decomposition of lauroyl peroxide, will rapidly and reversibly attack xanthate **42** in preference to abstracting a hydrogen from 2-propanol. If R is a secondary group, *irreversible* cleavage of the C–O bond in adduct **43** can now occur, in competition with the reverse of the initial addition which regenerates the less stable primary undecyl radical. The secondary radical R• thus produced has a choice between two principal pathways: (a) addition to the xanthate to give intermediate **45** and (b) abstraction of hydrogen from 2-propanol. The former is a faster process but highly reversible because, in contrast to the case with adduct radical **43**, the choice now is between two identical secondary radicals R•. The weaker C–S bond will break far more frequently so that the slower but irreversible abstraction of hydrogen from 2-propanol becomes competitive. Overall, the reduction requires stoichiometric amounts of initiator, the co-product being dithiocarbonate **44** derived from the peroxide. If 2-propanol is replaced by a solvent that does not easily relinquish its hydrogen atoms such as benzene, then the only remaining route becomes the slower,

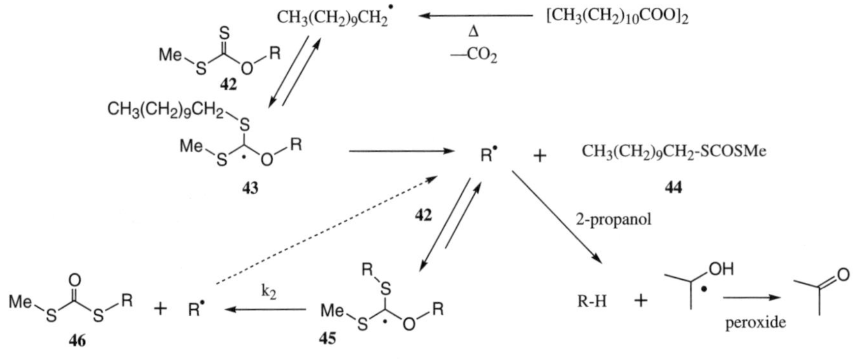

**Scheme 6.38** Reaction manifold of a xanthate with 2-propanol

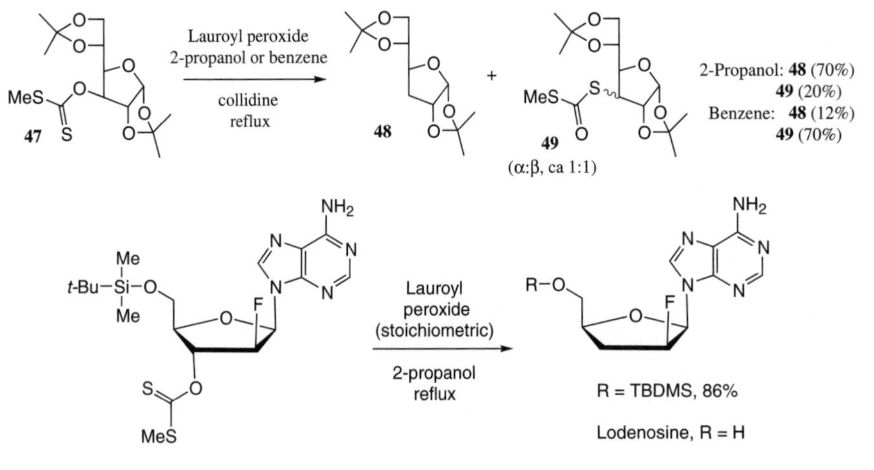

**Scheme 6.39** Examples of deoxygenation with lauroyl peroxide in 2-propanol

irreversible cleavage of the C–O bond in adduct **45** to give the *S-,S*-dithiocarbonate isomer **46** and initial radical R˙ which now propagates the chain. This represents a synthetically interesting *O*- to *S*-isomerization of the starting xanthate by a radical chain process. The conversion of the glucose derived xanthate **47** in Scheme 6.39 into either the deoxygenated product **48** or *S-,S*-dithiocarbonate **49** simply by choosing 2-propanol or benzene as solvent is a clear-cut illustration (Quiclet-Sire and Zard 1998*b*). Collidine is added to the medium as a protection for the isopropylidene groups against cleavage by adventitious acid. The second transformation in the same scheme is a more useful application of this tin-free variant of the Barton–McCombie deoxygenation for the synthesis of lodenosine, an anti-viral substance (Siddiqui *et al.* 2000). It must be noted, in passing, that the rearranged product **46** is often quite difficult to distinguish by thin-layer chromatography from the starting xanthate **42** and its formation as a side product in many Barton–McCombie deoxygenations has often gone unnoticed.

## 6.7   Further applications of degenerate processes: some chain reactions of sulfones

Controlling the selectivity by designing the reaction in such a way that the major unwanted side reactions are either degenerate and/or highly reversible so that the reactive species is for all practical purposes channelled into the desired direction can be extended to systems based on sulfones. This may be seen in the manifold outlined in Scheme 6.40 and employing allylethylsulfone to replace an iodide (X = I) or a xanthate (X = SCSOEt) by an allyl group.

In the presence of an initiator such as AIBN, an ethylsulfonyl radical can be generated by an addition–fragmentation process to the allyl ethylsulfone. The reaction of ethylsulfonyl radicals with the starting allyl sulfone is strictly degenerate (path **A**), whereas their reaction with either the iodide or the xanthate is endothermic and the equilibrium, if any, highly favours the reactants (path **B**). The only option left is extrusion of sulfur dioxide (path **C**), even though it is a reversible and comparatively slow reaction. However, the ethyl radicals thus generated are highly reactive and rapidly exchange an iodide or a xanthate from the starting material (path **D**) to give

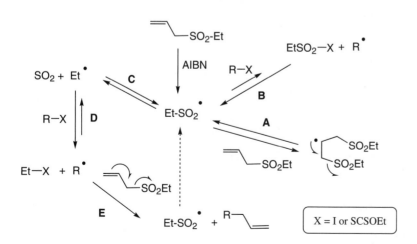

**Scheme 6.40**   Reaction manifold for the radical allylation of iodides and xanthates with ethyl allyl sulfone

radical R• which, in turn, has only the possibility of reacting with the allylethyl sulfone to give the desired allylated product and ethylsulfonyl radicals, thereby closing the chain (path **E**).

Two examples are given in Scheme 6.41. In the first, an iodide derived from arabinose is efficiently replaced by an allyl group but with no stereoselectivity (Le Guyader *et al.* 1997). In the second involving a xanthate, a 5-*exo* ring-closure step is inserted in the sequence, which leads only to the diastereoisomer drawn (Quiclet-Sire *et al.* 1998; PMB = *p*-methoxybenzyl). Allyl groups substituted in the 2 position (methallyl, 2-chloroallyl, etc.) can be introduced in the same manner by using the corresponding substituted allyl sulfones. Perhaps more importantly, it turns out that the mechanistic manifold displayed in Scheme 6.40 is applicable also to various vinyl sulfones (Bertrand *et al.* 1999). This is based on the fact that if appropriate substituents are present, attack of the carbon radical can occur on the carbon bearing the sulfone group resulting in an addition–fragmentation also generating a sulfonyl radical (Xiang and Fuchs 1996; Xiang *et al.* 1997). Three examples are given in Scheme 6.42; the first

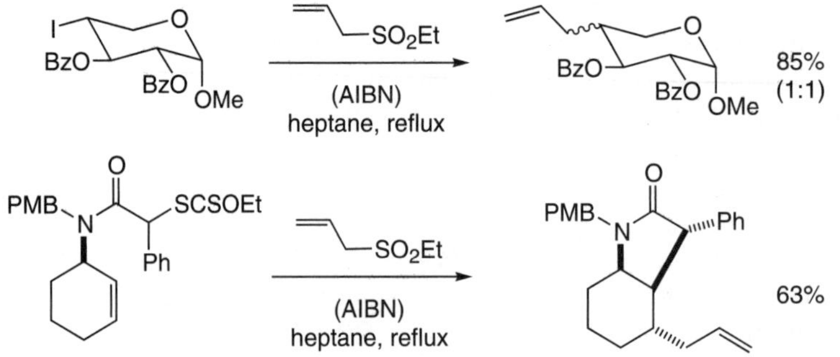

**Scheme 6.41**
Examples of radical
allylations

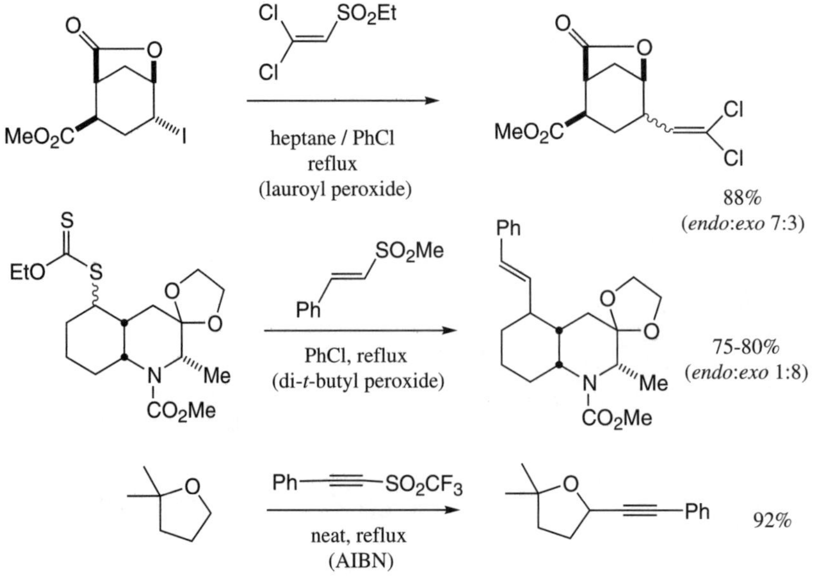

**Scheme 6.42**
Alkenylation and
alkynylation reactions

starts from an iodide conveniently obtained by an iodo-lactonization reaction (Bertrand *et al.* 1999) and the second from a xanthate, and corresponds to a key step in the synthesis of lepadin B (Kalaï *et al.* 2002). The use of trifluoromethylsulfones allows the vinylation or alkynylation of an alkane, through the formation of highly electrophilic trifluoromethyl radicals that are capable of abstracting a hydrogen from a hydrocarbon, as illustrated by the last example (Gong and Fuchs 1996).

These allylation and vinylation procedures are especially efficient with secondary and tertiary substrates, since the generation of the intermediate carbon radical is favoured (step **D** in Scheme 6.40). The process is less effective in the case of primary derivatives and fails with vinylic or aromatic iodides or xanthates since the equilibrium represented by step **D** becomes unfavourable. Nevertheless, the present approach complements transition-metal-mediated vinylations of iodides, which are easy to perform on vinylic and aromatic iodides but cannot normally be applied to aliphatic iodides, and especially secondary and tertiary derivatives, because of $\beta$-hydride elimination.

By using sulfonyl azides or sulfones derived from substituted oximes, it becomes possible to replace an iodide, a xanthate, and in some cases a telluride, by an azide (Ollivier and Renaud 2000*a*) or an oxime derivative (Kim *et al.* 2001) as illustrated by the examples pictured in Scheme 6.43. This tin-free approach to the creation of C–C or C–N bonds appears quite flexible and further variations will certainly emerge as it scope continues to be explored and expanded.

Another way to exploit the extrusion of sulfur dioxide as a source of radicals R$^{\bullet}$ is to attach the desired R group directly onto the allylsulfone, as outlined in Scheme 6.44. The reaction of sulfonyl radical, RSO$_2^{\bullet}$, with the starting alkylallyl sulfone is degenerate (step **A**) whereas loss of sulfur dioxide gives radical R$^{\bullet}$ (step **B**), which can add to the allyl sulfone to give the desired allyl derivative and the sulfonyl radical to propagate the chain (step **C**). This would represent a simple chain with two propagating steps, **B** and **C**. However, as the reaction progresses and the concentration of allylated product

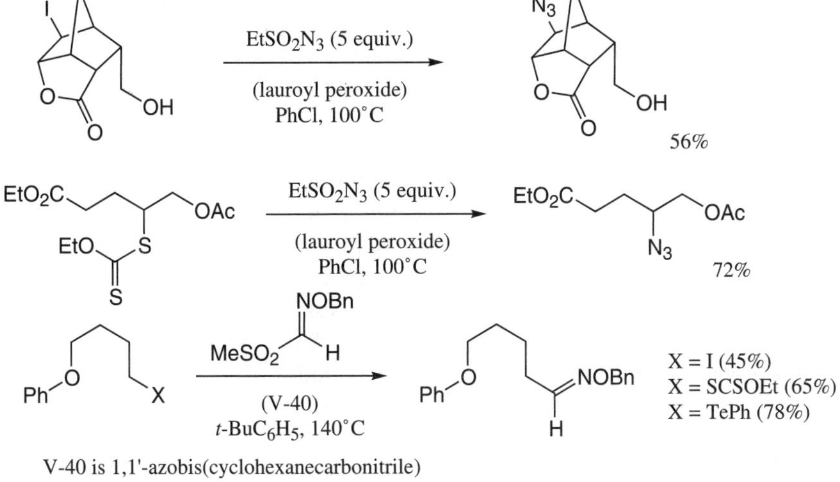

**Scheme 6.43** Azide and formyloxime transfers

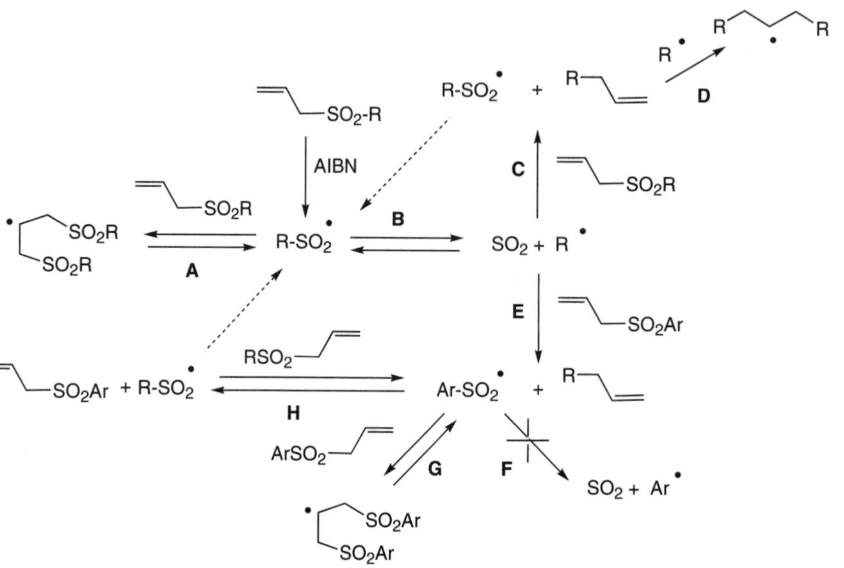

**Scheme 6.44** Reaction manifold of the radical desulfonylation of allyl sulfones

increases in the medium, the irreversible attack of radical R˙ on the olefinic terminus of the product becomes significant, leading to unwanted side products (step **D**).

From the outset, this would represent a serious limitation. One solution is to stop the reaction at low conversion and recycle the unreacted starting allyl sulfone. Another, more satisfying approach is to add to the medium an excess of an allyl*aryl* sulfone. As indicated in step **E**, this would react with radical R˙ to give the same allylated product and an *aryl*sulfonyl radical which, unlike its alkyl analogue, does not extrude sulfur dioxide (step **F**). This is a consequence of the high energy of an aryl radical. Furthermore, the addition of the *aryl*sulfonyl radical to the allyl*aryl* sulfone (step **G**) is degenerate. The only possibility left is the reversible addition to the alkylallyl sulfone (step **H**). This produces the alkylsulfonyl radical, thus propagating the chain, and regenerates at the same time the allyl*aryl* sulfone, which is in fact not consumed in the overall process and can be recovered at the end. By sheer concentration effect, the presence of an excess allyl*aryl* sulfone allows path **E** to dominate and therefore largely eliminate the formation of side products through undesired path **D**. In a sense, the allyl*aryl* sulfone acts as a relay for the introduction of the allyl group and shields the product from further reaction.

This procedure for the synthesis of allyl derivatives is illustrated by the two examples in Scheme 6.45 (Quiclet-Sire and Zard 1996*b*). In the first, the relay is allyltolyl sulfone; only one isomer is formed through reaction from the least hindered α-side of the nucleoside. The overall transformation is reminiscent of the Ramberg–Bäcklund reaction, except that a single C–C bond is created upon excision of the sulfone group instead of an olefin. In the second example, an external, electrophilic olefin has been incorporated into the sequence and there is no need to add a relay allyl*aryl* sulfone but simply to use the allylbenzyl sulfone in excess. The benzyl radicals reacts so much more rapidly with the maleimide that the selectivity issue is greatly simplified.

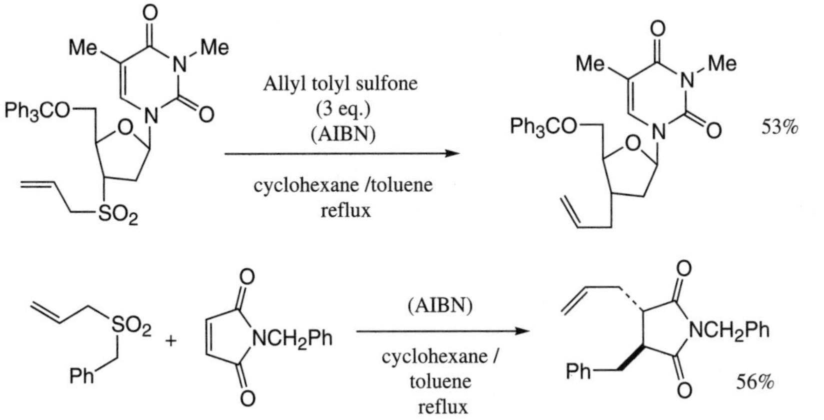

**Scheme 6.45**
Examples of
desulfonylative allylation

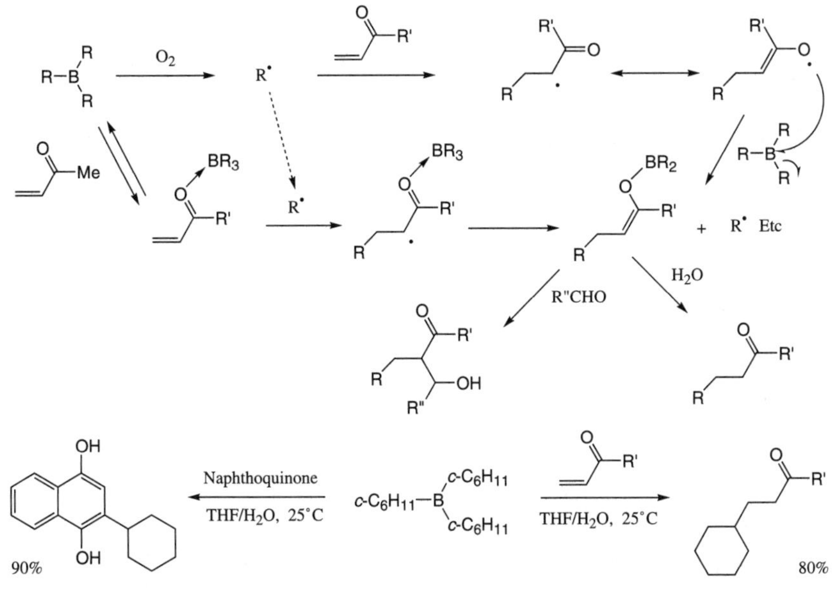

**Scheme 6.46**
Organoboron-mediated
radical reactions

## 6.8 Radicals from organoboron compounds

The autoxidation of triethyl borane liberates ethyl radicals and, as mentioned earlier in Chapter 2, this process has been cleverly exploited to initiate various radical reactions at temperatures as low as −78 °C. More generally, triorganoboranes represent sources of carbon-centred radicals, and these may be captured by an electrophilic olefin such as an enone in a chain reaction that gives rise ultimately to a boron enolate, as outlined in Scheme 6.46 (Suzuki *et al.* 1967; review: Ollivier and Renaud 2001). The triethylboron apparently also acts as a Lewis acid with respect to the carbonyl oxygen of the enone and enhances its electrophilicity (Béraud *et al.* 2000). Moreover, the resulting boron enolate may be captured by a subsequent ionic aldol type reaction leading to a considerable increase in the complexity of the products

(Nozaki *et al.* 1988). Two transformations starting with tricyclohexyl borane are shown in the bottom of the scheme, where methylvinyl ketone and naphthoquinone are used as the olefinic traps.

More complex organoboron precursors may be employed, the most useful perhaps being the readily accessible *B*-alkyl catechol boronates, since the C–B bond is selectively broken in the process (Ollivier and Renaud 1999). $\alpha,\beta$-Unsaturated ketones and aldehydes are suitable olefinic traps since the intermediate radical adduct has sufficient electron density on the oxygen atom to attack the starting organoboron derivative and thus propagate the chain. Other electrophilic olefins such as unsaturated esters, amides, and sulfones, even though inherently good radical traps, do not lead to intermediate adduct radicals with enough spin density on an oxygen atom to perpetuate the chain. This limitation has been circumvented in a clever manner by using a Barton carbonate or benzoate derivative as an external source of oxygen centred radicals to act as chain propagators (Cadot *et al.* 2000; Ollivier and Renaud 2000*b*). One example illustrating this contrivance is pictured in Scheme 6.47, starting with a catechol boronate derived from $\beta$-pinene. A further considerable advantage of this approach is the extra functionality provided by the pyridylsulfide group.

The formation of ethyl radicals from triethylborane can be also be exploited to produce and capture radicals from iodides, much in the same way as with the sulfone-based reagents described above. The example in Scheme 6.48 shows the addition of

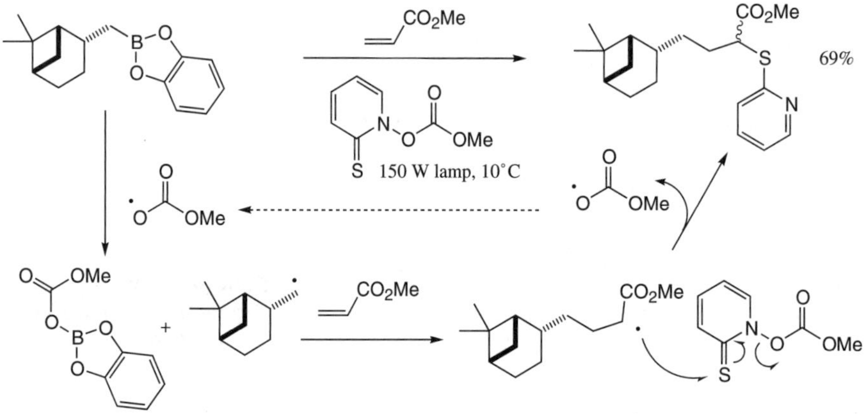

**Scheme 6.47** Barton-ester-mediated radical addition from organoboranes

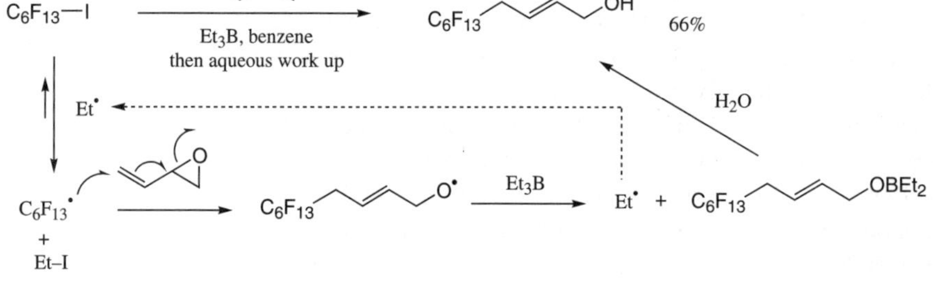

**Scheme 6.48** Triethylborane-mediated radical addition to butadiene epoxide

perfluorohexyl radical to butadiene epoxide, a representative of another family of alkene traps which leads to an oxygen-centred radical intermediate and therefore readily propagates the desired chain (Suzuki *et al.* 1971; Ichinose *et al.* 1988).

The present chapter provided a brief overview of Kharasch-type reactions. Although these processes have not been as extensively used in synthesis as tributyl-tin-hydride-based methods, they nevertheless offer several important advantages. They often surpass stannanes in terms of intermolecular additions to unactivated olefins and there is no need for high dilution or syringe pump techniques in most cases. They are therefore generally easier to scale up and do not present the toxicity hazards and purification problems associated with organotin derivatives.

## References

Abell, P. I. and Bohm, B. A. (1961). The stereochemistry of the free radical addition of hydrogen bromide to1-methylcycloheptene. *Journal of Organic Chemistry*, **26**, 252–254.

de Alvarenga, E. S. and Mann, J. (1993). Photocatalysed addition of pyrrolidines to buteno-lides: a concise synthesis of the pyrrolizidine alkaloid ring system. *Journal of the Chemical Society, Pertkin Transactions,* **1**, 2141–2142.

Axon, J., Boiteau, L., Boivin, J., Forbes, J. E., and Zard, S. Z. (1994). A new radical-based synthesis of lactams and indolones from dithiocarbonates (xanthates). *Tetrahedron Letters*, **35**, 1719–1722.

Baciocchi, E. and Muraglia, E. (1994). Synthesis of γ-haloesters and γ-ketoesters by homolytic addition of carbon radicals generated by α-haloesters and triethylborane to alkenes and silylenol ethers. *Tetrahedron Letters*, **35**, 2763–2766.

Bachi, M. D. and Khorshin, E. E. (1998). Thiol–oxygen cooxidation of monoterpenes. Synthesis of endoperoxides structurally related to yingzhaosu A. *Synlett*, 122–124.

Bachi, M. D., De Mesmaeker, A., and Stevenart-De Mesmaeker, N. (1987). Free-radical annelation in the synthesis of bicyclic β-lactams.5. Synthesis of carbapenems. *Tetrahedron Letters*, **28**, 2637–2640.

Back, T. and Collins, S. (1981). Selenosulfonation: boron trifluoride catalyzed or thermal addition of selenosulfonates to olefins. A novel regio- and stereo-controlled synthesis of vinyl sulfones. *Journal of Organic Chemistry*, **46**, 3249–3256.

Back, T., Collins, S., and Kerr, R. G. (1983). Selenosulfonation of acetylenes: preparation of novel β-(phenylseleno)vinyl sulfones and their conversion to acetylenic and β-functionalized sulfones. *Journal of Organic Chemistry*, **48**, 3077–3084.

Bacqué, E., Pautrat, F., and Zard, S. Z. (2003). A concise synthesis of the the tricyclic skele-ton of pleuromutilin and a new approach to cycloheptenes. *Organic Letters*, **5**, 325–328.

Barton, D. H. R., George, M. V., and Tomoeda, M. (1962). Photochemical transformations. Part XIII. A new method for the production of acyl radicals. *Journal of the Chemical Society*, 1967–1974.

Barton, D. H. R., Jang, D. O., and Jaszberenyi, J. Cs. (1992). Hypophosphorus acid and its salts: new reagents for radical chain deoxygenation, dehalogenation and deamination. *Tetrahedron Letters*, **33**, 5709–5712.

Barton, D. H. R., Jang, D. O., and Jaszberenyi, J. Cs. (1993). The invention of radical reac-tions.32. Radical deoxygenations, dehalogenations and deaminations with dialkylphosphites and hypophosphorus acid as hydrogen sources. *Journal of Organic Chemistry*, **58**, 6838–6842.

Batey, R. A., Harling, J. D., and Motherwell, W. B. (1992). Construction of bicyclic systems via a tandem free radical cyclopropylcarbinyl rearrangement–cyclisation strategy. *Tetrahedron*, **48**, 8031–8052.

Benko, Z., Fraser-Reid, B., Mariano, P. S., and Beckwith, A. L. J. (1988). Conjugate addition of methanol to $\alpha$-enones: photochemistry and stereochemical details. *Journal of Organic Chemistry*, **53**, 2066–2072.

Béraud, V., Gnanou, Y., Walton, J. C., and Maillard, B. (2000). New insight into the mechanism of the reaction between $\alpha,\beta$-unsaturated carbonyl compounds and triethyl borane (Brown's reaction). *Tetrahedron Letters*, **41**, 1195–1198.

Bertrand, M. (1994). Recent progress in the use of sulfonyl radicals in organic synthesis. *Organic Preparations and Procedures International*, **26**, 257–290.

Bertrand, F., Quiclet-Sire, B., and Zard, S. Z. (1999). A new radical vinylation reaction of iodides and dithiocarbonates. *Angewandte Chemie International Edition in English*, **38**, 1943–1946.

Biadatti, T., Quiclet-Sire, B., Saunier, J.-B., and Zard, S. Z. (1998). The tetrazolylmethyl and related radicals. A convenient access to tetrazoles and other heteroaromatic compounds. *Tetrahedron Letters*, **39**, 19–22.

Binot, G., Quiclet-Sire, B., Saleh, T., and Zard, S. Z. (2003). A convergent construction of quaternary centres and polycyclic structures. *Synlett*, 382–386.

Block, E., Aslam, M., Eswarakrishnan, V., Gebreyes, K., Hutchinson, J., Iyer, R., Laffitte, J.-A., and Wall, A. (1986). $\alpha$-Haloalkanesulfonyl bromides in organic synthesis. 5. Versatile reagents for the synthesis of conjugated polyenes, enones, and 1,3-oxathiol-1,1-dioxides. *Journal of the American Chemical Society*, **108**, 4568–4580.

Blomquist, A. T. and Wolinski, J. (1958). Addition of ethyl mercaptan to acetylenic compounds. *Journal of Organic Chemistry*, **23**, 551–554.

Boiteau, L., Boivin, J., Quiclet-Sire, B., Saunier, J.-B., and Zard, S. Z. (1998). Synthetic routes to $\beta$-lactams. Some unexpected hydrogen atom transfer reactions. *Tetrahedron*, **54**, 2087–2098.

Boivin, J., Pothier, J., Ramos, L., and Zard, S. Z. (1999*a*). An expeditious construction of 9-membered rings. *Tetrahedron Letters*, **40**, 9239–9241.

Boivin, J., Pothier, J., and Zard, S. Z. (1999*b*). A flexible, convergent approach to piperidines, pyridines, azepines, and related derivatives. *Tetrahedron Letters*, **40**, 3701–3704.

Boivin, J., Quiclet-Sire, B., Ramos, L., and Zard, S. Z. (1997). A new reductive deiodination by hydrogen atom transfer from cyclohexane. *Journal of the Chemical Society, Chemical Communications*, 353–354.

Byers, J. H. and Lane, G. C. (1993). Radical addition reactions of 2-(phenylseleno)propanedioates to alkenes and alkynes. *Journal of Organic Chemistry*, **58**, 3355–3360.

Cholleton, N. and Zard, S. Z. (1998) A convergent synthesis of 4-substituted 1,2,3,4-tetrahydroisoquinolin-1-ones. *Tetrahedron Letters*, **39**, 7295–7298.

Cholleton, N., Gillaizeau-Gauthier, I., Six, Y., and Zard, S. Z. (2000). A new approach to cyclohexenes and related structures. *Journal of the Chemical Society, Chemical Communications*, 535–536.

Cadot, C., Cossy, J., and Dalko, P. I. (2000). Radical carbon–carbon coupling reactions via organoboranes. *Chemical Communications*, 1017–1018.

Chatgilialoglu, C., Lunazzi, L., Macciantelli, D., and Placucci, G. (1984). Absolute rate constants for hydrogen abstraction from aldehydes and conformational studies of the corresponding aromatic acyl radicals. *Journal of the American Chemical Society*, **106**, 5252–5256.

Corey, E. J. and Shinoji, K. (1983). Total synthesis of the major human urinary metabolite of prostaglandin D2, a key diagnostic indicator. *Journal of the American Chemical Society*, **105**, 1662–1664.

Crich, D., Chen, C., Hwang, J.-T., Yuan, H., Papadatos, A., and Walter, R. I. (1994). Photoinduced free radical chemistry of the acyl tellurides: generation, inter- and intramolecular trapping, and ESR spectroscopic identification of acyl radicals. *Journal of the American Chemical Society*, **116**, 8937–8951.

Crimmins, M. T. and Mascarella, S. W. (1987). Radical Cleavage of cyclobutanes: alternative routes to (±)-silphinene. *Tetrahedron Letters*, **28**, 5063–5066.

Curran, D. P. and Chang, C.-T. (1989). Atom transfer cyclisation reactions of $\alpha$-iodo esters, ketones, and malonates: examples of selective 5-*exo*, 6-*endo*, 6-*exo*, and 7-*endo* ring closures. *Journal of Organic Chemistry*, **54**, 3140–3157.

Curran, D. P. and Chen, M.-H. (1987). Atom transfer cycloaddition. A facile preparation of functionalized (methylene)cyclopentanes. *Journal of the American Chemical Society*, **109**, 6558–6560.

Curran, D. P. and Ko, S.-B. (1998). Addition of electrophilic radicals to electron rich alkenes by the atom transfer method. Surmounting potentially reversible radical atom transfer reactions by irreversible ionic reactions. *Tetrahedron Letters*, **39**, 6629–6632.

Curran, D. P., Bosch, E., Kaplan, J., and Newcomb, M. (1989). Rate constants for halogen atom transfer from representative $\alpha$-halocarbonyl compounds to primary alkyl radicals. *Journal of Organic Chemistry*, **54**, 1826–1831.

Dang, H.-S. and Roberts, B. P. (1996). Homolytic aldol reactions: thiol catalysed radical chain addition of aldehydes to enol esters and to silyl enol ethers. *Journal of the Chemical Society, Chemical Communications*, 2201–2202.

De Riggi, I., Surzur, J.-M., and Bertrand, M. P. (1988). Additions radicalaires d'halogénures de sulfonyle. Cyclisation et bifonctionnalisation de cyclooctadiène 1,5 et de diènes 1,6. *Tetrahedron*, **44**, 7119–7125.

Delduc, P., Tailhan, C., and Zard, S. Z. (1988). A convenient source of alkyl and acyl radicals. *Journal of the Chemical Society, Chemical Communications*, 308–310.

Denieul, M.-P., Quiclet-Sire, B., and Zard, S. Z. (1996a). A synthetically useful source of propargyl radicals. *Tetrahedron Letters*, **37**, 5495–5498.

Denieul, M.-P., Quiclet-Sire, B., and Zard, S. Z. (1996b). Trifluoroacetonyl radicals: a versatile approach to trifluoromethyl ketones. *Journal of the Chemical Society, Chemical Communications*, 2511–2512.

Dingwall, J. G. and Tuck, B. (1986). Free radical catalysed additions to the double bond of diketene: a synthesis of novel oxetan-2-ones. *Journal of the Chemical Society, Perkin Transactions, 1*, 2081–2090.

Fang, J.-M. and Chen, M.-Y. (1987). Free radical type addition of toluenesulfonyl cyanide to unsaturated hydrocarbons. *Tetrahedron Letters*, **28**, 2853–2856.

Forbes, J. E., Saicic, R. N., and Zard, S. Z. (1999). Radical-mediated functional group transformations and C–C bond forming reactions with *S*-alkoxycarbonyl xanthates. Total synthesis of (±)-cinnamolide and (±)-methylenolactocin. *Tetrahedron*, **55**, 3791–3802.

Fraser-Reid, B., Holder, L. N., Hicks, D. R., and Walker, D. L. (1977). Synthetic applications of the photochemically induced addition of oxycarbinyl species to $\alpha$-enones. Part I. The addition of simple alcohols. *Canadian Journal of Chemistry*, **55**, 3978–3985.

Gomez, A. M., Montecon, S., Valverde, S., and Lopez, J. C. (1997). Photochemically induced addition of 2-propanol to hex-2-enono-$\delta$-lactones followed by radical cyclisations: a novel entry to branched cyclohexanes and cyclopentanes from carbohydrates. *Journal of Organic Chemistry*, **62**, 6612–6614.

Gancarz, R. A. and Kice, J. L. (1981). Se–phenyl arenesulfonates: their facile formation and striking chemistry. *Journal of Organic Chemistry*, **46**, 4899–4906.

Gong, J. and Fuchs, P. L. (1996). Alkynylation of of C–H bonds via reaction with acetylenic triflones. *Journal of the American Chemical Society*, **118**, 4486–4487.

Gonzalez Martin, C., Murphy, J. A., and Smith, C. R. (2000). Replacing tin in radical chemistry: *N*-ethylpiperidine hypophosphite in cyclisation reactions of aryl radicals. *Tetrahedron Letters*, **41**, 1833–1836.

Graham, S. R., Murphy, J. A., and Coates, D. (1999*a*). Hypophosphite-mediated carbon–carbon bond formation: a clean approach to radical methodology. *Tetrahedron Letters*, **40**, 2415–2416.

Graham, S. R., Murphy, J. A., and Kennedy, A. R. (1999*b*). Hypophosphite-mediated carbon–carbon bond formation: total synthesis of epialboatrin and structural revision of alboatrin. *Journal of the Chemical Society, Perkin Transactions 1*, 3071–33073.

Hoffman, F. W., Ess, R. J., Simmons, T. C., and Hanzel, R. S. (1956). The desulfurization of mercaptans with trialkyl phosphites. *Journal of the American Chemical Society*, **78**, 6414.

Hudlicky, M. (1992). *Chemistry of organic fluorine compounds*. Ellis Horwood, New York.

Ichinose, Y., Oshima, K., and Utimoto, K. (1988). Et$_3$B-induced radical reaction of 1,3-diene monoxides with C$_6$F$_{13}$I, PhSH, or Ph$_3$GeH. *Chemistry Letters*, 1437–1440.

Ichinose, Y., Matsunaga, S., Fugami, K., Oshima, K., and Utimoto, K. (1989). Triethylborane-induced radical addition of alkyl iodides to acetylenes. *Tetrahedron Letters*, **30**, 3155–3158.

Israelashvili, S. and Diamant, E. (1952). The peroxide-catalyzed addition of carbon tetrachloride to *d*-limonene. *Journal of the American Chemical Society*, **74**, 3185–3186.

Jang, D. O. (1996). Hypophosphorus acid dehalogenation in water. *Tetrahedron Letters*, **37**, 5367–5368.

Julia, M. (1971). Free radical cyclization. *Accounts of Chemical Research*, **4**, 386–392.

Julia, M. and Maumy, M. (1976). Free radical cyclization. Ethyl 1-cyano-2-methylcyclohexanecarboxylate. *Organic Synthesis*, **55**, 55–62.

Kalaï, C., Tate, E., and Zard, S. Z. (2002). A short formal route to (±)-lepadin B using a xanthate-mediated free radical cyclisation/vinylation sequence. *Journal of Chemical Society, Chemical Communications*, 1430–1431.

Kaoudi, T., Quiclet-Sire, B., Seguin, S., and Zard, S. Z. (2000). An expedient construction of 7-membered rings adjoining aromatic systems. *Angewandte Chemie International Edition in English*, **39**, 731–733.

Kenney, R. and Fisher, G. S. (1974). Reaction of terpenes with diethyl phosphite under free radical conditions. *Journal of Organic Chemistry*, **39**, 682–686.

Kharasch, M. S., Freiman, M., and Urry, W. H. (1948*a*). Reaction of atoms and free radicals in solution. XIV. Addition of polyhalomethanes to butadiene sulfone. *Journal of Organic Chemistry*, **13**, 570–575.

Kharasch, M. S., Skell, P. S., and Fisher, P. (1948*b*). Reaction of atoms and free radicals in solution. XII. The addition of bromo esters to olefins. *Journal of the American Chemical Society*, **70**, 1055–1059.

Kharasch, M. S., Jensen, E. V., and Urry, W. H. (1945). Addition of carbon tetrachloride and chloroform to olefins. *Science*, **102**, 128.

Kim, S., Song, H.-J., Choi, T.-L., and Yoon, J.-Y. (2001). Tin-free radical acylation reactions with methanesulfonyl oxime ethers. *Angewandte Chemie International Edition in English*, **40**, 2524–2526.

Le Guyader, F., Quiclet-Sire, B., Seguin, S., and Zard, S. Z. (1997). A new radical allylation reaction of iodides. *Journal of the American Chemical Society*, **119**, 7410–7411.

Liard, A., Quiclet-Sire, B., and Zard, S. Z. (1996). A practical method for the reductive cleavage of the sulfide bond in xanthates. *Tetrahedron Letters*, **37**, 5877–5880.

Liard, A., Quiclet-Sire, B., Saicic, R. N., and Zard, S. Z. (1997). A new synthesis of α-tetralones. *Tetrahedron Letters*, **38**, 1759–1762.

Liu, L. K., Chi, Y., and Jen, K.-Y. (1980). Copper-catalyzed additions of sulfonyl iodides to simple and cyclic alkenes. *Journal of Organic Chemistry*, **45**, 406–410.

Lopez-Ruiz, H. and Zard, S. Z. (2001). A flexible access to highly functionalised boronates. *Journal of Chemical Society, Chemical Communications*, 2618–2619.

Ly, T.-M., Quiclet-Sire, B., Sortais, B., and Zard, S. Z. (1999). A convergent approach to indolines and indanes. *Tetrahedron Letters*, **40**, 2533–2536.

Mayo, F. R. and Walling, C. (1940). The peroxide effect in the addition of reagents to unsaturated compounds and in rearrangement reactions. *Chemical Reviews*, **27**, 351–412.

Miura, K., Fugami, K., Oshima, K., and Utimoto, K. (1988). Synthesis of vinylcyclopentanes from vinylcyclopropanes and alkenes promoted by benzenethiyl radicals. *Tetrahedron Letters*, **29**, 5135–5138.

Miyata, O., Ogawa, Y., Ninomiya, I., and Naito, T. (1997). An enantioselective synthesis of (−)-α-kainic acid via thiyl radical addition–cyclisation–elimination reaction. *Synlett*, 275–276.

Nagahara, K., Ryu, I., Komatsu, M., and Sonoda, N. (1997). Radical carbonylation: ester synthesis from alkyl iodides, carbon monoxide, and alcohols under irradiation conditions. *Journal of the American Chemical Society*, **119**, 5465–5466.

Najéra, C., Baldó, B., and Yus, M. (1988). Regio- and stereo-selective synthesis of β-sulfonyl-α,β-unsaturated carbonyl compounds via iodosulfonylation–dehydroiodination reaction. *Journal of the Chemical Society, Perkin Transactions 1*, 1029–1032.

Newcomb, M. (1993). Competition methods and scales for alkyl radical reaction kinetics. *Tetrahedron*, **49**, 1151–1176.

Nozaki, K., Oshima, K., and Utimoto, K. (1988). Facile route to boron enolates. Et$_3$B-mediated Reformatsky-type reaction and three components coupling reaction of alkyl iodides, methylvinyl ketone, and carbonyl compounds. *Tetrahedron Letters*, **29**, 1041–1044.

Ollivier, C. and Renaud, P. (1999). *B*-Alkyl catecholboranes as a source of radicals for efficient conjugate additions to unsaturated ketones and aldehydes. *Chemistry A European Journal*, **5**, 1468–1473.

Ollivier, C. and Renaud, P. (2000a). Formation of carbon–nitrogen bonds via a novel radical azidation process. *Journal of the American Chemical Society*, **122**, 6496–6497.

Ollivier, C. and Renaud, P. (2000b). A convenient and general tin-free procedure for radical conjugate additions. *Angewandte Chemie International Edition in English*, **39**, 925–928.

Ollivier, C. and Renaud, P. (2001). Organoboranes as a source of radicals. *Chemical Reviews*, **101**, 3415–3434.

Ollivier, C., Bak, T., and Renaud, P. (2000). An efficient and tin free procedure for radical iodine atom transfer reactions. *Synthesis*, 1598–1602.

Oswald, A. A. (1960). Organic sulfur compounds. II. Synthesis of indanyl aryl sulfides, sulfoxides, and sulfones. *Journal of Organic Chemistry*, **25**, 467–469.

Oswald, A. A., Griesbaum, K., Thaler, W. A., and Hudson, B. E. Jr. (1962). Organic sulfur compounds. VIII. Addition of thiols to conjugated olefins. *Journal of the American Chemical Society*, **84**, 3897–3904.

Parsons, P. J. and Cowell, J. K. (2000). A rapid synthesis of (−)-tetrahydrolipstatin. *Synlett*, 107–109.

Patrick, T. M. Jr. (1952). The free radical addition of aldehydes to unsaturated polycarboxylic esters. *Journal of Organic Chemistry*, **17**, 1009–1016.

Pombrowski, G. W., Dinnocenzo, J. P., Farid, S., Goodman, J. L., and Gould, I. R. (1999). α-CH bond dissociation energies of some tertiary amines. *Journal of Organic Chemistry*, **64**, 427–431.

Qiu, Z.-M. and Burton, D. L. (1994). UV initiated addition of iododifluoromethylketones to electron-deficient olefins. *Tetrahedron Letters*, **35**, 1813–1616.

Quiclet-Sire, B. and Zard, S. Z. (1996a). An unusual route to deoxysugars by hydrogen atom transfer from cyclohexane. A possible manifestation of polar effects in a radical process. *Journal of the American Chemical Society*, **118**, 9190–9191.

Quiclet-Sire, B. and Zard, S. Z. (1996b). New allylation reaction, *Journal of the American Chemical Society*, **118**, 1209–1210.

Quiclet-Sire, B. and Zard, S. Z. (1998a). A concise access to a new class of selective anti-inflammatory steroid derivatives. *Tetrahedron Letters*, **39**, 1073–1074.

Quiclet-Sire, B. and Zard, S. Z. (1998b). A practical modification of the Barton–McCombie reaction and radical O- to S- rearrangement of xanthates. *Tetrahedron Letters*, **39**, 9435–9438.

Quiclet-Sire, B. and Zard, S. Z. (1999). Some new radical reactions for organic synthesis. *Phosphorus, Sulfur, and Silicon*, **153–154**, 137–154.

Quiclet-Sire, B., Seguin, S., and Zard, S. Z. (1998). A new radical allylation reaction of dithiocarbonates. *Angewandte Chemie International Edition in English*, **37**, 2864–2867.

Renaud, P. and Vionnet, J.-P. (1993). Radical additions to 7-oxabicyclo[2.2.1]hept-5en-2-one. Facile preparation of all-*cis*-Corey lactone. *Journal of Organic Chemistry*, **58**, 5895–5896.

Roberts, B. P. and Winter, J. N. (1999). Polarity-reversal catalysis of hydrogen-atom abstraction reactions: concepts and applications in organic chemistry. *Chemical Society Reviews*, 25–35.

Schenck, G. O., Koltzenburg, G., and Grossmann, H. (1957). Durch Benzophenon photosensibilisierte Synthese der Terebinsaüre. *Angewandte Chemie*, **69**, 177–178.

Schiesser, C. H. and Wild, L. M. (1996). Free-radical homolytic substitution: new methods for formation of bonds to heteroatoms. *Tetrahedron*, **52**, 13265–13314.

Schmitt, K. D. (1995). Surfactant-mediated phase transfer as an alternative to propanesultone alkylation. Formation of a new class of zwitterionic surfactants. *Journal of Organic Chemistry*, **60**, 5474–5479.

Siddiqui, M. A., Driscoll, J. S., Abushanab, E., Kelley, J. A., Barchi, J. J., and Marquez, V. E. (2000). The 'beta-fluorine effect' in the non-metal hydride radical deoxygenation of fluorine-containing nucleoside xanthate. *Nucleosides, Nucleotides and Nucleic acids*, **19**, 1–12.

Smith, T. A. K. and Whitham, G. H. (1989). Radical induced 1,3-rearrangement-cyclisations of some unsaturated allylic sulfones. *Journal of the Chemical Society, Pertkin Transactions 1*, 319–325.

Smith, A. B. III, Rano, T. A., Chida, N., and Sulikowski, G. A. (1990). Total synthesis of (+)-hitachimycin. *Journal of Organic Chemistry*, **55**, 1136–1138.

Sosnovsky, G. (1964). *Free radical reactions in preparative organic chemistry*. Macmillan, New York.

Stacey, F. W. and Harris, J. F. Jr. (1963). Formation of carbon–heteroatom bonds by free radical chain additions to carbon–carbon multiple bonds. *Organic Reactions*, **13**, 150–376.

Suzuki, A., Matsumoto, H., Itoh, M., Brown, H. C., Rogic, M. M., and Rathke, M. W. (1967). A facile reaction of organoboranes with methylvinyl ketones. A convenient new ketone synthesis via hydroboration. *Journal of the American Chemical Society*, **89**, 5708–5709.

Suzuki, A., Miyaura, N., Itoh, M., Brown, H. C., Holland, G. W., and Negishi, E. (1971). A new four-carbon atom homologation involving the free radical chain reaction of 1,3-butadiene monoxide with organoboranes. Synthesis of 4-alkyl-2-buten-1-ols from olefins via hydroboration. *Journal of the American Chemical Society*, **93**, 2792–2793.

Thalman, A., Oertle, K., and Gerlach, H. (1984). Ricinoleic acid lactone. *Organic Syntheses*, **63**, 193–197.

Udding, J. H., Giesselink, J. P. M., Hiemstra, H., and Speckamp, W. N. (1994*a*). Xanthate transfer cyclisation of glycolic acid-derived radicals. Synthesis of five- to eight-membered ring ethers. *Journal of Organic Chemistry*, **59**, 6671–6682.

Udding, J. H., Hiemstra, H., and Speckamp, W. N. (1994*b*). Xanthate transfer addition of a glycine radical equivalent to alkenes. A novel route to α-amino acid derivatives. *Journal of Organic Chemistry*, **59**, 3721–3725.

Urry, W. H. and Juveland, O. O. (1958). Free radical additions of amines to olefins. *Journal of the American Chemical Society*, **80**, 3323–3328.

Wagner, P. J., Sedon, J. H., and Lindstrom, M. J. (1978). Rates of radical b-cleavages in photogenerated diradicals. *Journal of the American Chemical Society*, **100**, 2579–2580.

Walling, C. and Huyser, E. S. (1963). Free radical additions to olefins to form carbon–carbon bonds. *Organic Reactions*, **13**, 91–149.

Xiang, J. and Fuchs, P. L. (1996). Alkynylation of of C–H bonds via reaction with vinyl and dienyl triflones. Stereospecific synthesis of vinyl triflones via organocopper addition to acetylenic triflones. *Journal of the American Chemical Society*, **118**, 11986–11987.

Xiang, J., Jiang, W., Gong, J., and Fuchs, P. L. (1997). Stereospecific alkenylation of of C–H bonds via reaction with β-heteroatom-functionalized trisubstituted vinylic triflones. *Journal of the American Chemical Society*, **119**, 4123–4129.

Yamago, S., Mizoe, H., and Yoshida, J. (1999*a*). Synthesis of vinylic *C*-glycosides from telluroglycosides. Addition of photochemically and thermally generated glycosyl radicals to alkenes. *Tetrahedron Letters*, **40**, 2343–2346.

Yamago, S., Mizoe, H., Goto, R., and Yoshida, J. (1999*b*). Radical-mediated imidoylation of telluroglycosides. Insertion of isonitriles into the glycosidic carbon–tellurium bond. *Tetrahedron Letters*, **40**, 2347–2350.

Yorimitsu, H., Nakamura, T., Shinokubo, H., and Oshima, K. (1998). Triethylborane-mediated atom transfer radical cyclization reaction in water. *Journal of Organic Chemistry*, **63**, 8604–8605.

Zard, S. Z. (1997). On the trail of xanthates: some new reactions from an old functional group. *Angewandte Chemie International Edition in English*, **36**, 672–685.

# 7 The persistent radical effect: non-chain processes

## 7.1 Selective suppression of fast modes: the Fischer–Ingold effect

In the previous chapters, we dealt mostly with chain reactions where the desired end product is elaborated in the propagation steps. The faster these were, the greater the efficiency: less initiator is needed; fewer unwanted side reactions can compete; and radical–radical interactions constituting the termination become negligible. Indeed, the whole approach was geared to reduce radical–radical interactions by keeping the steady-state concentration of the intermediate radical species as low as possible. Since these interactions are generally diffusion controlled and notoriously unselective, it might seem, at first sight, that domesticating inter-radical reactions for synthetic purposes is doomed from the outset. It turns out that there is an ingenious solution to this problem based on the use of the persistent radical effect, now called the Fischer–Ingold effect. This phenomenon, which has only recently been well understood (Fischer 1986, 2001; Studer 2001), underlies several radical reactions that occur in nature or that have been discovered by man. Much of the radical chemistry of nitric oxide and various nitroso derivatives, of hypoiodites, $N$-iodoamides, of vitamin $B_{12}$, cobalamines, and simpler organocobalt compounds is governed by the persistent radical effect. Another, industrially important recent development that will also be discussed briefly, concerns the control of radical polymerizations by the application of this principle.

The Fischer–Ingold effect can be apprehended without the need for an elaborate—but more rigorous—mathematical modelling. Let us consider a compound **A** (say a hypothetical diazo derivative **X–N=N–Y**), that can be decomposed thermally or photochemically into two radicals **X•** and **Y•** which will then undergo recombination. If we neglect eventual cage effects and assume that the rate of these recombinations are diffusion controlled (and therefore comparable) then, statistically, from the three possible reactions shown by eqns (1–3) of Fig. 7.1, one would expect roughly a yield of 25% of each of the symmetrical dimers **X–X** and **Y–Y**, and a 50% yield of the cross-product **X–Y**. There is therefore already some statistical selectivity in favour of the latter.

Now what will the effect on the selectivity be if **Y•** is a persistent radical, that is, if reaction (2) does not occur ? We are now left with two reactions (1) and (3) as shown in Fig. 7.2 and the first answer that comes to mind is that the selectivity in favour of **X–Y** would increase to around 66%, again on statistical grounds. In fact, the selectivity for the cross-product becomes almost total.

The reason is that we are not dealing with a hypothetical mathematical situation where all **X•** and **Y•** are generated instantly and react instantly. In this real-life chemical transformation, compound **A** is being decomposed over a certain period of time (minutes, hours) which is vastly greater than the lifetime of the intermediate radicals

$$A \xrightarrow{\Delta \text{ or } h\nu} X^\bullet + Y^\bullet$$

$$X^\bullet + X^\bullet \longrightarrow X–X \quad (1)$$

$$Y^\bullet + Y^\bullet \longrightarrow Y–Y \quad (2)$$

$$X^\bullet + Y^\bullet \longrightarrow X–Y \quad (3)$$

**Fig. 7.1** Dimerization reactions between two transient radicals

$$A \xrightarrow{\Delta \text{ or } h\nu} X^\bullet + Y^\bullet$$

$$X^\bullet + X^\bullet \longrightarrow X–X \quad (1)$$

$$X^\bullet + Y^\bullet \longrightarrow X–Y \quad (3)$$

**Fig. 7.2** Dimerization reactions between a transient and a persistent radical

(micro- to nanoseconds), whose concentration in the medium remains small throughout. Now, while the decomposition of **A** and the cross-coupling reaction (3) do not modify the relative concentration of $X^\bullet$ and $Y^\bullet$ (either one of each is produced or one of each is consumed), the formation of **X–X** in eqn (1) consumes in contrast only $X^\bullet$ radicals. The consequence is that very soon after the radicals start being generated, the *relative* concentration becomes tilted greatly in favour of radicals $Y^\bullet$, even if the *absolute*, steady-state concentration of both $X^\bullet$ and $Y^\bullet$ remains small (in other words, $[Y^\bullet]_{\text{steady state}} \gg [X^\bullet]_{\text{steady state}}$). Thus, every time a radical $X^\bullet$ is created by the decomposition of **A**, its chances of capturing a radical $Y^\bullet$ are much greater than capturing another $X^\bullet$; of course, radicals $Y^\bullet$ being by definition persistent can only react with $X^\bullet$. The formation of the cross-coupling product **X–Y** will therefore rapidly dominate and the selectivity, for all practical purposes, will be almost complete. The same result will qualitatively obtain if reaction (2) is (comparatively) slow or easily reversible under the experimental conditions (as for example with Gomberg's triphenylmethyl radical we discussed in the introduction; see Schemes 1.1 and 1.2).

## 7.2 Dissociation of nitroso derivatives

The effective and selective suppression of fast termination steps by the presence of a persistent radical explains nicely why, for example, photolysis in an inert atmosphere of *N*-nitroso dimethylamine does not lead to apparent change (Scheme 7.1), even though the photolytic cleavage of the N–N bond to give dimethylaminyl radicals and nitric oxide occurs with a quantum yield of close to unity and that the

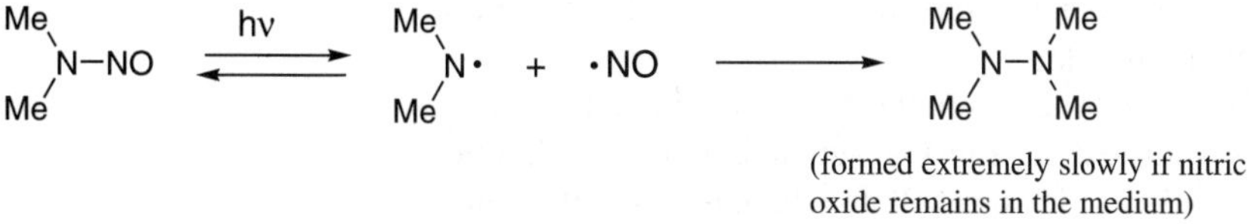

$$\underset{\text{Me}}{\overset{\text{Me}}{\diagdown}}N–NO \; \underset{\longleftarrow}{\overset{h\nu}{\rightleftharpoons}} \; \underset{\text{Me}}{\overset{\text{Me}}{\diagdown}}N^\bullet \; + \; ^\bullet NO \; \longrightarrow \; \underset{\text{Me}\quad\text{Me}}{\overset{\text{Me}\quad\text{Me}}{\diagdown\diagup}}N–N$$

(formed extremely slowly if nitric oxide remains in the medium)

**Scheme 7.1** Photolysis of *N*-nitroso-dimethylamine

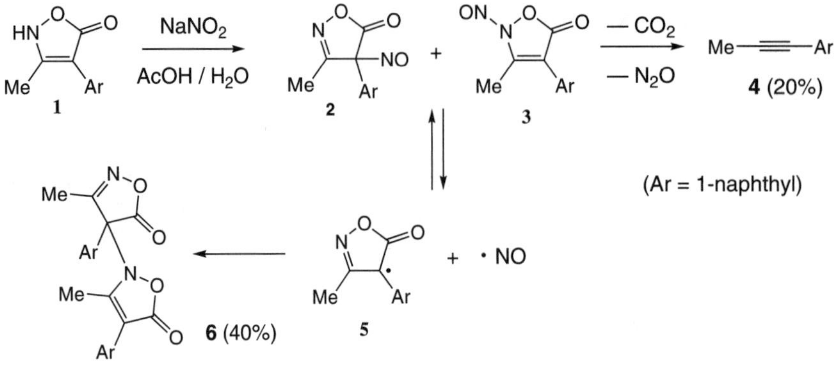

**Scheme 7.2** Formation of dimers in the synthesis of alkynes from isoxazolinones

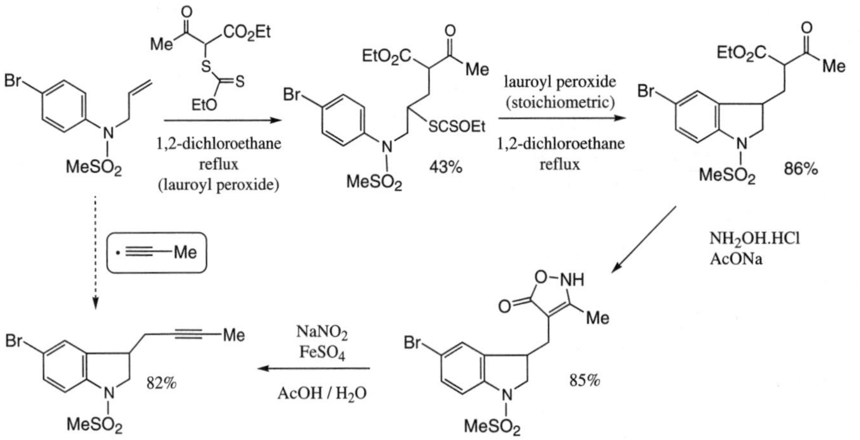

**Scheme 7.3** Synthetic equivalents of alkynyl radicals

former recombine to give tetramethylhydrazine at diffusion-controlled rates (Fischer 1986). The self reaction of dimethylaminyl radicals is in fact effectively suppressed by the presence of the co-product, nitric oxide, which is a persistent radical; and the cross-coupling mode gives back starting material; hence the apparent, macroscopic inertness to irradiation.

This phenomenon has been exploited in a recent synthesis of acetylenes (Boivin *et al.* 1991). Treatment of isoxazolinone **1** with sodium nitrite/acetic acid unexpectedly furnished dimer **6** as the major product (40%) and only 20% of the desired alkyne **4** (Scheme 7.2). It turns out that nitrosation produces both *C*- and *N*-nitroso intermediates, **2** and **3**, but only the latter can collapse into the alkyne. The *C*-nitroso isomer appears to be more favoured; however, it apparently dissociates spontaneously and reversibly into nitric oxide and the stabilized isoxazolinyl radical **5**. As the gaseous nitric oxide escapes from the medium, the isoxazolinyl radicals combine to give the unsymmetrical dimer **6**.

To suppress this unwanted radical side reaction and to favour the ionic process (incongruous in this book!), clearly nitric oxide must be kept in the medium. In practice, this was accomplished by addition of ferrous sulfate which reacts with sodium nitrite and acetic acid to generate nitric oxide *in situ*. The path leading to dimer is effectively curtailed because the isoxazolinyl radical is continuously captured by the nitric oxide to give back both the *C*- and *N*-nitroso derivatives, and thus only alkyne is ultimately produced. The example in Scheme 7.3 highlights one application, where the xanthate-transfer process discussed in the previous chapter and the present alkyne synthesis were combined to provide the synthetic equivalent of propynyl radicals (Boutillier and Zard 2001). Alkynyl radicals are even higher in energy than vinyl and aryl radicals and are very difficult to generate and capture in a useful manner.

## 7.3  Photolysis of nitrite esters: the Barton reaction

The photolysis of nitrite esters, also known as the Barton reaction (Barton *et al.* 1961) is, from a synthetic perspective, perhaps the most important radical transformation of nitroso derivatives. Irradiation of a nitrite ester **7** (easily made from the reaction of an alcohol and nitrosyl chloride in the presence of an organic base) produces reversibly an alkoxy radical and nitric oxide (Scheme 7.4). Alkoxy radicals are highly reactive and readily abstract a suitably located hydrogen atom to form a carbon radical. Combination with nitric oxide leads to a *C*-nitroso derivative **8** which can be isolated, usually as its more stable dimer **9**. In many applications, the crude photolysis product is directly converted to the oxime tautomer by heating in a protic solvent or, under harsher conditions, directly to the corresponding aldehyde or ketone. 1,5- Hydrogen migration is the most favoured especially in rigid systems where the alkoxy radical and the migrating hydrogen are in close proximity, but rare cases of 1,4- and 1,6-transfers have been reported (for a review, see: Majetich and Wehless 1995).

The Barton reaction allows the intramolecular functionalization of unactivated alkyl groups, a task that is generally very difficult to accomplish by other methods. Originally, the reaction was conceived to solve a specific but vexing problem in steroid chemistry, namely the partial synthesis of aldosterone outlined in Scheme 7.5 (Barton and Beaton 1961). This substance is an extremely potent mineralo-corticoid

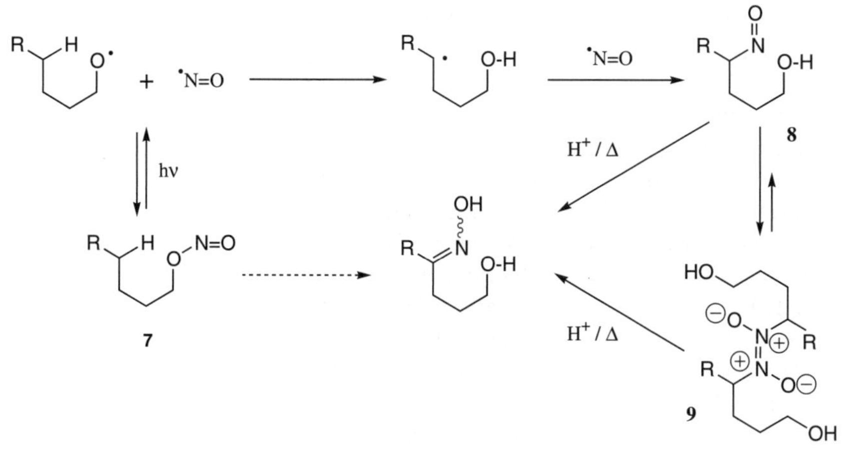

**Scheme 7.4**  The Barton nitrite photolysis reaction

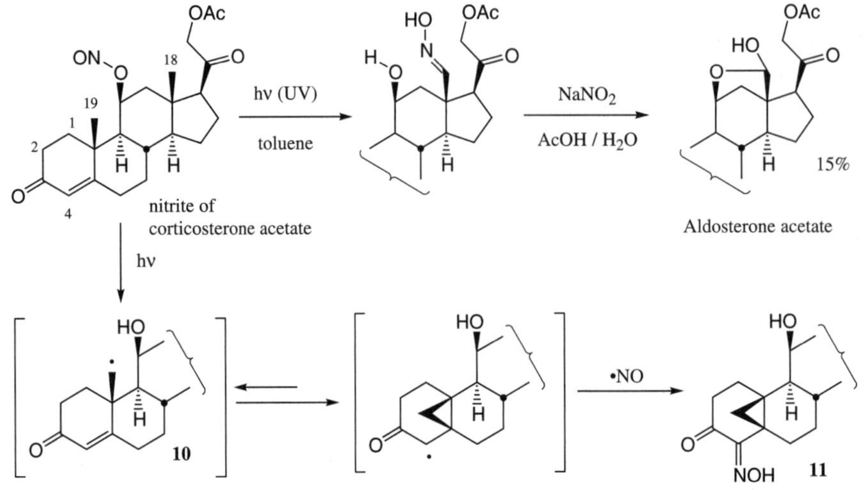

**Scheme 7.5** Synthesis of aldosterone acetate using the Barton nitrite photolysis

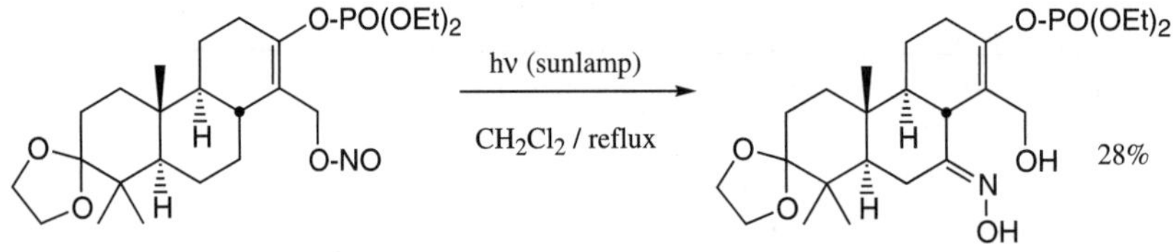

**Scheme 7.6** Further synthetic application of the Barton nitrite photolysis

that could only be obtained by a tedious extraction in vanishingly small quantities not sufficient for a thorough clinical study of its properties. Moreover, its structure contained a modified 18-methyl group which made its partial synthesis especially daunting since none of the available potential steroid precursors had a substituent on C-18. It may be interesting to note that it took nearly 20 years before a crystalline sample of aldosterone could be obtained.

Photolysis of the easily accessible nitrite of corticosterone acetate gave the oxime which was cleaved with sodium nitrite and aqueous acetic acid into aldosterone acetate; this compound exists as the hemiacetal. The overall yield is low (15%) because the intermediate alkoxy radical is equidistant to both the 18- and 19-methyl groups, and hydrogen abstraction from the latter cannot be avoided. Nevertheless, this ingenious and spectacularly short synthesis is still by far the most effective route to aldosterone; it allowed the preparation of a 70 g sample of the precious natural product when the world stock was only a few milligrams. It is interesting to note that the radical on C-19 **10** is in equilibrium with its cyclopropyl isomer, and it is the latter that is captured by the nitric oxide to furnish oxime **11** after tautomerization of the nitroso intermediate. This is one of the earliest examples of a cyclopropane being formed by a radical addition to an olefin. As we have seen in earlier chapters, it is the reverse transformation (i.e. opening of a cyclopropyl ring) that usually prevails; presumably, the shape of the steroid backbone and the greater stability of the radical on C-4 drive this process forward. It is also worth noting that if an unsaturation is introduced between carbons 1,2 in the A ring, the steroid structure bends in such a way as to place the 19-methyl a little further from the intermediate alkoxy radical: H-abstraction now only occurs from the 18-methyl. Exploiting the sensitivity of the hydrogen migration to geometrical factors has thus yielded an improved synthesis of aldosterone (Barton *et al.* 1975).

A key step in a synthesis of the limonoid azadiradione (Corey and Hahl 1989) is a more recent application of the Barton nitrite photolysis (Scheme 7.6). The yield in this case is also a low 28% but, again, it would be difficult to imagine a more expedient way of introducing a functionality in such an unactivated position of the molecule.

## 7.4 Amidyl radicals by photolysis of nitrosamides

Photolytic rupture of the N–N bond in *N*-nitroso amides constitutes a simple and convenient, but surprisingly little used, route to amidyl radicals, and these are sufficiently reactive to pluck off a hydrogen atom from an unactivated C–H bond. Amidyl radicals also add readily to suitably located internal olefins and both of these synthetic possibilities are exemplified by the reactions depicted in Scheme 7.7. The first (Edwards and Rosich 1967) shows that, as with alkoxy radicals, a 1,5-disposition of the hydrogen atom is the preferred mode.

The second transformation (Chow and Perry 1985) illustrates an efficient construction of a functionalized aza-bicyclic [3.2.1] skeleton. In both examples, the resulting carbon-centred radical is rapidly captured by the nitric oxide and tautomerization provides the corresponding oxime which may be cleaved to the ketone.

## 7.5 Photolysis of hypoiodites, *N*-iodoamides, and sulfenates

The photolysis or thermolysis of hypoiodites is conceptually similar to that of nitrite esters (Scheme 7.8). The fact that iodine atoms, unlike nitric oxide, do recombine

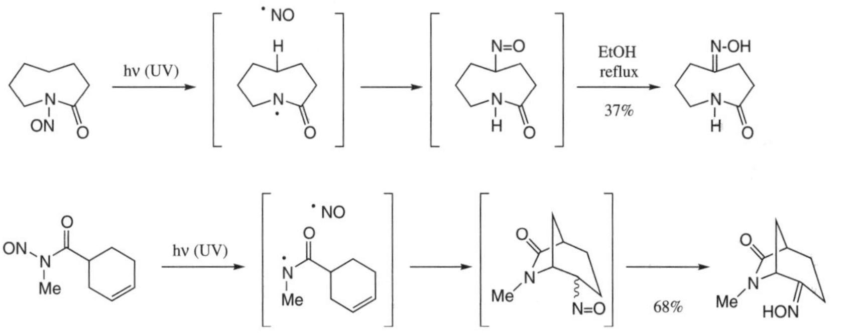

**Scheme 7.7** Photolysis of *N*-nitrosoamides

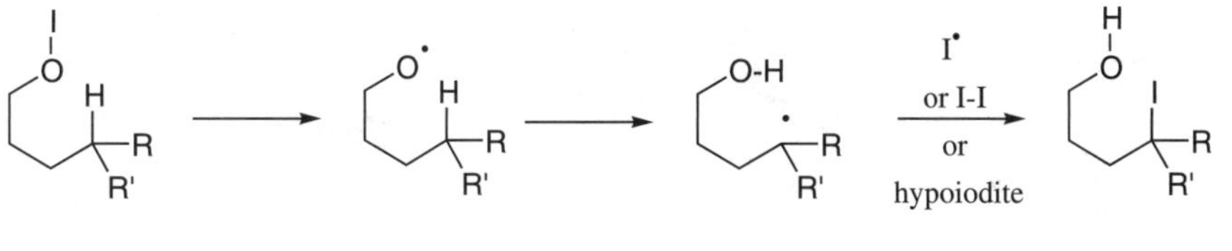

**Scheme 7.8** Alkoxy radicals from hypoiodites

(reversibly under irradiation) to give molecular iodine does not really affect the out-come since both $I^{\bullet}$ and $I_2$ react extremely rapidly with alkyl radicals to furnish the same iodide. A persistent radical effect thus prevails. It is of course also quite possi-ble that some iodine transfer takes place directly from the starting hypoiodite itself, especially at the beginning of the reaction when its concentration is still high. In this case, the mechanism would be an iodine transfer of the Kharasch-type as discussed in the preceding chapter.

All three possibilities converge on the same product and the effectiveness of the process is demonstrated by the clean conversion displayed in Scheme 7.9 of the $6\beta$-hypoiodite **12** into the C-19 iodide **13**, which in this case evolves further into the tetrahydrofuran (Kalvoda and Heusler 1971). The hypoiodite is made *in situ* from the corresponding alcohol by the action of iodine and lead tetraacetate on the alco-hol. There is no competition in this instance with other abstractable hydrogens and the yield is high.

The use of hypoiodites as a source of the highly reactive alkoxy radicals is becoming more popular (Majetich and Wehless 1995). Lead tetraacetate may be replaced by hypervalent iodine or selenium oxidants such as iodobenzene diacetate [PhI(OAc)$_2$] or diphenylselenium diacetate [Ph$_2$Se(OAc)$_2$], or their corresponding di-trifluoroacetate analogues. The example in Scheme 7.10, taken from the extensive work of Suarez and his group (Dorta *et al.* 1988), illustrates a remote functionaliza-tion applied to cedrol. In this case, a significant amount of ring opening occurred to give a ketone, reflecting once again the high reactivity of alkoxy radicals and their general lack of discrimination.

In the same way as the aforementioned *N*-nitroso amides, *N*-iodoamides can be used as a source of amidyl radicals. The difference lies in the last step where an iodine is transferred instead of the nitroso group. The radical sequence is often followed by an intramolecular ionic substitution of the iodide by the amide oxygen to give the corresponding lactone upon work up. The reasonably efficient conversion

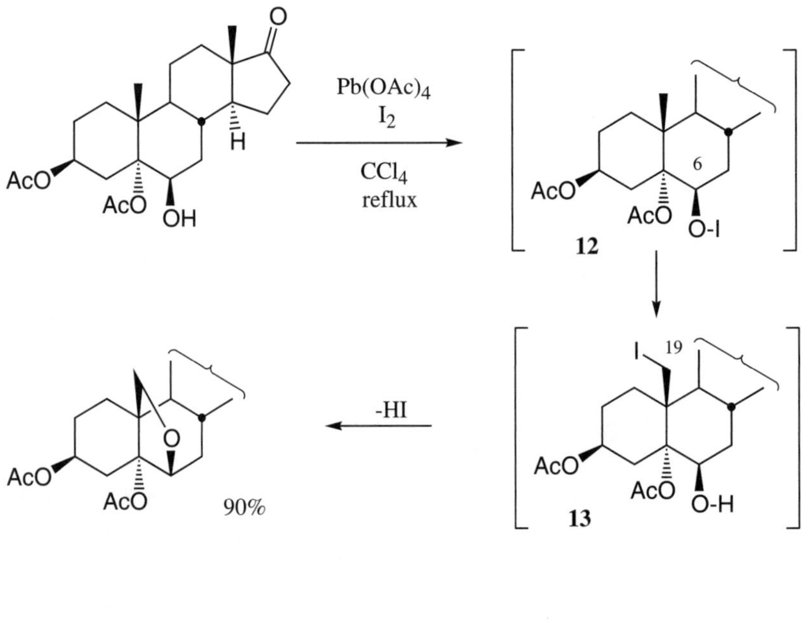

**Scheme 7.9**  C-19 Functionalization in steroids via hypoiodites

displayed in Scheme 7.11 of the steroid amide into a lactone through a 1,5 transfer of hydrogen is one example of application (Barton *et al.* 1965; Baldwin *et al.* 1968). The rate of hydrogen abstraction by the amidyl radical has been measured more recently; it depends somewhat on whether abstraction is taking place on *N*-alkyl or on the acyl side chain of the amide, being $1 \times 10^5$ and $4 \times 10^4$ s$^{-1}$, respectively (Sutcliffe and Ingold 1982).

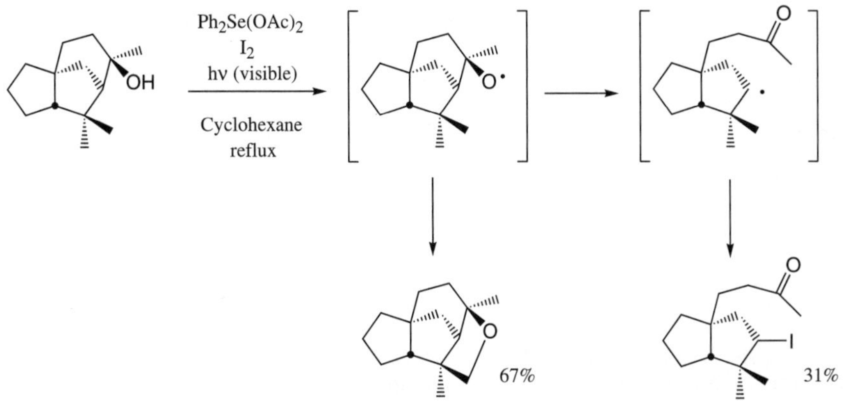

**Scheme 7.10** Further application of the hypoiodite photolysis

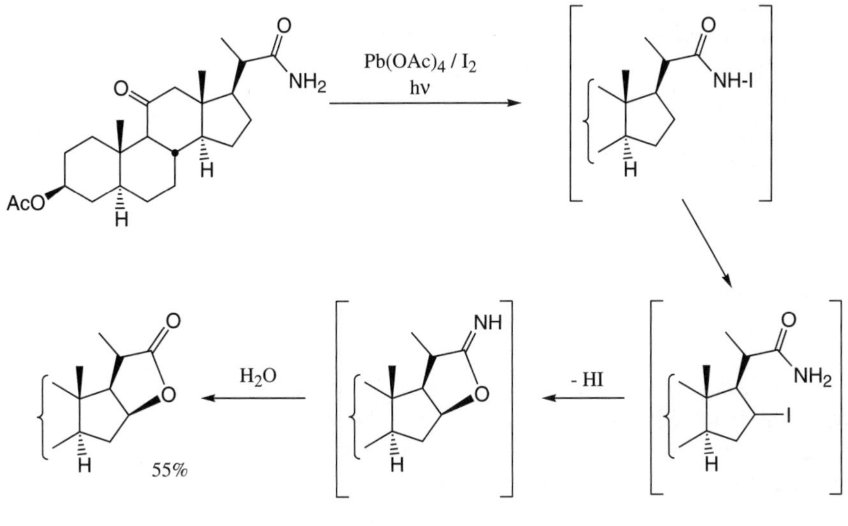

**Scheme 7.11** Photolysis of *N*-iodoamides

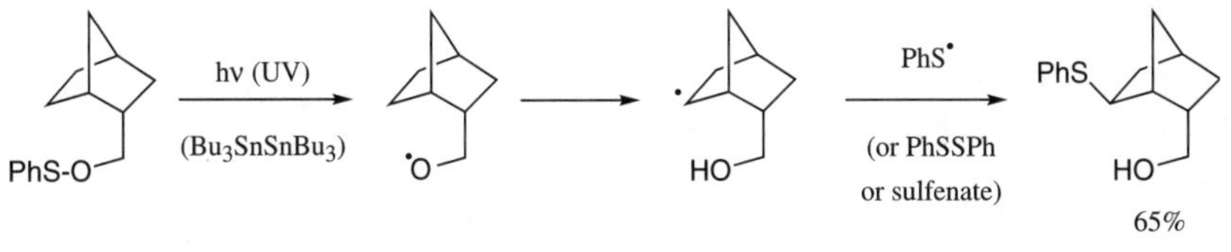

**Scheme 7.12** Alkoxy radicals by photolysis of sufenates

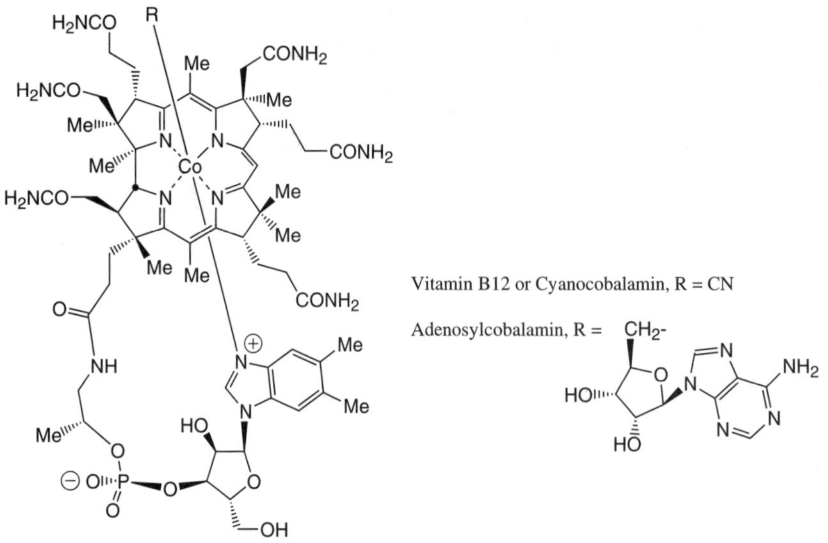

Vitamin B12 or Cyanocobalamin, R = CN

Adenosylcobalamin, R = CH₂-

**Fig. 7.3**  Vitamin $B_{12}$

We shall end up this section with a brief look at phenyl sulfenates which can also be decomposed by irradiation with UV light to furnish alkoxy and phenylthiyl radicals; the latter can dimerise to diphenyl disulfide reversibly under irradiation thus ensuring again a Fischer–Ingold type control of the selectivity. Diphenyl disulfide and the starting sulfenate are both capable of reacting with the carbon radical to afford the same sulfide, and it is possible that some of the product arises by this route; however, this homolytic substitution is much less rapid than that with molecular iodine. Sulfenates are therefore generally less effective than hypoiodites. Addition of catalytic amounts of hexabutylditin shortens irradiation times and can improve the yield. The example pictured in Scheme 7.12 is typical (Petrovic *et al.* 1997).

## 7.6   Mimicking vitamin $B_{12}$: the radical chemistry of organocobalt derivatives

The cobalt containing vitamin $B_{12}$ or cyano-cobalamine and its biologically active form, coenzyme $B_{12}$ or adenosyl-cobalamine (Fig. 7.3), represent some of the most complex organometallic compounds found in nature. Coenzyme $B_{12}$ mediates a number of important biochemical transformations where the first step appears to be the homolytic rupture of the carbon to cobalt bond of the adenosyl moiety to give a primary 5′-adenosyl radical (Frey 1990).

The rich and versatile chemistry of cobalamines has inspired a vast amount of work, both mechanistic and synthetic. The key feature is the weakness of the C–Co bond (20–30 Kcal mole$^{-1}$) which upon thermolysis or, more commonly, photolysis (visible light) gives a carbon radical and a persistent Co(II)-centred radical. Fischer–Ingold control of the selectivity operates, allowing numerous useful synthetic transformations (Scheffold *et al.* 1983; Pattenden 1988; for kinetic modelling

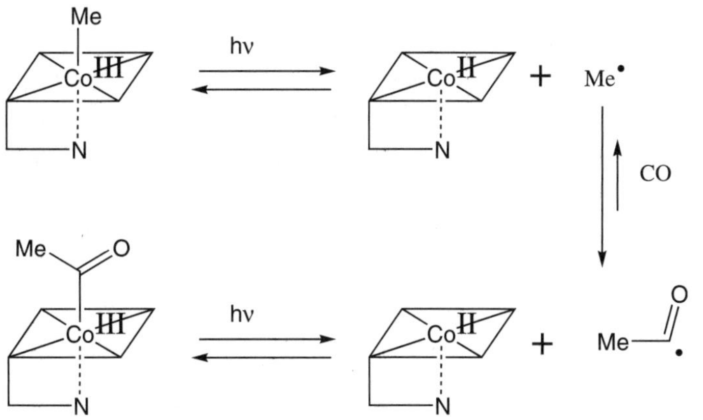

**Scheme 7.13**   Vitamin B12-mediated carbonylation of methyl radicals

studies, see: Daikh and Finke 1992; Waddington and Finke 1993). The effect of the persistent radical effect manifests itself, like the *N*-nitroso-amine in Scheme 7.1, by a surprising stability to irradiation in the absence of a trap. Thus, prolonged exposure to light under anaerobic conditions of methyl-cobalamine, drawn in a simplified form in Scheme 7.13, causes hardly any modification. Yet, if the irradiation is performed under a blanket of carbon monoxide, then a good yield of acetyl-cobalamine is rapidly obtained (Kräutler 1984). In an inert medium, the methyl radicals produced by homolysis are continuously captured by the persistent Co(II) radical to give back starting methyl-cobalamine. This degeneracy is broken by the presence of carbon monoxide, which reacts almost irreversibly with the methyl radical, leading finally to the product. This sequence is thought to be used by bacteria for the biosynthesis of acetic acid or its equivalent, a coenzyme A bound acetyl group.

It is possible to accomplish a similar type of chemistry using simpler organocobalt complexes, the most common being cobaloximes, salens, and salophens; these are shown in Scheme 7.14, along with a typical preparation of cobaloximes derived from dimethylglyoxime (Pratt and Craig 1973; Scheffold *et al.* 1983). The Co(I) anion obtained by reduction of the Co(II) complex is a powerful nucleophile capable of reacting with a large variety of alkylating agents (the reaction in some cases may not be a simple $S_N2$ process and could involve initial electron transfer; cf. $S_{RN}1$ reactions in Chapter 8). Beautiful colour changes accompany the sequence reflecting changes in the oxidation level of the cobalt as well as the nature of the ligands around it.

The transformations that follow radical generation by thermal or photochemical homolysis of an organocobalt complex depend on the type of traps that are placed in the medium (internal or external olefins, atom or group transfer reagents, other persistent radical species). Some of these possibilities are summarized in the reaction manifold depicted in Scheme 7.15 ($L_n$ standing for *n*-ligands around the cobalt). It is thus possible to capture the first radical R$^\bullet$ directly by a persistent radical X$^\bullet$ (e.g. TEMPO or nitric oxide) or by a reagent Y–Z (e.g. BrCCl$_3$ or PhSeSePh) to give respectively R–X or R–Y. In the latter case, the Z$^\bullet$ fragment can either dimerize or combine with $^\bullet$Co$^{II}$L$_n$ depending on its structure. Both of these possibilities are simple functional group interconversions and have seldom been used, even though they

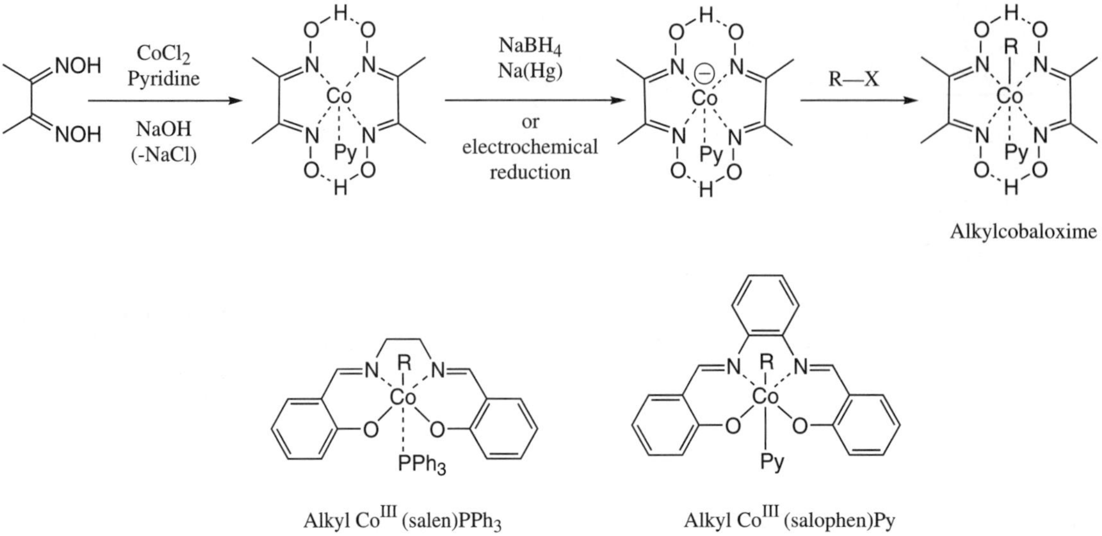

Alkylcobaloxime

Alkyl Co$^{III}$ (salen)PPh$_3$

Alkyl Co$^{III}$ (salophen)Py

**Scheme 7.14**   Alkylcobalt complexes

may provide a solution to specific problems in organic synthesis. A more generally interesting variation is the inter- or intramolecular interception of R$^{\bullet}$ by an olefin to give an intermediate **14** which is in equilibrium with its cobalt complex **15**. If the adduct radical **14** is not stabilized (e.g. G = H), then it is often possible to isolate the corresponding complex **15**; otherwise the reaction evolves to give either olefin **16** through dismutation or the saturated analogue **17** by protonolysis of organocobalt complex **15**. The former process is favoured when G is a stabilizing but not electron-withdrawing group such as a phenyl, whereas the latter usually dominates when G is an electron-withdrawing group such as a nitrile or an ester, and the medium contains a proton source such as methanol. If radical **14** is especially stabilized, as when two stabilizing groups are present, the formation of complex **15** becomes strongly disfavoured (also for steric reasons) with respect to radical **14**, then dismutation is the preferred route, even if one of the groups is electron-withdrawing and the medium contains a proton source. Finally, the adduct radical **14** can be captured by reagents X$^{\bullet}$ or Y–Z as before to give the corresponding products **18** or **19**, respectively. Some synthetic applications of these various alternative pathways are displayed in the schemes that follow.

The effect of the nature of the substituent on the olefinic trap and the experimental conditions are illustrated by the transformations in Scheme 7.16 starting with mannosyl-cobaloxime (Ghosez *et al.* 1988). Irradiation with a sun lamp for several hours of a benzene solution containing excess styrene leads to the olefinic 1α–adduct in good yield and high selectivity (98:2) with respect to the reduced analogue. In contrast, irradiation in the presence of acrylonitrile in a refluxing mixture of benzene and methanol gives cleanly the reduced 1α–derivative. That this is the result of protonolysis was easily shown by running the reaction in deuterated methanol (MeOD) and finding that the product contained one deuterium atom next to the

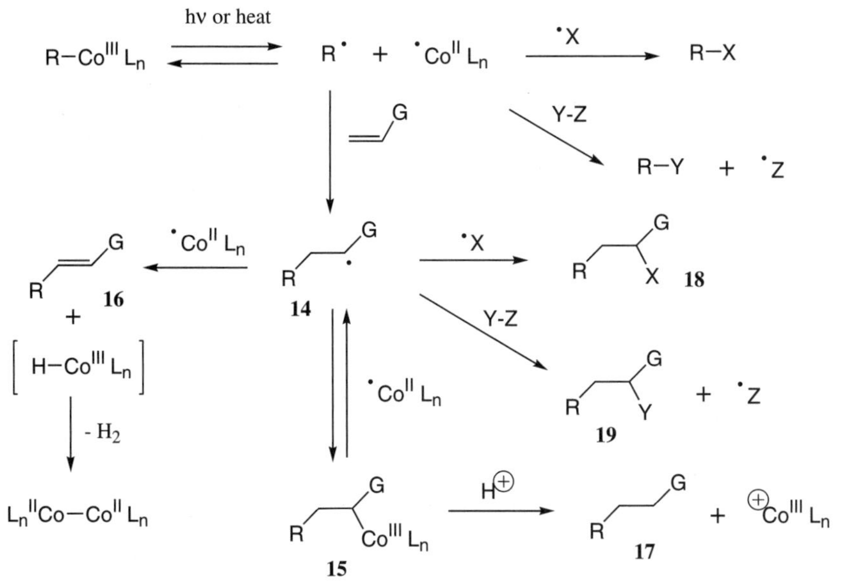

**Scheme 7.15** Reaction manifold of alkyl cobalt complexes

nitrile group. These are two clear-cut cases; in general, varying proportions of the two types are obtained.

In the example displayed in Scheme 7.17, the mannose unit is elongated on carbon-6, in an elegant approach to KDO derivatives (Branchaud and Meier 1988). This conversion illustrates the case where two stabilizing groups on the olefinic trap result in the formation of the olefin, despite the presence of a nitrile group and the use of aqueous ethanol as solvent.

In Scheme 7.18, the intramolecular alternative is exemplified by the synthesis of a large variety of benzofuran and dihydrobenzofuran derivatives simply by changing the trap and experimental conditions (Patel and Pattenden 1987). Addition of the

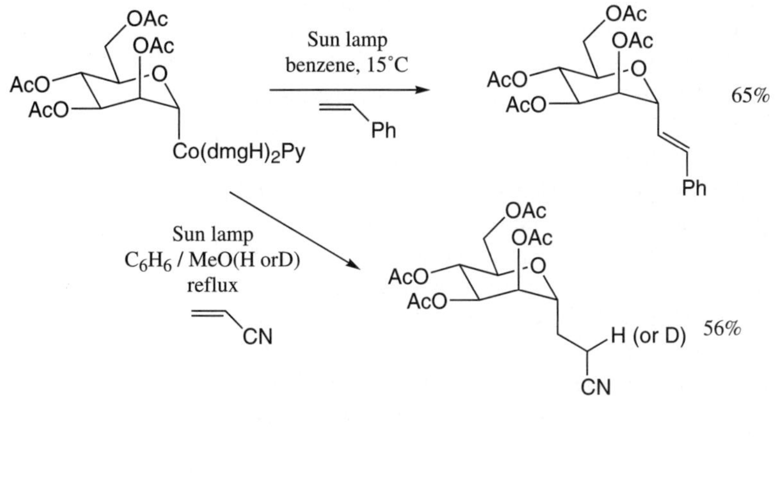

**Scheme 7.16** Intermolecular radical additions involving organocobalt complexes

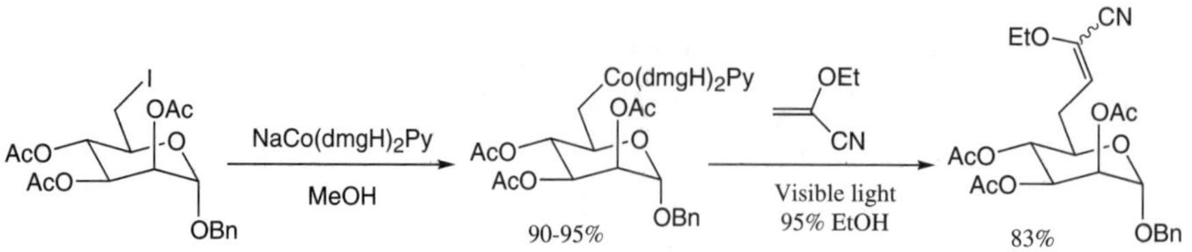

**Scheme 7.17** Organocobalt-mediated radical addition to ethoxy-acrylonitrile

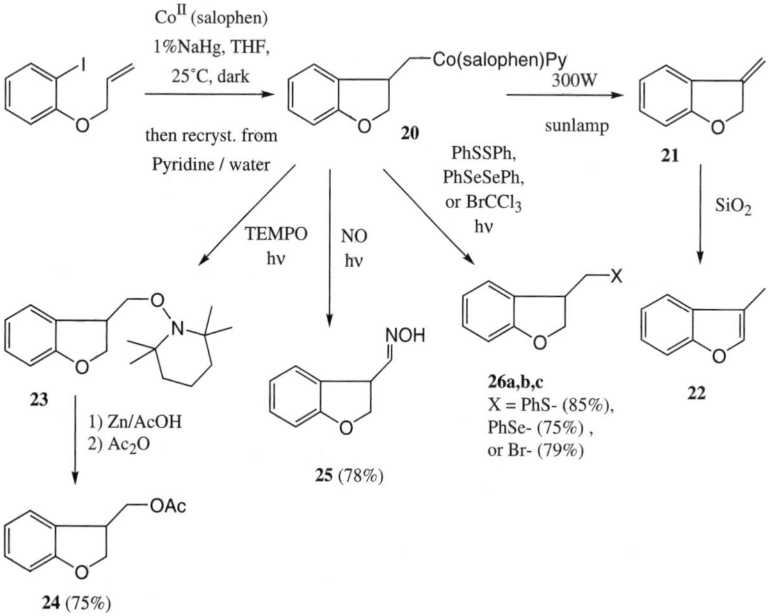

**Scheme 7.18** Synthetic transformations based on organocobalt derivatives

*in situ* prepared emerald-green cobalt (I) salophen reagent to *O*-allyl-iodophenol in THF caused the rapid discharge of the green colour and the formation of another cobalt complex corresponding to cyclized derivative **20**, which can be isolated as black crystals. It is interesting to note that with this aromatic substrate, the more common but less nucleophilic cobaloxime (I) reagent (Scheme 7.14) or even reduced vitamin $B_{12}$ itself failed to induce the radical transformation. Irradiation of a dichloromethane solution of **20** under nitrogen afforded the olefinic product **21** which, upon purification by silica gel chromatography, was converted into its more stable benzofuran isomer **22**. Alternatively, irradiation in the presence of TEMPO gave **23**, a compound easily transformed into acetoxy derivative **24** by reductive acetylation. Capture of the intermediate radical with nitric oxide produced ultimately the corresponding oxime **25**. These are two examples of persistent free radical traps ($X^{\bullet}$ in Scheme 7.15). Finally, the corresponding phenyl sulfide, phenyl selenide or bromide (**26a,b,c**) can be obtained by using the appropriate atom or group transfer reagent.

## 7.7 Living free-radical polymerizations

To complete this chapter, we shall discuss briefly a rapidly emerging area in radical chemistry, namely living free-radical polymerizations. Notwithstanding their enormous importance, radical polymerizations in general are not within the scope of this book; nevertheless, since much of the recent work in the special field of living-radical polymerization involves persistent nitroxyl radicals and the application of the Fischer–Ingold effect, a short description is probably useful (Hawker 1997; Fischer 2001; Hawker *et al.* 2001).

When a monomer such as styrene or methyl acrylate is polymerized with a radical initiator, one obtains a polymer where the chains have varying lengths and the average molecular weight depends (among other factors) on the ratio of monomer to initiator. Since the polymerization terminates irreversibly through dimerization or disproportionation, the distribution of molecular weights is fairly large (wide polydispersity) and it is not generally easy to build block copolymers where a segment made of a given monomer is followed by one or more segments made from other monomers. And even if such block macromolecules were constructed, the wide polydispersity in each of the constituting segments would translate into a poor control over the mechanical and other properties of the end product. These difficulties can sometimes be overcome by employing living ionic polymerizations but the technical hurdles on scale-up are quite challenging: very pure monomers and solvents, rigorous absence of water and oxygen, and low compatibility with many functional groups.

An ingenious solution to this problem hinges on the fact that persistent nitroxyl radicals such as TEMPO react *reversibly* with stabilized radicals if the temperature is raised sufficiently (Scheme 7.19). Thus, heating a suitable *N*-alkoxy tetra–methylpiperidinyl derivative **27**, where the R group represents a stabilized radical (e.g. substituted benzyl), to around 120 °C causes homolysis into R• and TEMPO

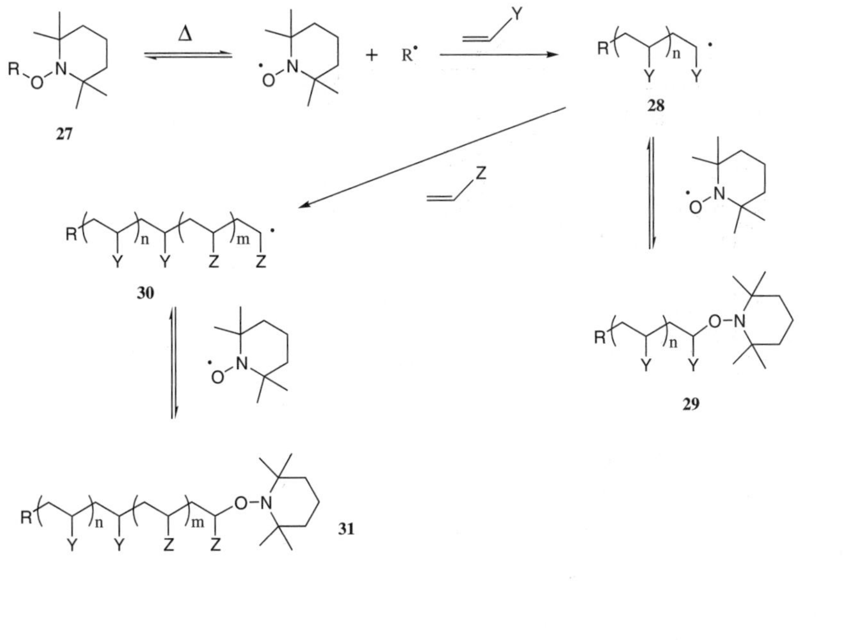

**Scheme 7.19** TEMPO-controlled radical polymerization

radicals. If this is done in the presence of an excess of a monomer such as styrene (Y = Ph), then the R• radicals can start the polymerization process. The nitroxyl radicals in contrast are incapable of acting as initiators but combine readily but *reversibly* with the end of the growing polymer **28** to give what is called a 'dormant' species **29**. The constant presence in the medium of the persistent nitroxyl radical eliminates most of the radical–radical interactions between two polymeric radicals through operation of the Fischer–Ingold effect, preventing hence premature termination of the polymerization process. Polymer growth becomes quite uniform leading to a macromolecule (**29**) with a narrow polydispersity and, because cross-coupling with the persistent nitroxyl radical is dominant, essentially all the polymer chains are 'capped' by the TEMPO unit. If another monomer is now incorporated into the medium and the heating resumed, radical **28** is regenerated from the 'dormant' intermediate polymer **29** (whence the term 'living polymerization') and the new monomer adds to give, via the growing polymer **30**, a structure **31** made of two distinct polymeric blocks. The procedure can be repeated in principle and many modifications can be envisaged to create high molecular weight material with a wide variety of finely tuned physical and mechanical properties. Both the R and TEMPO end units can be chemically modified, a rigid segment can be placed between two flexible arms and vice versa, or the chain may be made to grow in two or more directions at the same time to produce hyperbranched and even dendrimeric macro-molecular architectures. The compatibility of radical processes with many functional Y and Z groups opens tremendous opportunities. Radical chemistry based on the aforementioned cobalt derivatives as well as on atom (halides) and group (xanthates, dithioesters, etc.) transfer reactions described in the previous and following chapters are currently also under intense scrutiny in the context of living polymerization (Gridnev and Ittel 2001; Kamigaito *et al.* 2001). This exciting and rapidly evolving domain certainly holds much promise for future development.

## References

Baldwin, J. E., Barton, D. H. R., Dainis, I., and Pereira, J. L. C. (1968). Photochemical transformations. Part XXIV. The synthesis of 18-hydroxyestrone. *Journal of the Chemical Society (C)*, 2283–2289.

Barton, D. H. R., Beaton, J. M., Geller, L. E., and Pechet, M. M. (1961). A new photochemical reaction. *Journal of the American Chemical Society*, **83**, 4076–4080.

Barton, D. H. R. and Beaton, J. M. (1961). A synthesis of aldosterone acetate. *Journal of the American Chemical Society*, **83**, 4083–4089.

Barton, D. H. R., Beckwith, A. L. J., and Goosen, A. (1965). Photochemical transformations. Part XVI. A novel synthesis of lactones. *Journal of the Chemical Society*, 181–190.

Barton, D. H. R., Basu, N. K., Day, M. J., Hesse, R. H., Pechet, M. M., and Starratt, A. N. (1975). Improved syntheses of aldosterone. *Journal of the Chemical Society, Perkin Transactions 1*, 2243–2247.

Boivin, J., Elkaim, L., Ferro, P. G., and Zard, S. Z. (1991). A new practical method for the synthesis of acetylenes. *Tetrahedron Letters*, **32**, 5321–5324.

Boutillier and Zard, S. Z. (2001). Synthetic equivalents of alkynyl and propargyl radicals. *Chemical Communications*, 1304–1305.

Branchaud, B. P. and Meier, M. S. (1988). A synthesis of ammonium 3-deoxy-*D*-manno-2-octulosonate (ammonium KDO) from *D*-mannose via cobaloxime-mediated radical alkyl–alkenyl cross-coupling. *Tetrahedron Letters*, **29**, 3191–3194.

Chow, Y. L. and Perry, R. A. (1985). Chemistry of amidyl radicals: intramolecular reactivities of alkenylamidyl radicals. *Canadian Journal of Chemistry*, **63**, 2203–2210.

Corey, E. J. and Hahl, R. W. (1989). Synthesis of a limonoid, azadiradione. *Tetrahedron Letters*, **30**, 3023–3026.

Daikh, B. E. and Finke, R. G. (1992). The persistent radical effect: a prototype example of the extreme, $10^5$ to 1, product selectivity in a free radical reaction involving persistent $^{\bullet}Co^{II}$[macrocycle] and alkyl free radicals. *Journal of the American Chemical Society*, **114**, 2938–2943.

Dorta, R. L., Francisco, C. G., Freire, R., and Suarez, E. (1988). Intramolecular hydrogen abstraction. The use of organoselenium reagents for the generation of alkoxy radicals. *Tetrahedron Letters*, **29**, 5429–5432.

Edwards, O. E. and Rosich, R. S. (1967). Photolysis of a medium ring nitrosamide. *Canadian Journal of Chemistry*, **45**, 1287–1290.

Fischer, H. (1986). Unusual selectivities of radical reactions by internal suppression of fast modes. *Journal of the American Chemical Society*, **108**, 3925–3927.

Fischer, H. (2001). The persistent radical effect: a principle for selective radical reactions and living radical poymerizations. *Chemical Reviews*, **101**, 3581–3610.

Frey, P. A. (1990). Importance of organic radicals in enzymatic cleavage of unactivated C–H bonds. *Chemical Reviews*, **90**, 1343–1357.

Ghosez, A., Göbel, T., and Giese, B. (1988). Syntheses and reactions of glycosylcobaloximes. *Chemische Berichte*, **121**, 1807–1811.

Gridnev, A. A. and Ittel, S. D. (2001). New polymer synthesis by nitroxide-mediated living radical poymerizations. *Chemical Reviews*, **101**, 3611–3660.

Hawker, C. J. (1997). Living free radical polymerisation: a unique technique for the preparation of controlled macromolecular architectures. *Accounts of Chemical Research*, **30**, 373–382.

Hawker, C. J., Bosman, A. W., and Harth, E. (2001). New polymer synthesis by nitroxide-mediated living radical poymerizations. *Chemical Reviews*, **101**, 3661–3688.

Kalvoda, J. and Heusler, K. (1971). Die Hypojodit-Reaktion (Verfahren zur Intramolecularen Substitution an nicht-activierten C-Atomen). *Synthesis*, 501–526.

Kamigaito, M., Ando, T., and Sawamoto, M. (2001). Metal-catalyzed living radical poymerization. *Chemical Reviews*, **101**, 3689–3745.

Kräutler, B. (1984). Acetyl-cobalamin from photoinduced carbonylation of methyl-cobalamin. *Helvetica Chimica Acta*, **67**, 1053–1059.

Majetich, G. and Wheless, K. (1995). Remote intramolecular free radical functionalizations: an update. *Tetrahedron*, **51**, 7095–7129.

Patel, V. F. and Pattenden, G. (1987). Radical reactions in synthesis. Homolysis of alkylcobalt salophens in the presence of radical-trapping agents. *Tetrahedron Letters*, **28**, 1451–1454.

Pattenden, G. (1988). Cobalt-mediated radical reactions in organic synthesis. *Chemical Society Reviews*, **17**, 361–382.

Petrovic, G., Saicic, R. N., and Cekovic, Z. (1997). Free radical phenylthio group transfer to nonactivated δ-carbon atom in the photolysis reactions of alkylbenzene sulfenates. *Tetrahedron Letters*, **38**, 7107–7110.

Pratt, J. M. and Craig, P. J. (1973). Preparation and reactions of organocobalt (III) complexes. *Advances in Organometallic Chemistry*, **11**, 331–446.

Scheffold, R., Rytz, G., and Walder, L. (1983). Vitamin $B_{12}$ and related complexes as catalysts in organic synthesis. In *Modern synthetic methods*, Vol. 3 (ed. R. Scheffold), pp. 355–440. Verlag Sauerlander, Aarau.

Studer, A. (2001). The persistent radical effect in synthesis. *Chemistry A European Journal*, **7**, 1159–1164.

Sutcliffe, R. and Ingold, K. U. (1982). Kinetic applications of electron paramagnetic resonance spectroscopy. Intramolecular reactions of some *N*-alkyl carboxamidyl radicals. *Journal of the American Chemical Society*, **104**, 6071–6075.

Waddington, M. D. and Finke, R. G. (1993). Neopentylcobalamin (neopentyl $B_{12}$) cobalt–carbon bond thermolysis products, kinetics, activation parameters, and bond dissociation energy: a chemical model exhibiting $10^6$ of the $10^{12}$ enzymic activation of coenzyme $B_{12}$'s cobalt–carbon bond. *Journal of the American Chemical Society*, **115**, 4629–4640.

# 8 Redox processes

## 8.1 Electron-transfer processes

In essentially all the reactions we have explored so far, the radicals were generated by thermal or photochemical homolysis. There is another way to produce radicals, shown in a formal manner in Scheme 8.1. It consists in removing an electron from an electron-rich species, represented by an anion, or by adding one electron to an electron deficient entity, now represented by a cation. It is of course possible to oxidize a radical to the cation level or reduce it to the anion. These constitute alternative ways of 'destroying' the radical character, in addition to recombination and disproportionation. Such transformations are referred to as redox processes; they are exceedingly important in radical chemistry, and their impact on organic synthesis can hardly be overstated.

Electron transfer can be accomplished through the agency of various transition metals and metal salts or certain organic substances; it can also be electrochemically or photochemically induced. This is a vast field of study and only a tiny fraction can be covered in this chapter (reviews: Kochi 1973; Eberson 1987; Iqbal *et al.* 1994; Dalko 1995). The emphasis will be on transition metal-mediated redox processes and an attempt will be made to give as much a coherent—yet simple—picture as possible. In the great majority of cases, the redox reaction occurs at the beginning to generate the radical species, which then behaves as a true free radical, undergoing the various transformations typical of free radicals. Also quite frequently, a redox process intervenes to 'terminate' the sequence by converting the last radical in the sequence into an ionic species. Thus, most of the systems we shall examine will not be of the chain type; nevertheless, it must be remembered that the radical concentration must be kept low at all times (by controlling the rate of generation by the redox system), otherwise unwanted competition from radical–radical interactions can complicate the outcome.

Generation of radicals by oxidation involves removing one electron from an anionic species (the Kolbe anodic oxidation of carboxylate salts is one such case; cf. Scheme 5.1) or from a neutral electron-rich substance such as an enol. In the latter case, an electron is taken away from the $\pi$-orbital (the HOMO) to give a radical cation, which then looses a proton to give the free radical. An overall equivalent sequence consists in first removing the proton to make the enolate before the electron transfer. Both of these routes are displayed in Scheme 8.2. Which path will prevail in a given situation will depend on the substrate, the oxidant, and exact experimental parameters. As for producing radicals by reduction, most examples involve neutral substrates (represented in the same Scheme by R–X) since true organic

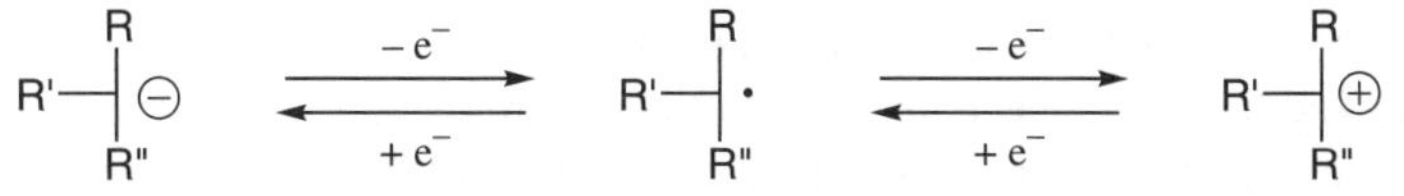

**Scheme 8.1**
Generation of radicals by electron transfer

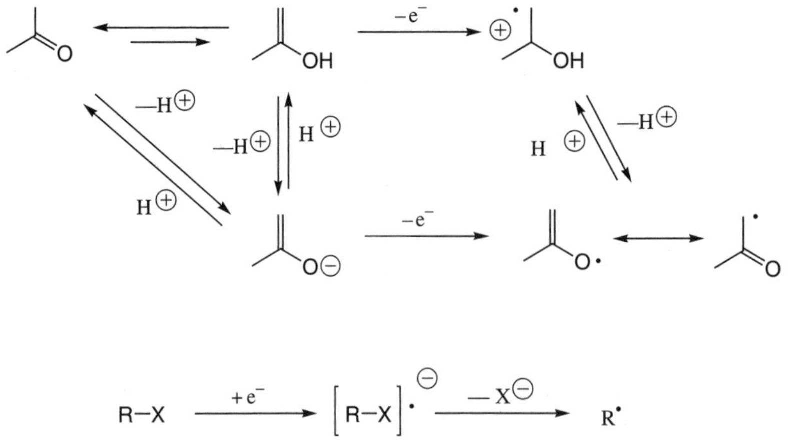

**Scheme 8.2** Oxidation of enols and enolates and fragmentation of radical anions

cationic species, with the notable exception of diazonium and iminium salts, are less commonly encountered. Addition of an electron to the anti-bonding orbital of R–X gives a radical anion, which rapidly collapses into a free radical $R^{\bullet}$ and an anion $X^{-}$ (Savéant 1993; Costentin *et al.* 2003).

Chemists have classified electron-transfer processes either as outer sphere, when the two partners are not attached together by some bond while the transfer is occurring, or inner sphere, when the transfer happens within some complex structure containing the two interacting species. For our purposes, unless the evidence points otherwise, we shall for convenience assume that the electron transfer takes place in an outer-sphere fashion.

## 8.2 Formation of radicals by oxidation with transition metal salts: general reaction scheme

Transition metals in a high oxidation state are often capable of extracting one electron from electron-rich organic substances. Ketones, esters, nitriles, and various other 'carbon acids' that can form enols, enolates, and structures related to them (e.g. enamines) are by far the most commonly used substrates. Their oxidation can lead to a free radical, which then follows one or more of the paths deployed in Scheme 8.3. Before discussing specific examples, it is perhaps useful to explore this seemingly complicated manifold in some detail in order to gain an overall picture of the various possibilities. For clarity, a ketone (possibly containing another electron-withdrawing group E) has been used as a typical substrate and its oxidation is shown to occur from the corresponding enolate, even though, as we have seen in Scheme 8.2, the same radical **1** can be obtained from the enol. The rate of radical production will depend on the exact structure of the substrate, its propensity to exists as the corresponding enol or enolate in the medium, the pH, the solvent, the temperature, and of course the redox potential of the metallic salt (which can be strongly affected by the nature of the ligands around the metal) and the exact mechanism by which electron transfer actually occurs (i.e. inner or outer sphere). Upon transfer of an electron to

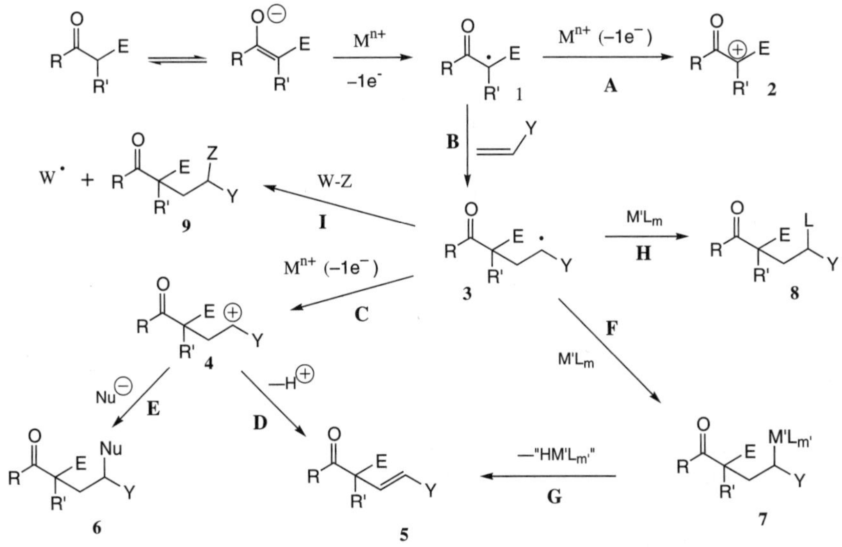

**Scheme 8.3** Oxidation manifold

the metallic salt (denoted as $M^{n+}$), the oxidation level of the metal formally decreases by one unit, to $M^{(n-1)+}$ (not shown in the Scheme).

Once created, radical **1** can be oxidized by the metallic oxidant into cation **2** (path **A**). This is a high-energy route because the positive charge is next to at least one electron-withdrawing group. Nevertheless, such oxidations can and do occur occasionally. The formation of cation **2** (and products therefrom) is in competition with the desired radical process, for example, an addition to an olefin, leading to a new radical **3** (path **B**). Again for purposes of clarity, only the intermolecular case is shown, even though most of the examples that will be discussed will involve cyclizations (and sometimes an elegant combination of several radical steps). To ensure polarity matching and improve the rate of addition, the Y substituent on the olefin will usually be a donating group so that, in contrast to the initial radical **1**, adduct radical **3** is now easier to oxidize by the metal salt into a cation equivalent **4** (path **C**), which then either undergoes elimination of a proton to give olefin **5** (path **D**) or simply adds a nucleophile to give **6** (path **E**). Of course, other reactions typical of cations, such as Wagner–Meerwein rearrangements, can intervene. Another route to olefin **5** involves the intermediacy of an organometallic species **7** arising from the combination of radical **3** with a metal salt or complex, followed by elimination (paths **F, G**). Alternatively, a ligand transfer can occur to furnish **8** via path **H** (the ligand in this case is usually a halide, $L = Cl$ or $Br$). The metal ($M'$) involved in the last two transformations need not be the same as M, used for the initial oxidation. As we shall see shortly, combinations of metal salts (especially those where $M'$ is copper) often result in cleaner reactions, since each of the two metals plays the role for which it is best suited. Finally, the medium may contain a radical trap W–Z compatible with the oxidizing system (path **I**). Transfer of the Z-group gives rise to **9** and leaves behind a radical $W^{\bullet}$, whose fate (oxidation, dimerization etc.) does not interfere with the desired process. Path **I** may for example correspond to hydrogen abstraction from the solvent (Z = H) or homolytic substitution on a disulfide or diselenide (Z = R''S or R''Se).

The metal salts that are most commonly employed include Mn(III), Cu(II), Fe(III), Ce(IV), Ag(II), and Pb(IV). We shall examine them in turn trying, as much as possible, to relate each particular transformation with the general reaction manifold in Scheme 8.3.

## 8.3   Oxidations involving Mn(III) and Cu(II)

Radical generation through oxidation of enolizable substrates using Mn(III) salts is by far the most common, and the field is expanding rapidly (Melikyan 1993; Iqbal *et al.* 1994; Snider 1996). Manganese triacetate, $Mn(OAc)_3$. $2H_2O$, is the usual oxidant; it is actually a trimer made up of an oxo-centred triangle of Mn(III) ions bridged by acetate units. For simplicity and convenience, we shall use the simplified formula $Mn(OAc)_3$ throughout. In Scheme 8.4, two additions are displayed, involving in one case the radical generated from acetic acid itself (often the solvent for these reactions) and, in the second, the radical produced from cyanoacetic acid (Heiba *et al.* 1968, 1974). The overall transformation for both follows path **B–C–E** in Scheme 8.3, with the carboxylate group acting as an internal nucleophile. The formation of the thermodynamically more stable *trans*-lactone in the case of methylstyrene is due to the reversibility of the ionic ring-closure process.

The difference in the reaction conditions is dramatic, reflecting the large difference in p$K$a (for the weakly acidic C–H bond) between acetic acid and cyanoacetic acid. The latter is estimated to react about $4 \times 10^5$ faster (Fristad *et al.* 1985). In the

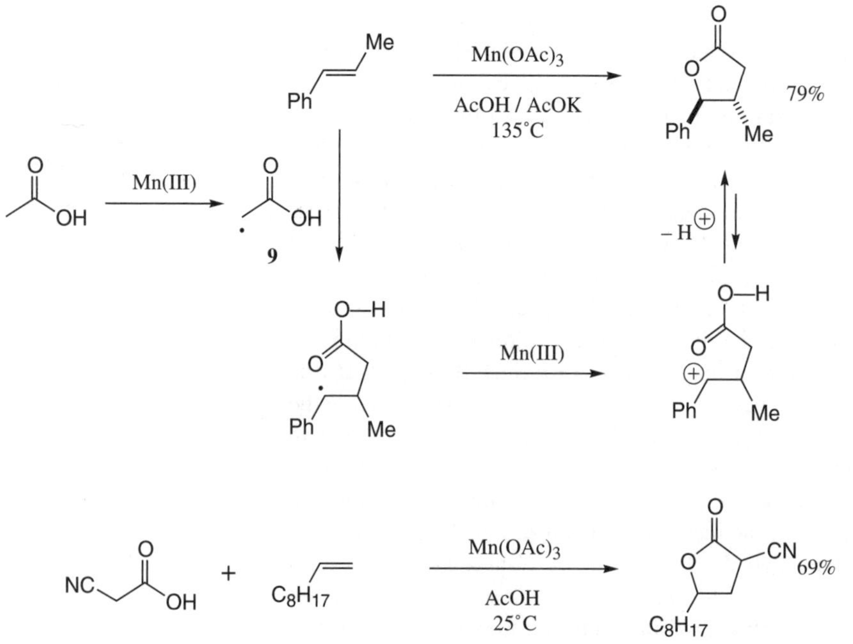

**Scheme 8.4**  Mn(III)-mediated synthesis of lactones

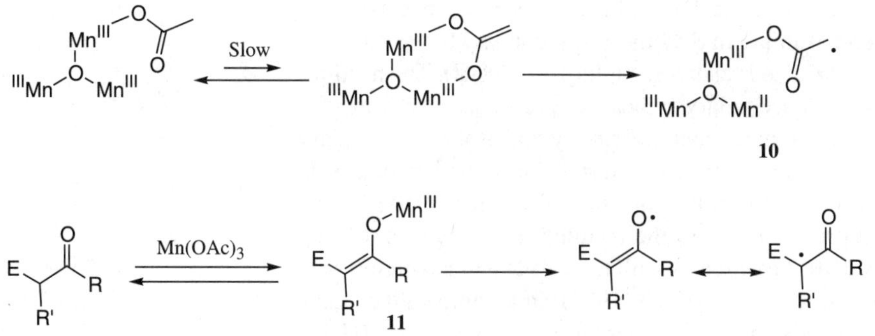

**10**

**Scheme 8.5** Mn(III)-oxidation of carboxylic acids and enols

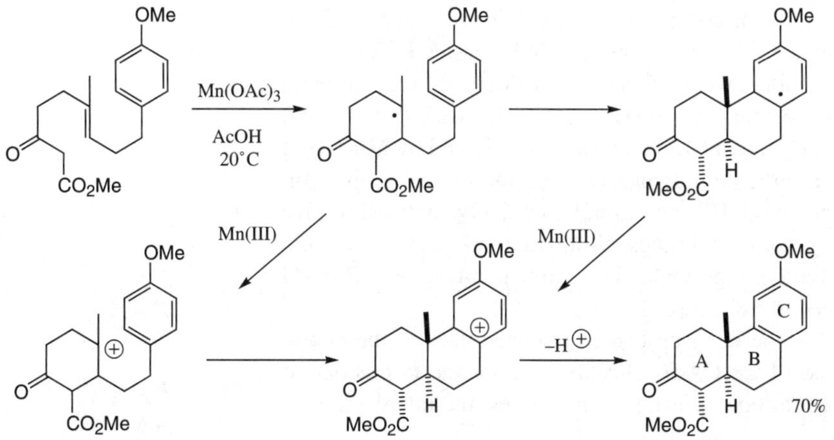

**Scheme 8.6** Mn(III)-mediated tandem cyclizations

case of acetic acid, the slow step is the formation of the manganese enolate, which then undergoes presumably an inner-sphere electron transfer to give the radical **10**, where the carboxylic function may in fact not be free but bound to the manganese (Scheme 8.5). Substrates containing more than one electron-withdrawing group enolize more readily but the ease of electron transfer from the intermediate manganese complex will depend on the exact nature of the substituents. Which of the two steps is rate determining may thus vary from one substrate to another. From extensive studies by the group of Snider (1996), it appears that for R = Me and R' = H, the electron transfer is rate determining, whereas for R = R' = Me, it is the formation of the manganese complex **11** that is the slower step (E = ester group in both cases).

One interesting synthetic application concerns an approach to podocarpic acid (Scheme 8.6), where a cascade was used to construct rings A and B (Snider *et al.* 1985). The ring closure onto the aromatic ring may in principle follow two routes: either the intermediate tertiary radical closes before it is oxidized or it is first converted into a cation equivalent, which then participates in a Friedel–Crafts reaction. The ester group in the product easily equilibrates under the reaction conditions and ends up in the least hindered equatorial position.

In this and the preceding examples, the last radical in the sequence is either benzylic, tertiary, or as in second example in Scheme 8.4, secondary but $\gamma$- to a carboxylic group. Such radicals are oxidized relatively rapidly by Mn(III). The need for a carboxylic group in the case of a secondary radicals is presumably to bring the Mn(III) into close proximity to the radical centre. Primary and isolated secondary radicals are not oxidized quickly enough to avoid further unwanted radical side reactions, usually hydrogen abstraction from the solvent. This difficulty may be circumvented by incorporating cupric acetate into the medium. In terms of oxidizing power, Cu(II) is weaker than Mn(III) but it reacts with radicals at a much higher rate, estimated to be of the order of $10^6$ $M^{-1}$ $s^{-1}$ (Kochi 1973). In a comparative study, $Cu(OAc)_2$ oxidized secondary radicals about 350 times faster than $Mn(OAc)_3$ (Heiba and Dassau 1971). Thus, Mn(III) is used to generate the initial radical and Cu(II) to terminate the sequence.

The high rate of capture and ultimate oxidation of the adduct radical by Cu(II) ensures a kinetic control over the system and generally results in much cleaner reactions. This is illustrated in Scheme 8.7 by the conversion of ketoester **12** into **14** in poor yield (24%) with $Mn(OAc)_3$ alone, while alkene **16** is formed in excellent yield if a combination of manganic acetate and cupric acetate is used (Snider and McCarthy 1993a). Because primary radical **13** is not rapidly oxidized by Mn(III), it eventually follows path **B–I** by abstracting a hydrogen from the solvent to give **14**, albeit inefficiently. In the presence of Cu(II), however, it is quickly captured to give a formal Cu(III) complex **15**, which undergoes elimination of acetic acid and cuprous acetate to afford the alkene in high yield. This corresponds to path **B–F–G** in the reaction manifold displayed in Scheme 8.3.

The cuprous acetate released in the last step can be oxidized back to the cupric level by Mn(III), allowing the use of $Cu(OAc)_2$ in catalytic amounts. In practice, it is better not to reduce the concentration of Cu(II) too much, lest the interception of

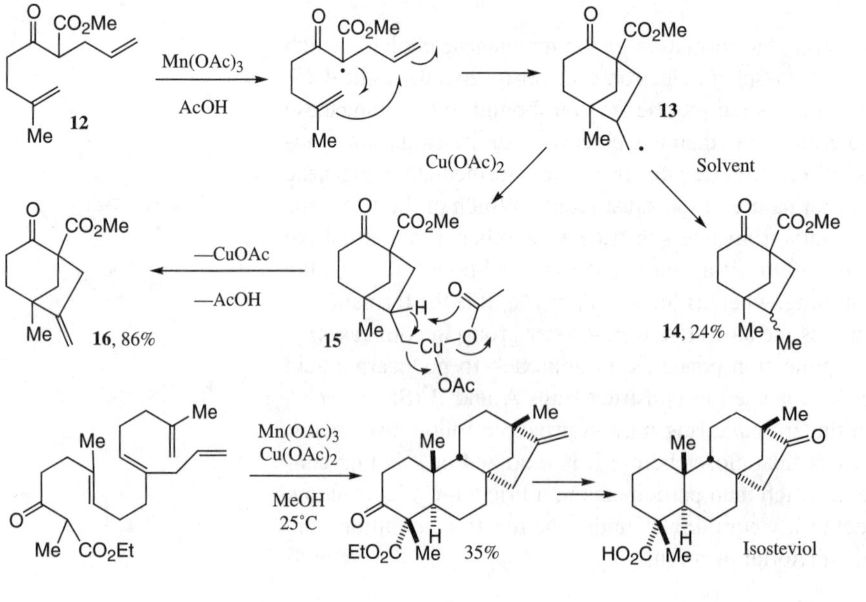

**Scheme 8.7** Mn(III)–Cu(II)-mediated tandem cyclizations–olefin formation

radical **13** becomes too slow to compete with hydrogen abstraction from the solvent. The formation of **15** may be considered as a case of the persistent radical phenomenon (the Fischer–Ingold effect discussed in the preceding chapter), since Cu(II) is a $d^7$ species and therefore paramagnetic.

The Mn(III)/Cu(II) combination is exceedingly powerful, allowing the construction of complex, functionalized structures starting from simple, readily available substrates. The synthesis of isosteviol, for example, could be accomplished by a spectacular, quadruple cyclization sequence displayed in the bottom part of Scheme 8.7 (Snider *et al.* 1998). The control of relative stereochemistry is provided by the chair–chair folding of the chain during the cyclization process.

A further important aspect concerns the formation of the olefin. The collapse of the putative Cu(III) intermediate **15** is subject to steric effects: it is normally the most accessible hydrogen that is removed to form acetic acid. Thus, when a choice is available, the reaction usually leads to the least substituted alkene. In this respect, this step is akin to the Hofmann elimination. This preference is seen in the first example in Scheme 8.8 where two sequential oxidative cyclizations have been performed (Kates *et al.* 1990). At room temperature, only the easily enolizable ketoester reacts to give compound **17**, with the final elimination step occurring away from the cyclohexanone ring to produce the di-substituted distal olefin. At a higher temperature, the enol on the other side of the ketone can now be converted into the corresponding radical and then to the bicyclic derivative **18** with the least substituted terminal olefin. The possibility of oxidizing cleanly simple ketones by working at around 80 °C is further highlighted by the second example involving a 6-*endo* cyclization from a tetralone precursor (Snider and Cole 1995; the 5-*exo* mode of cyclisation is disfavoured in this case because of the resulting strain in the structure). Although both possible olefins are disubstituted, the major product arises from the elimination proceeding through the least congested transition-state.

The intermediate Cu(III) complex can also collapse to give the equivalent of a cation, if the latter is especially stabilized (benzylic, allylic, or tertiary), or react with a neighbouring nucleophile. The latter possibility is illustrated by the formation of cyclopropane **20** in modest yield upon oxidative cyclization of malonate **19a** (Scheme 8.9). Owing to steric encumbrance, the elimination route leading to **21** is

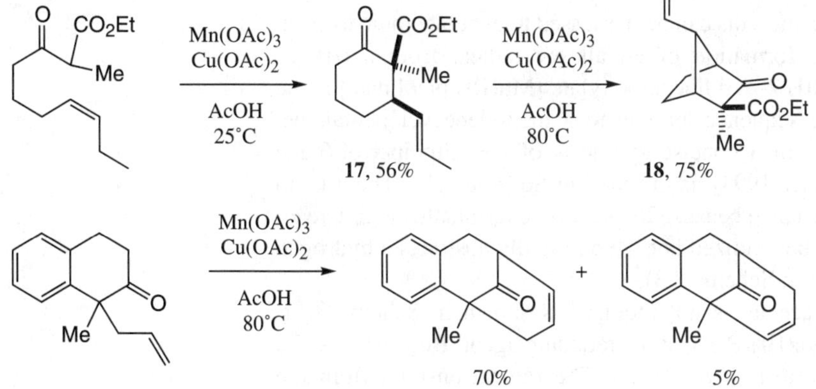

**Scheme 8.8**
Mn(III)–Cu(II) mediated cyclizations-olefin formation from simple ketones

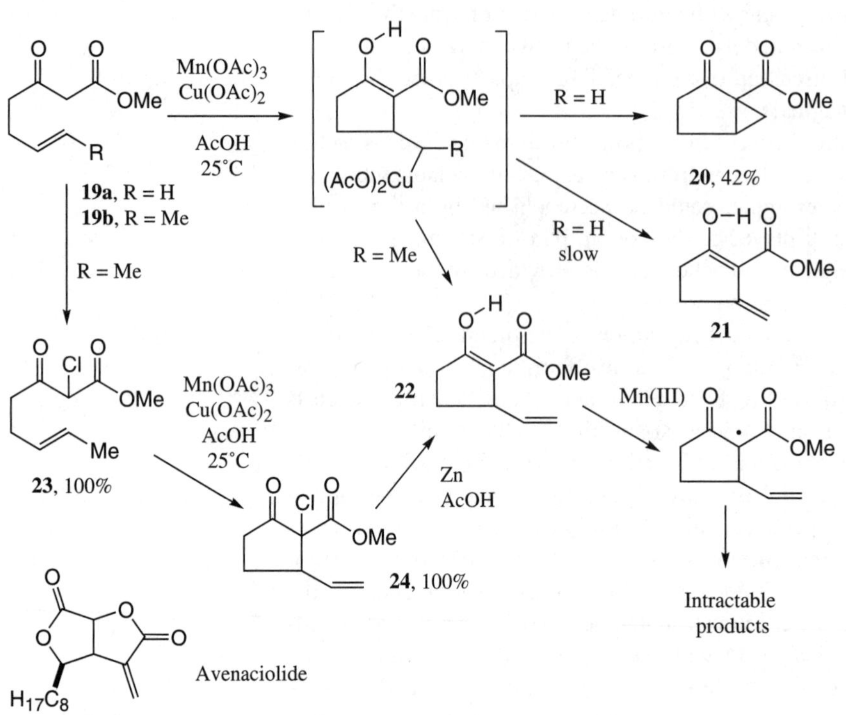

**Scheme 8.9**
Mn(III)–Cu(II)-mediated
cyclization of
chloroketoesters

comparatively slow and substitution can therefore take place. With the crotyl-derived substrate **19b**, elimination proceeds readily to give **22**, but this compound still contains an acidic hydrogen and is oxidized further by Mn(III) into intractable material (Snider 1996). This serious limitation, caused by the presence in the product of a site that is of comparable acidity and oxidizability to the one in the substrate, can be overcome by blocking further enolization with a chlorine atom. Starting from **23**, the radical sequence stops cleanly at **24**; reductive removal of the chlorine atom finally provides **22** in essentially quantitative yield. The chlorine may also be exploited as a leaving group. This ingenious contrivance has been applied in a short synthesis of avenaciolide (Snider and McCarthy 1993*b*).

The oxidizing power of manganic salts can be harnessed to generate radicals from various functional groups. The formation of an alkoxy radical from a tertiary alkynylcyclobutanol using Mn(III) 2-pyridinecarboxylate [Mn(III) picolinate] is the starting point of the interesting sequence leading to a methylenecyclopentanone. This transformation, a key step in a concise synthesis of (−)-silphiperfol-6-ene [Snider *et al.* 1993; Vo and Snider 1994], is outlined in Scheme 8.10. There is no need for cupric acetate in this instance because the sequence terminates with a reactive vinyl radical, which cannot be oxidized but which rapidly abstracts a hydrogen atom from the solvent (path **B–I** in Scheme 8.3).

Another case where cupric acetate is not required is shown in Scheme 8.11. Interestingly, both an oxidant, Mn(III)(Pic)$_3$, and a reducing agent, Bu$_3$SnH, may be present in the same flask! (Iwasawa *et al.* 1993). The radical ensuing from the

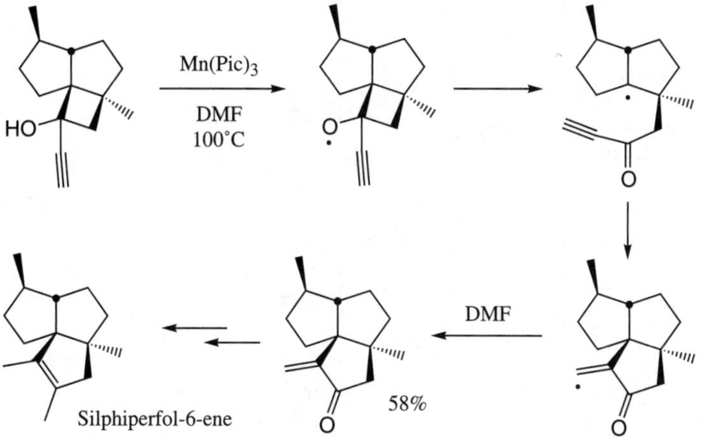

**Scheme 8.10** Ring expansion of alkynyl cyclobutanol derivatives

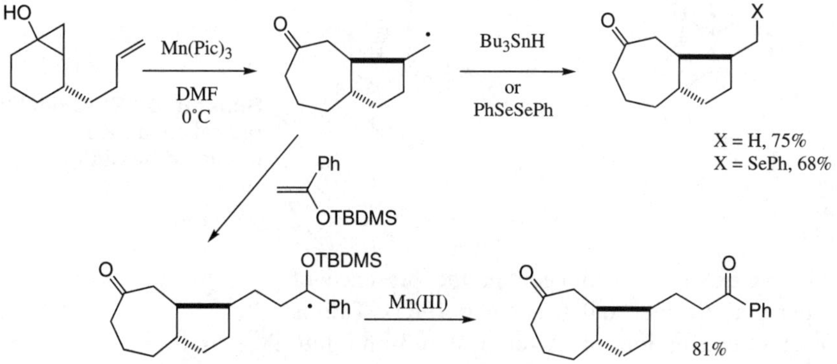

**Scheme 8.11** Sequential reactions from cyclopropanols mediated by Mn(III) picolinate

cyclization can thus be either reduced with the stannane or captured with diphenyl diselenide, depending on which reagent is present in the medium (path **B–I**). Alternatively, an intermolecular C–C bond formation can be achieved using a suitable external olefin such as a silyl enol ether: the adduct radical is oxidized by Mn(III) to the cation, which then rapidly collapses to the ketone with loss of the silicon appendage (path **B–C** in Scheme 8.3).

Copper(II) is not a particularly strong oxidant, but its rapid interception of carbon radicals allows a better regulation of the course of the reaction started by another, more powerful (but slower) oxidizing agent. In some cases, with systems that are especially prone to oxidation, Cu(II) can be used alone. For example, dimedone can be made to add to butadiene as in the first example in Scheme 8.12 (Vinogradov *et al.* 1988). This reaction is mechanistically similar to the aforementioned Mn(III)-mediated addition of acetic acid to styrene (cf. Scheme 8.4). Analogous transformations can be accomplished using Co(II) salts (Iqbal *et al.* 1994).

The fact that cuprous ions can be oxidized back to the cupric level by molecular oxygen or related hydroperoxy radicals has been exploited in various catalytic oxidative transformations. In this manner, $\alpha,\beta$ and $\beta,\gamma$-unsaturated ketones

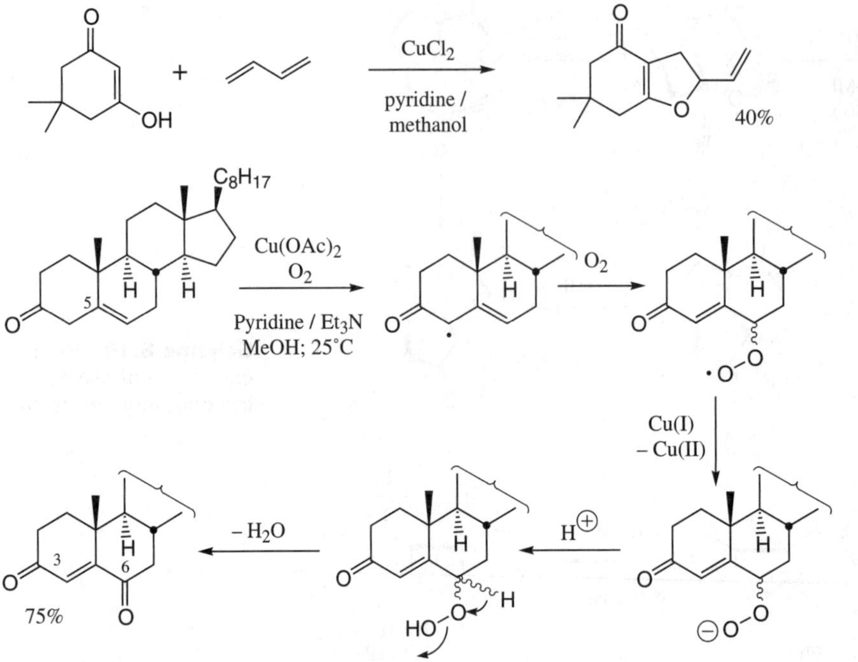

**Scheme 8.12** Oxidative radical sequences mediated by Cu(II)

were converted into enediones upon exposure to oxygen in the presence of catalytic amounts of Cu(OAc)$_2$/pyridine (Volger and Brackman 1965). This is illustrated in Scheme 8.12 by the conversion of $\Delta^5$-cholesten-3-one into $\Delta^4$-cholesten-3,6-dione.

The use of an oxygen-centred radical, usually benzoyloxyl [Ph(C = O)O$^\bullet$] derived from the thermal decomposition of dibenzoyl peroxide or *t*-butyl perbenzoate, to abstract an allylic hydrogen and oxidation of the ensuing allylic radical with cupric ions has been used to prepare allylic benzoates from olefins (Andrus and Lashley 2002). In this transformation, known as the Kharasch–Sosnovsky reaction, the copper is used catalytically and the peroxide is the stoichiometric oxidant. By use of a chiral ligand around the copper, a useful asymmetric induction can sometimes be achieved. This lends support to the intermediacy of a Cu(III) complex, as in Scheme 8.7 above, and an internal delivery of the benzoyloxy group.

In another application, inspired by an earlier work of van Rheenen (1969*a*), cupric-acetate-mediated oxidation with molecular oxygen was used to efficiently degrade the bile acid side chain into a 20-keto-pregnane (Barton *et al.* 1989). As depicted in Scheme 8.13, hydroperoxidation of the readily enolizable $\alpha$-ketoester **25**, obtained from the bile acid precursor in two steps, follows the same route as that for $\Delta^5$-cholesten-3-one, except that the corresponding intermediate hydroperoxide closes onto the ester group, leading presumably to the unstable derivative **26**. This intermediate collapses rapidly into aldehyde **27** and further Cu(II)/O$_2$ hydroperoxidation ultimately leads to the desired 20-ketone **28**, which is immune to this mildly

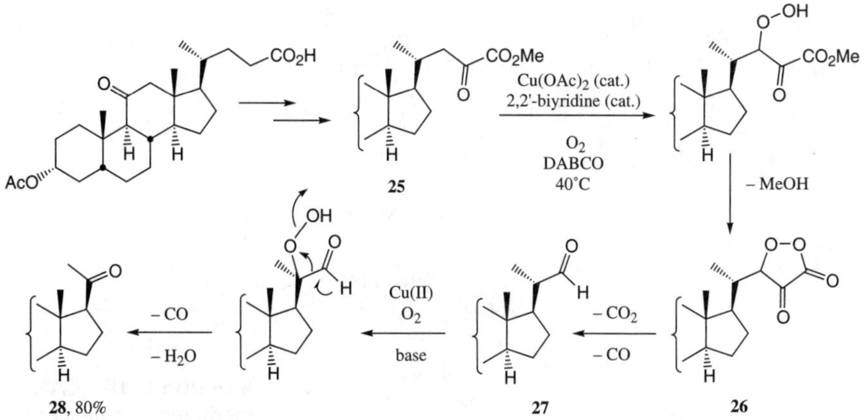

**Scheme 8.13** Degradation of the bile acid side chain

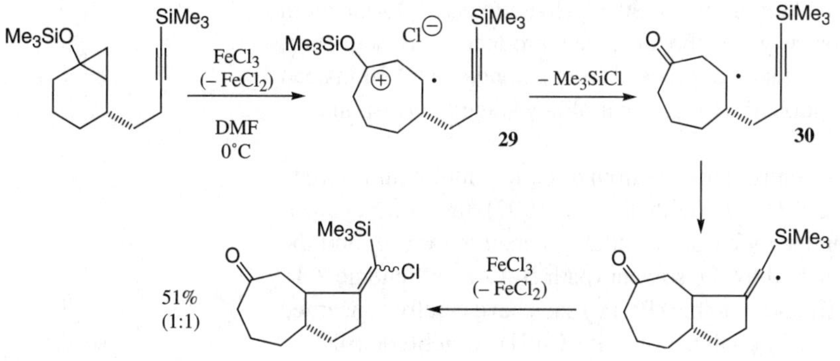

**Scheme 8.14** Fe(III)-mediated oxidative cleavage of silylated cyclopropylcarbinols

oxidizing system. All three extra carbons can thus be removed in one operation. By using CuCl$_2$ in chloroform, the degradation can be stopped cleanly at the level of aldehyde **27**. This gentler oxidizing combination is nevertheless capable of usefully cleaving enamines (van Rheenen 1969*b*).

## 8.4   Oxidations involving Fe(III), Ce(IV), Ag(II), and Pb(IV)

The use of other metal salts such as Fe(III), Ce(IV), Ag(II), and Pb(IV), for the oxidative generation of radical has remained relatively limited in comparison with the Mn(III)/Cu(II) combination. The Fe(III) system certainly deserves a more thorough study, in view of cheapness and general non-toxicity of iron salts (Citterio *et al.* 1989). The transformation shown in Scheme 8.14 (Booker-Milburn 1992) uses ferric chloride as the oxidant and is akin to the one in Scheme 8.11. Electron

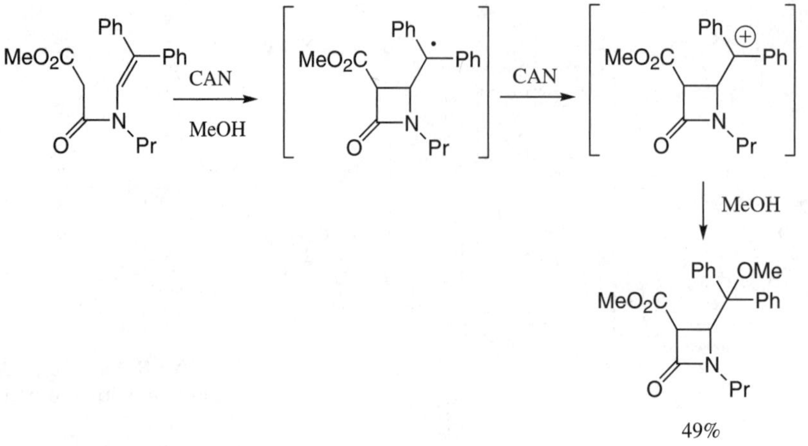

**Scheme 8.15** CAN-mediated synthesis of a β-lactam

transfer from the electron-rich bond of the trimethylsiloxycyclopropane produces radical cation **29**, a synthetic equivalent of radical **30**. Cyclization and chlorine atom transfer from ferric chloride finally gives the observed product. This sequence, which for convenience and clarity is taken to proceed from ketone radical **30** instead of **29** (loss of the trimethysilyl group can in fact occur at any stage), corresponds to path **B–H** in Scheme 8.3.

Reactions with Ce(IV) salts centre mostly around ceric ammonium nitrate (CAN). In the example in Scheme 8.15 (D'Annibale *et al.* 1997), the 4-*exo* cyclization to the β-lactam ring gives an easily oxidized, doubly benzylic radical, and the resulting cation is ultimately quenched by the solvent (path **B–C–E** in Scheme 8.3).

Oxidations with argentic (Ag(II) and plumbic (Pb(IV)) salts have chiefly concerned carboxylic acids. Both can be used in conjunction with Cu(II), which controls the termination part of the sequence. Various possibilities are displayed in Schemes 8.16. In the first example (Doll 1999), the acyl radical arising from the decarboxylation closes onto the pyridine ring. Such Ag(II)-mediated decarboxylative additions to activated pyridines and pyridinium salts have been studied extensively by Minisci and his group (Minisci 1975; Minisci *et al.* 1983). Decarboxylative additions to quinones using similar conditions are also known (Goldman *et al.* 1974). There is no need for cupric ions since the oxidative rearomatization of the intermediate adduct radical occurs easily. This is not the case in the second transformation, where the addition of cupric acetate is needed to convert the secondary carbon radical cleanly into an olefin (Goldman *et al.* 1974). In both examples, cheap ammonium persulfate is the stoichiometric oxidant and both the silver and cupric sulfates are used in catalytic amounts. The last example is an application of a lead tetraacetate oxidation in combination with cupric acetate (Sheldon and Kochi 1972). If the cupric acetate is replaced by lithium chloride, lithium bromide, or iodine, then a decarboxylative halogenation is observed, leading respectively to the nor-chloride, -bromide, or -iodide. Thus, depending on the additive, either a decarboxylative elimination or a halodecarboxylation may be accomplished. The strong oxidizing power of Ag(II) and Pb(IV) limits however the use of these reactions to robust substrates.

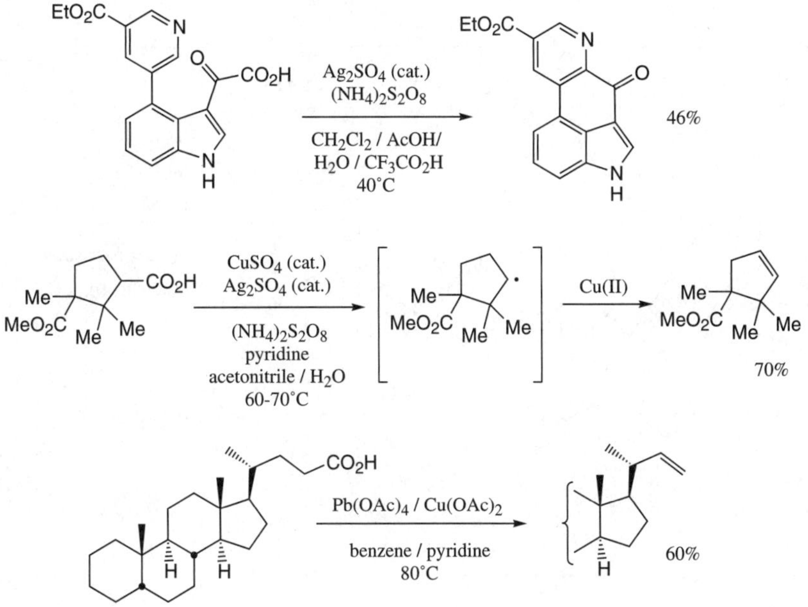

**Scheme 8.16** Ag(II)- and Pb(IV)-induced oxidative decarbovylations

## 8.5   Formation of radicals by reduction with transition metal salts: general reaction scheme

The generation of radicals by electron transfer from a low-valent transition metal is perhaps of even greater synthetic potential than by oxidation of an electron-rich substrate, because a wider variety of precursors, reducing systems, and radical traps may be used. In some cases, the reductant is sufficiently mild to be compatible with the presence of an oxidizing agent and one or more oxidation steps can be incorporated into the sequence. Such combinations are especially powerful and translate into a vast number of synthetically useful transformations. A general reaction manifold is displayed in Scheme 8.17, outlining the various pathways open to the radical species. R–X represents any easily reducible functional group such as a halide, a carbonyl derivative, an enone, an epoxide, a hydroperoxide, a diazonium salt, and a nitrogen or sulfur containing functionality (imine, iminium, chloroamine, nitro, oxime, hydroxylamine, sulfone, sulfonium, etc.).

Electron transfer to R–X results in the formation a radical anion, which rapidly collapses into a radical R$^\bullet$ by expelling an anion (X$^-$). If the reducing agent is powerful or if R$^\bullet$ is especially easily reduced, a second electron transfer may occur (path **J**) leading to an anion equivalent (R$^-$) **31**, which then undergoes the usual ionic reactions associated with anions (path **K**). Many organometallic reagents are obtained in this way (e.g. RLi or RMgX from RX and lithium or magnesium metal; organochromium derivatives by reaction of R–X and a chromous salt, etc.). It has been shown, for example, that the formation of Grignard reagents using highly reactive, vaporized magnesium can be inhibited by addition of typical radical inhibitors (Péralez *et al.* 1994). Other well-known reactions such as the Birch and related

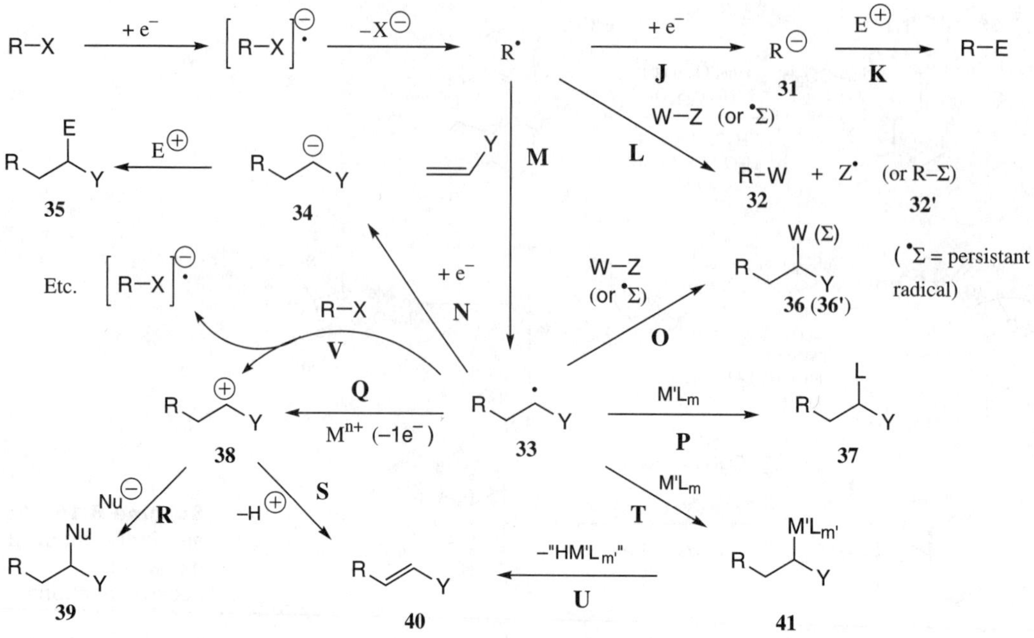

**Scheme 8.17**   Reduction manifold

dissolving metal reductions (e.g. the Bouveault–Blanc reaction) also proceed by a similar electron-transfer pathway. All of these classical reactions are of enormous importance in organic chemistry but will not be further discussed here because the radical intermediate is often difficult to capture under the usual reaction conditions.

It is possible to avoid the premature reduction step by using various combinations of substrate/reductant/experimental conditions. The intermediate radical $R^\bullet$ may be intercepted through a homolytic substitution reaction using various W–Z type radicophiles (e.g. a thiol) to give compound R–Y (**32**) and radical $Z^\bullet$, which may in turn undergo reduction or dimerization. Radical $R^\bullet$ may be intercepted by a persistent radical $\Sigma^\bullet$ to give **32′** (path **L**; one classical example is Gomberg's experiment, outlined in Scheme 1.1, where the triphenylmethyl radical, generated by zinc reduction of chlorotriphenylmethane is trapped either by oxygen or by another triphenylmethyl radical). Addition to an olefin (or fragmentation, translocation, or any combination thereof) is also possible and leads to a new radical **33** (path **M**), whose fate depends on the substituent Y, the reagents present, and the exact experimental conditions. Further reduction generates an anionic species **34** that can be quenched by a proton or by a more elaborate electrophile $E^+$ (path **N**). Reaction of $R^\bullet$ with a radicophile W–Z or with a persistent radical $\Sigma^\bullet$ provides **36** or **36′** (path **O**). It is also possible to form **37** by incorporating a ligand-transfer step from the initial transition metal M or from an auxiliary complex $M'L_n$ (path **P**). Finally, if substituent Y is an electron-releasing group, an oxidation step may intervene, for example, through the agency of a mild oxidizing reagent compatible with the initial reducing system (e.g. Cu(I)/Cu(II) or Fe(III)/Cu(II) redox systems), to produce a cation equivalent **38** (path **Q**) which can be quenched by a nucleophile to give **39** (path **R**) or loose a proton to furnish an olefin **40** (path **S**). The same olefin may be reached by a reductive elimination of a metallic complex **41** (path **T–U**). One interesting variant is an oxidation of the radical by electron transfer to the substrate R–X to give the same cationic species **38** and the initial radical anion which in turn regenerates radical R• (path **V**). In the latter case, a chain reaction (**M** → **V**→ **M** → **V** etc.) may operate in principle, whereby the reducing agent acts simply as the initiator. In practice, such redox chains are often relatively inefficient, requiring extensive initiation and/or the presence of a relay.

The intricacy of the manifold in Scheme 8.17, which overlaps to a certain extent with the oxidation manifold displayed in Scheme 8.3 because of the possible co-existence of reducing and oxidizing species, reflects the richness and versatility of this approach for generating and harnessing radicals for organic synthesis. Despite the apparent daunting complexity, it is possible to navigate through the numerous competing reaction pathways, favouring one route and disfavouring the other by a judicious choice of the various parameters. The examples detailed below and arranged according to the reducing system employed will illustrate some of the synthetic applications, without really doing justice to the vast amount of work accomplished in this area. We shall thus examine reactions based on chromous salts, Sm(II) iodide, Ti(III), Fe(II), Cu(I), and Ru(I) salts and complexes, as well as combinations of a low-valent metal salt reductant and Cu(II) oxidant. Dissolving metal reductions, involving especially alkali metals, zinc, magnesium, nickel, and reactions mediated by non-metallic reducing agents will be briefly discussed. Finally, a short presentation of electrochemical and photochemical methods and $S_{RN}1$ processes will complete this chapter.

## 8.6   Reductions with Cr(II) and Ti(III) salts

Chromium(II) and titanium(III) salts are strong reducing agents, capable of generating radicals by electron transfer from a variety of precursors. In one early and elegant application (Scheme 8.18), the problem of selectively removing the bromine from the $9\alpha$-position of steroids, while keeping the important $11\beta$-hydroxy group, was solved by a combination of chromous acetate and a thiol (Barton *et al.* 1966). Electron transfer from chromous acetate to the C–Br anti-bonding orbital gives a radical anion intermediate, which rapidly collapses into a radical **42** by expelling a bromide anion. The radical is quenched by the thiol to provide the desired reduced steroid **43** in high yield. The route we have followed corresponds to path **L** in Scheme 8.17, where the W–Z reagent represents the thiol. In the absence of thiol, the products observed reflect the behaviour of the organochromium anion equivalent **44**, produced by further reduction of radical **42** : $\beta$–elimination of hydroxide to form alkene **45** and ring closure to cyclopropylsteroid **47** (i.e. path **J–K**). The latter may also arise by a reversible radical cyclization followed by reduction to give chromium enolate **46** and protonation. The relative yield of **45** and **47** is variable and depends strongly on the exact experimental conditions.

The radical generated from a halide by reduction with a chromous salt may be captured by an internal olefin as in the example in Scheme 8.19 (Lübbers and

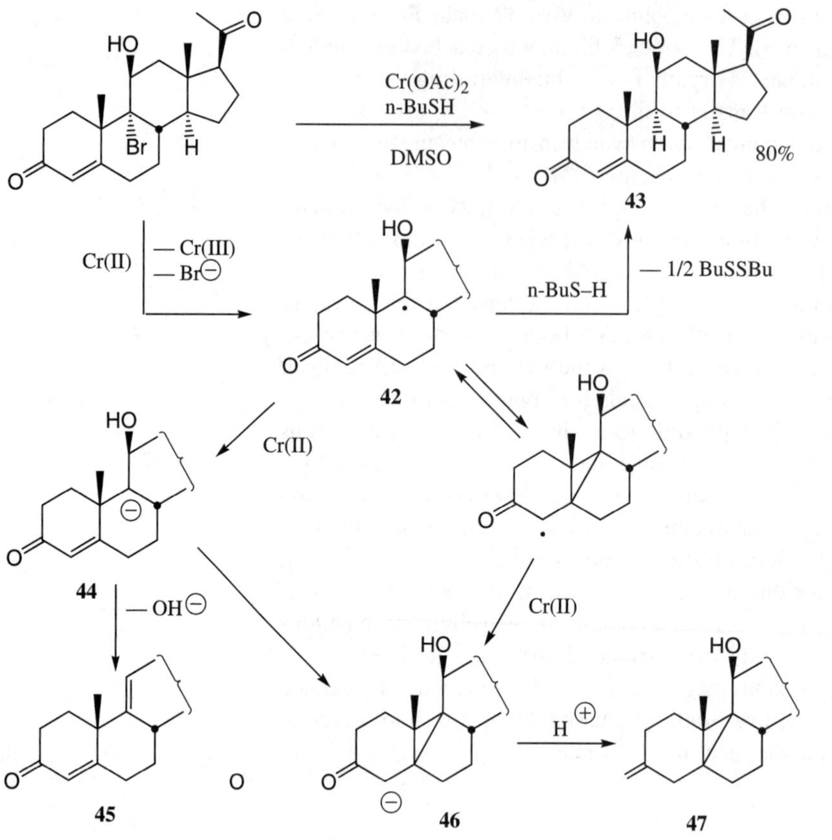

**Scheme 8.18**
Reductive debromination of a corticosteroid with a Cr(II) acetate/thiol combination

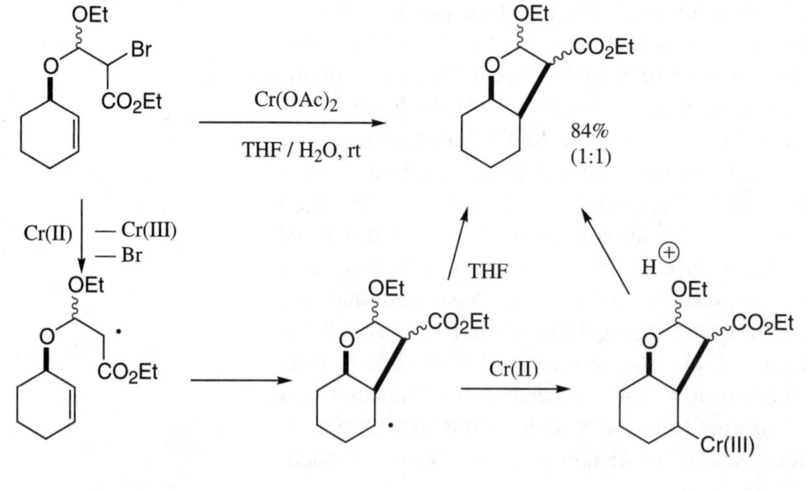

**Scheme 8.19**
Cr(II)-acetate-mediated reductive cyclization

Schäfer 1990). Since the reaction was performed in THF, a reasonably good hydrogen atom donor, the last step may involve hydrogen abstraction from the solvent rather than reduction to an organochromium intermediate followed by protonolysis. Other, similar reactions have been described (Takai *et al.* 1989; Hackmann and Schäfer 1993), but the potential of chromous salts in synthetic radical chemistry has certainly not been exploited to the extent it deserves.

Titanium(III) salts and complexes have been more extensively studied. Early applications concerned mainly the reduction of hydroperoxides, chloramines, and diazonium salts with aqueous titanium trichloride. The last process is a variation on the Meerwein arylation reaction, normally performed with copper salts (see below). In the first transformation in Scheme 8.20 (Citterio and Vismara 1980; Citterio 1984), the radical generated by extrusion of nitrogen from the diazonium chloride is captured with methylvinyl ketone, and the resulting adduct is in turn intercepted to give a titanium enolate which suffers a rapid protonolysis in the acidic medium. The path followed corresponds to **M–N** in Scheme 8.17 with $E^+$ representing a proton.

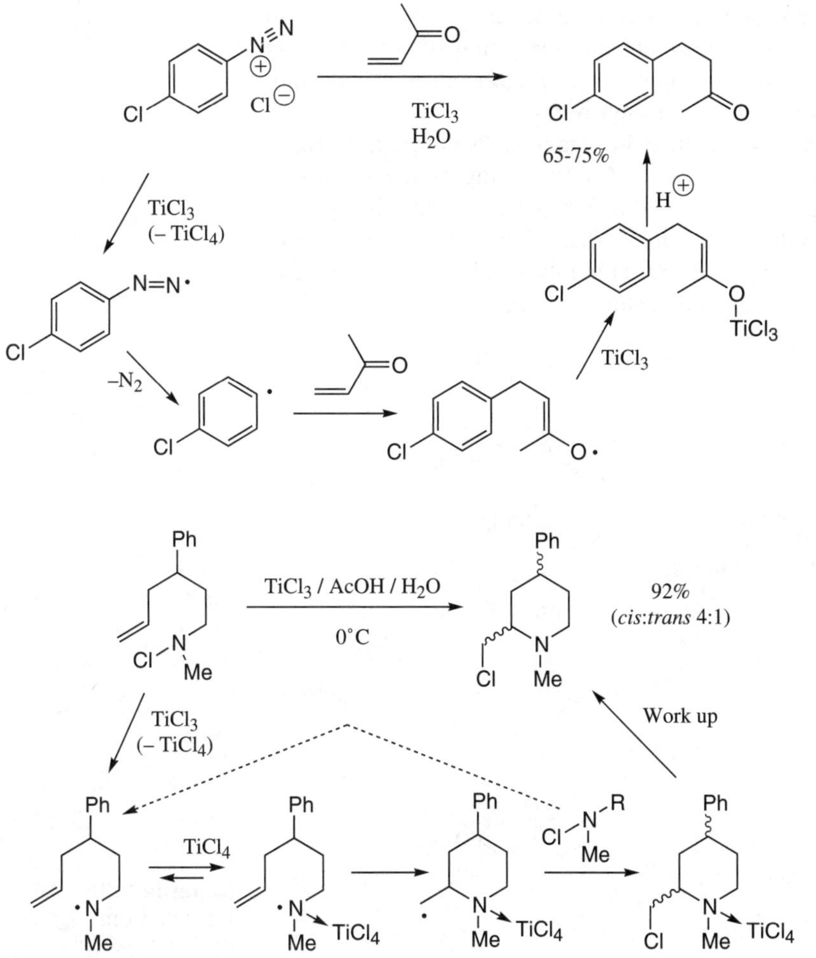

**Scheme 8.20** Radical generation from diazonium salts and N-chloroamines by reduction with Ti(III) chloride

Electron transfer to diazonium salts constitutes perhaps the most practical, tin-free source of the highly energetic aromatic radicals (Galli 1988).

*N*-Chloroamines and *N*-chloroamides are other nitrogen functions that can be reduced with Ti(III) salts leading to the formation of the respective aminyl and amidyl radicals (Stella 1983). Cuprous and ferrous salts have also been used in this regard. In the case of aminyls, complexation with the titanium metal enhances their reactivity towards addition to an olefin. The second reaction in Scheme 8.20 illustrates a 6-*exo*-cyclization of an aminyl radical (Stella *et al.* 1977). Transfer of the chlorine is shown for convenience to take place from the starting chloroamine making the overall process similar to a Kharasch-type chlorine atom transfer. The actual mechanism may be more complicated than the one presented due to complexation of the various nitrogen-containing species with titanium salts.

Hydroperoxides are also easily reduced with titanium trichloride to give alkoxy radicals, as in the second example outlined in Scheme 8.21 (Karim and Sampson 1988). In this case, the hydroperoxide precursor is produced by reversible addition of hydrogen peroxide to cyclohexanone in methanol. The corresponding alkoxy radical undergoes fast ring opening to a primary carbon radical, in turn captured by the external electrophilic olefin. The sequence is once again terminated by reduction to the titanium enolate and protonolysis. It is possible that the rupture of the weak O–O bond is preceded by the formation of a titanium peroxide intermediate, the electron transfer occurring as a consequence in an inner-sphere fashion.

Organotitanium complexes have been applied in the reductive opening of epoxides. Using dicyclopentadienyltitanium chloride (Cp$_2$TiCl), Nugent and Rajanbabu (1988) performed the reductive cyclization outlined in Scheme 8.22. The epoxide rings cleaves regioselectively to the more stable tertiary radical **48**, and the subsequent *endo*-selective ring closure (*endo* : *exo* 9 : 1) is followed by further reduction with Cp$_2$TiCl. The resulting bicyclic organotitanium (IV) derivative may be cleaved

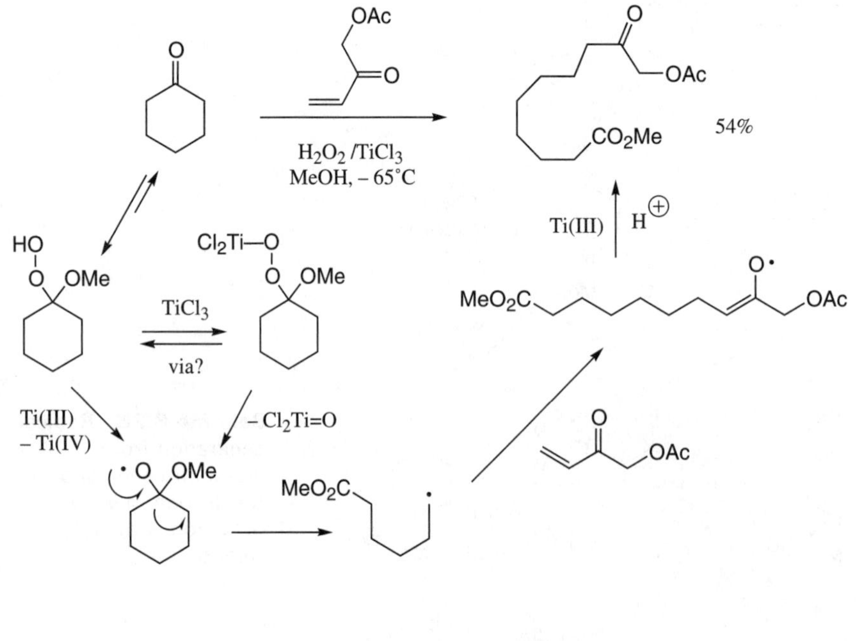

**Scheme 8.21** Alkoxy radicals from hydroperoxides

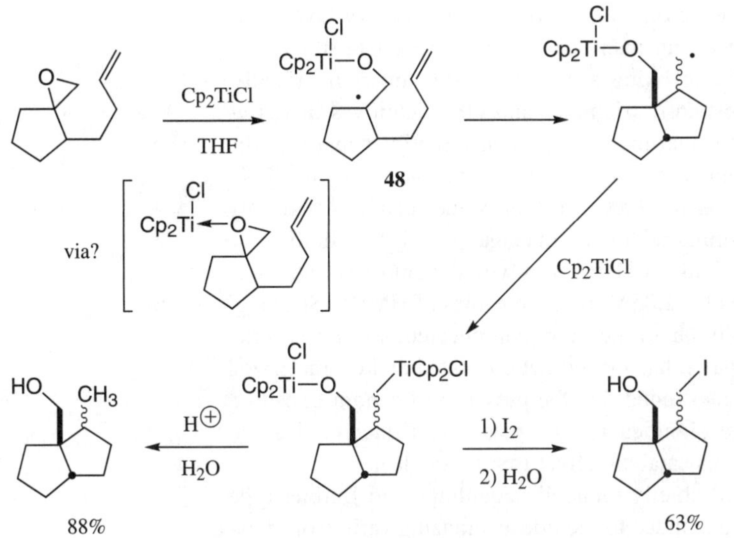

**Scheme 8.22**
Regioselective reductive opening of epoxides and cyclopentane ring formation

with acid or converted into the iodide by exposure to iodine. In this case, too, it is quite likely that the electron transfer step is an inner-sphere process, taking place within a complex around the titanium atom and implicating the oxirane oxygen.

Interception of the intermediate radical (e.g. **48**) with a hydrogen atom donor such as 1,4-cyclohexadiene allows an overall reductive ring opening of the epoxide (Rajanbabu and Nugent 1994). The regiochemistry is often opposite to that of an 'ionic' hydride reduction. The reductive ring opening can be made catalytic in Cp$_2$TiCl by using zinc or manganese as the stoichiometric reductants (Gansäuer *et al.* 1998; Gansäuer and Bluhm 2000; Gansäuer and Pierobon 2000). Recently, ring closure of the incipient radical intermediate onto an aldehyde or a ketone has also been reported, but the possibility that this step may be occurring after further reduction to the titanate anion level has not been ruled out completely (Fernández-Mateos *et al.* 1999).

## 8.7   Reductions with Sm(II) iodide

Perhaps the most popular and synthetically versatile one-electron reducing agent is samarium diiodide, SmI$_2$. Its reducing properties were first explored by the group of Kagan (Girard *et al.* 1980), who also reported a convenient and mild preparation based on the reaction of metallic samarium with iodine, diiodomethane, or 1,2-diiodoethane. SmI$_2$ is normally handled in solution, and blue solutions of SmI$_2$ in THF are now commercially available. The reduction potential of the Sm(III)/Sm(II) couple in THF has recently been measured: $E° = -1.41$ V (Enemaerke *et al.* 1999). Samarium iodide is thus a powerful reducing agent, capable of interacting usefully with a large variety of functional groups. Interestingly, electron transfer to halides (e.g. benzyl bromide) appears to be an outer sphere process, whereas with ketones (e.g. acetophenone) it occurs in an inner-sphere fashion (Enemaerke *et al.* 1999).

The latter indicates that prior complexation with the oxygen atom (very favourable with lanthanides) is important. Indeed, in many cases, the stereochemistry of the products can only be understood by invoking a complex with samarium. Finally, addition of HMPA (hexamethylphosphoric triamide) and other additives increases significantly the reduction power of samarium diiodide and sometimes modifies the mode of action, that is, whether inner or outer sphere (Enemaerke *et al.* 2000; Kuhlman and Flowers 2000; Miller *et al.* 2000); in fact, some reactions will only proceed in the presence of this or similar additives (Inanaga *et al.* 1987). In the case of HMPA, the reduction potential of $SmI_2$ in THF varies with the amount of additive, reaching a maximum negative value of $-2.05$ V for 4 molecules of HMPA (Shabangi *et al.* 1998; the situation may actually be a little more complicated, see: Enemaerke *et al.* 2000). A complex, $[SmI_2(hmpa)_4]$, has indeed been isolated and characterized (Hou and Wakatsuki 1994). It was also found that the presence of certain transition metal salts, such as $NiI_2$, enhanced significantly the reactivity (Machrouhi *et al.* 1996), but the mechanistic basis of this catalytic effect is still not clear.

Initial work with $SmI_2$ concerned chiefly pinacolic couplings and Barbier-type additions, but the field dramatically expanded to include an amazing variety of transformations (Kagan and Namy 1986; Molander 1992, 1994; Molander and Harris 1996*a*, 1998*a*). The rich and unique chemistry of this reagent is due to the exquisite combinations and sequences of radical and anionic transformations that can be implemented, often with spectacular results.

Pinacolic couplings mediated by $SmI_2$ exhibit a remarkable efficiency. This is probably due to the rapid and clean formation of a samarium-complexed ketyl radical anion from an aldehyde or a ketone and its subsequent coupling with another such species, presumably within a further complex. Aromatic aldehydes and ketones react within minutes, aliphatic aldehydes within hours, and aliphatic ketones require a day or so for complete reaction. The intramolecular version is especially useful for constructing rings containing a vicinal diol group, as shown by the example in Scheme 8. 23. In this case, one samarium atom straddles the two oxygens of the carbonyl groups resulting in the formation of a diol with a *cis*-stereochemistry (Chiara *et al.* 1991).

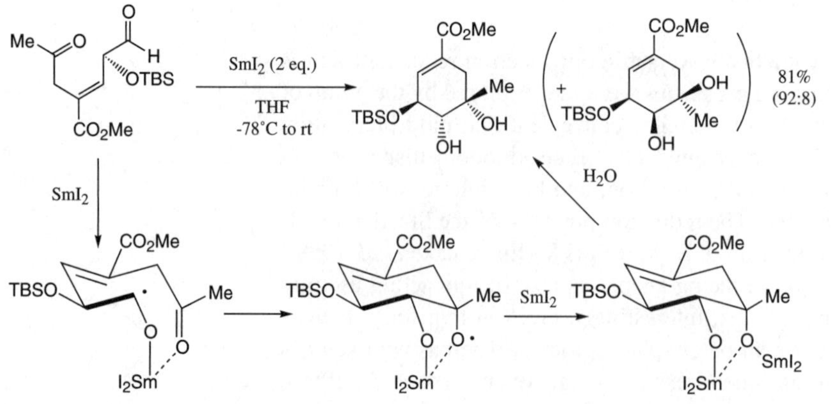

**Scheme 8.23**
Stereoselective
Sm(II)-mediated
6-membered ring
formation

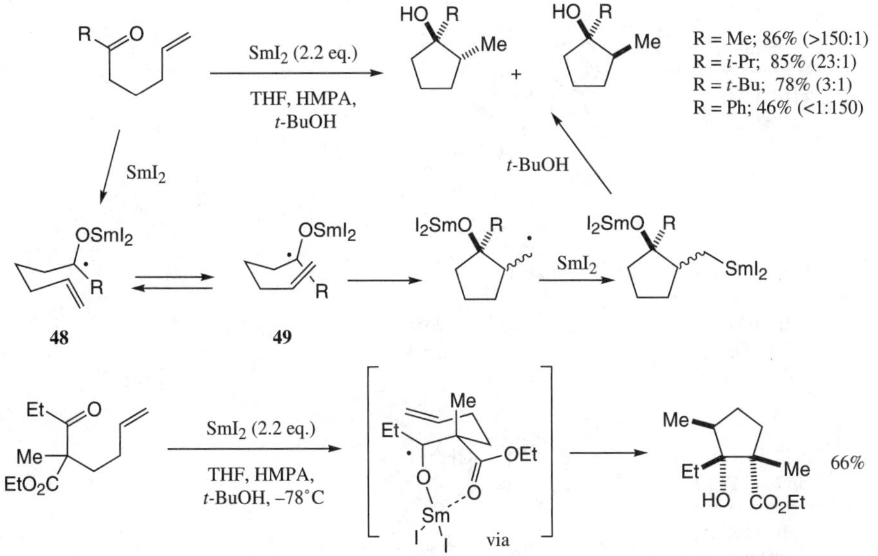

**Scheme 8.24**
Substituent effect on
stereoselectivity of
Sm(II)-mediated 5-mem-
bered ring formation

The ketyl radical obtained by one-electron transfer to an aldehyde or a ketone is relatively persistent in the medium, since the transfer of a second electron would lead to a high-energy dianion. It may thus be easily captured with an internal or external olefin. In the case of a simple cyclization, the stereochemistry depends on the substituents and on the possibility of chelation, which introduces an element of rigidity into the transition-state. This is illustrated in Scheme 8.24 for the reductive ring closure of a simple ketone (Molander and McKie 1992) and of a β-ketoester (Molander and Kenny 1987, 1989). In the first case, the olefin may face either the R (as in **48**) or the ketyl group (as in **49**). There is a preference for the former, presumably because of electronic repulsion between the lone pairs of the oxygen and the π-system of the olefin; this preference diminishes as the size of R increases and is reversed when R is a phenyl group because, in addition to steric encumbrance, there is now a repulsion between the two π-systems. With the ketoester, chelation and the preference for an anti-disposition of the olefin with respect to the oxygen of the ketyl dictate the shape of the transition state. The last step involves reduction of the primary carbon radical to the organosamarium intermediate, which then undergoes protonolysis by *t*-butanol. The electron transfer from $SmI_2$ to primary radicals appears to be an outer-sphere process (Shabangi *et al.* 1999) and its rate constant is of the order of $5 \times 10^5$–$7 \times 10^6$ $M^{-1}$ $s^{-1}$ depending on the amount of HMPA present (Hagesawa and Curran 1993). Not unexpectedly, electron transfer to benzylic radicals is faster: $k = 5.3 \times 10^7$ $M^{-1}$ $s^{-1}$ (Skene *et al.* 1996). In all of these reactions, the THF or HMPA ligands around the samarium atom have been omitted for clarity.

Structures of much greater complexity may be obtained, as in the first transformation in Scheme 8.25. This highly diastereoselective ring closure (diastereomeric ratio 24 : 1; only the major isomer is drawn) is a key step in a beautiful total synthesis of grayanotoxin III, which also features two other $SmI_2$-promoted reactions: a

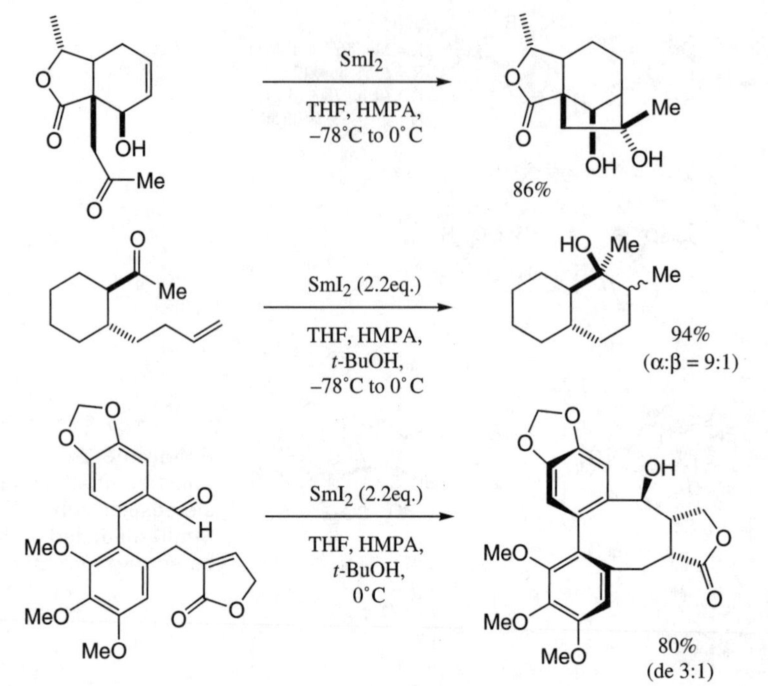

**Scheme 8.25**
Examples of Sm(II)-
mediated ring closures

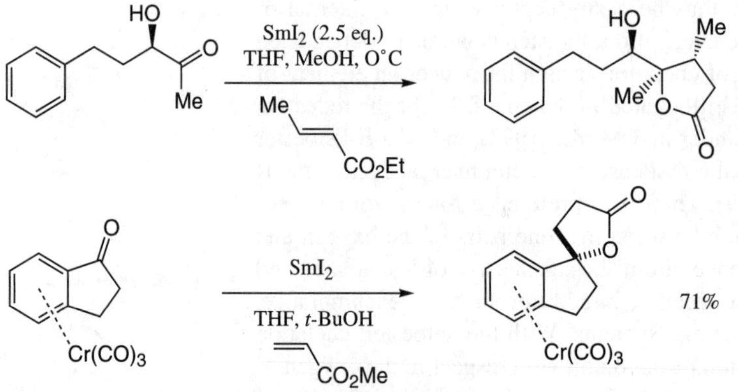

**Scheme 8.26**  Sm(II)-
mediated intermolecular
additions and lactone
formation

second ketyl–olefin cyclization and an intramolecular pinacolic coupling leading to
a 7-membered ring (Kan *et al.* 1994). The construction of a 6-membered ring by a
*6-exo* ring closure is illustrated by the second example (Molander and Mckie 1995).
Abstraction of the allylic hydrogen, often observed with alkyl radicals, is not normally a problem with ketyls in view of their relative stability and persistence. The
formation of an 8-membered ring spanning a biphenyl system, by addition of a ketyl
radical to an activated olefin, has recently been exploited in an elegant total synthesis of (−)-steganone (Monovich *et al.* 2000).

Examples of intermolecular additions to alkenes, dienes, and alkynes are less
numerous but can also be quite efficient, especially if the unsaturation is unhindered

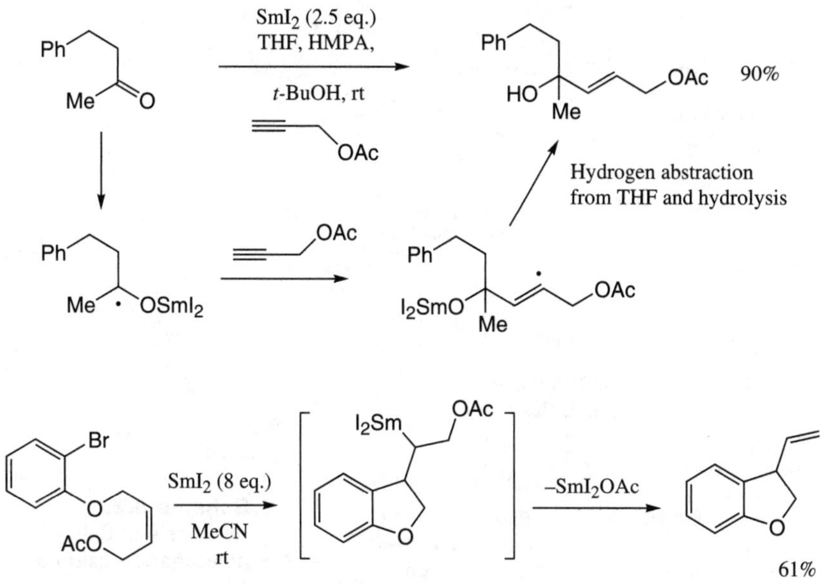

**Scheme 8.27**
Contrasting additions to alkenyl and alkynyl acetates

(preferably terminal) and activated by a suitable group. In the first example in Scheme 8.26, a complex structure is obtained with remarkable stereochemical control—99 : 1, only the major isomer is shown—upon addition of the ketyl radical anion to ethyl crotonate (Kawatsura *et al.* 1994). Again, chelation with the hydroxy group is responsible for the high diastereoselectivity. The THF solution of SmI$_2$ is added to a mixture of the ketone and olefin in methanol; the diol initially produced closes onto the ester group to give the lactone. The second example represents a relatively rare instance of a radical addition reaction starting with an organometallic substrate (Merlic and Walsh 1998). The corresponding ketyl radical adds to methyl acrylate from the least hindered β-face, opposite to the chromium tricarbonyl group. Lactonization is also observed in this case.

When alkynes traps are used, as in the first example in Scheme 8.27 (Inanaga *et al.* 1991*b*), the highly reactive vinylic radical arising from the addition abstracts a hydrogen from the solvent before undergoing reduction with SmI$_2$. This is why no elimination of the acetoxy group to give an allene is observed. Such eliminations readily occur with similar additions to alkenes, where reduction of the ensuing secondary radical is faster than hydrogen abstraction from the solvent, as demonstrated by the last transformation (Inanaga *et al.* 1991*a*). The initial radical is generated by electron transfer to the aromatic bromide then β-elimination of the acetoxy group readily occurs from the organosamarium intermediate.

The possibility of capturing the intermediate carbon radical by an internal alkyne has been exploited to link two carbohydrate units via the Stork silicon tether. The elegant construction of an α-C-disaccharide displayed in Scheme 8.28 (Skrydstrup *et al.* 1997) also illustrates the use of a pyridylsulfone as the substrate. After the radical cyclization, the silicon bridge is removed with fluoride anion; catalytic reduction/debenzylation and peracetylation finally deliver the desired disaccharide. Pyridyl

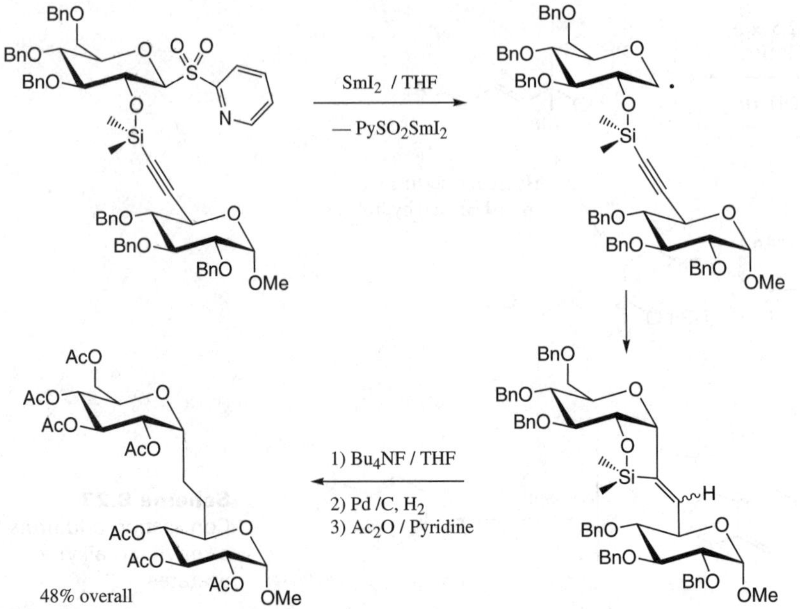

**Scheme 8.28**
Synthesis of C–C disaccharide using a silicon tether

sulfones are much more easily reduced in comparison to simple phenylsulfones, and this enhanced reactivity often obviates the need for HMPA.

The oragnosamarium species behaves in fact as the synthetic equivalent of an anion in a way typical of a classical organometallic reagent. The organosamarium intermediate exhibits roughly the reactivity of a Grignard reagent: it is thus capable of undergoing reactions with a great variety of electrophiles and the overall route corresponds to the **M–N** sequence in Scheme 8.17. The transformations deployed in Schemes 8.29 and 8.30 give an idea of the synthetic scope. The former is centred around o-iodophenol (Curran and Totleben 1992), whereas the latter involves a simple δ,ε-enone (Molander and McKie 1992). In these experiments, the substrate is added to the solution of SmI$_2$ followed, after a few minutes, by the electrophile. The hydrolysis step, where applicable, has been omitted for clarity.

The irreversible and relatively fast reduction step may be used to drive forward an otherwise unfavourable equilibrium. One such case is detailed in Scheme 8.31 (Katritzky et al. 1999), where electron transfer to an α-aminobenzotriazole results in a transient radical anion which expels a benzotriazolyl anion to give an α-aminoalkyl radical. This fairly stabilized tertiary radical adds reversibly to the internal double bond, with the equilibrium apparently lying towards the open form. However, the primary, ring-closed radical is easier to reduce to the level of the anion by an extra equivalent of SmI$_2$ and this irreversible step pushes the process in the desired direction. Finally, the organosamarium species is quenched with methylethyl ketone.

The possibility of combining radical and ionic reactions in various order allows the conception of some spectacular cascades. For instance, starting with the acid chloride derived from O-allyl salicylic acid, the sequence leads to the formation of a

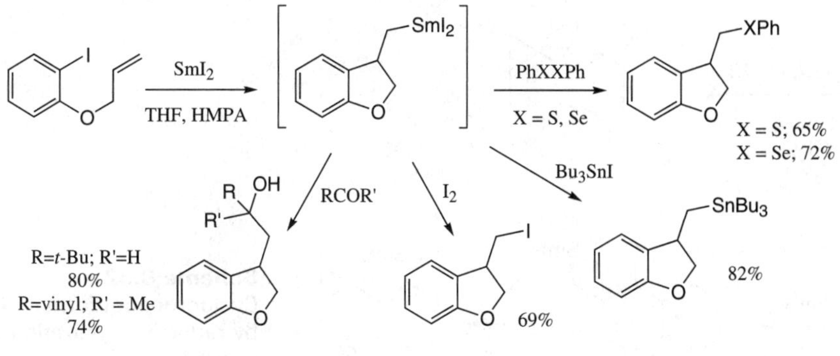

**Scheme 8.29** Reaction manifold of organosamarium cyclization product

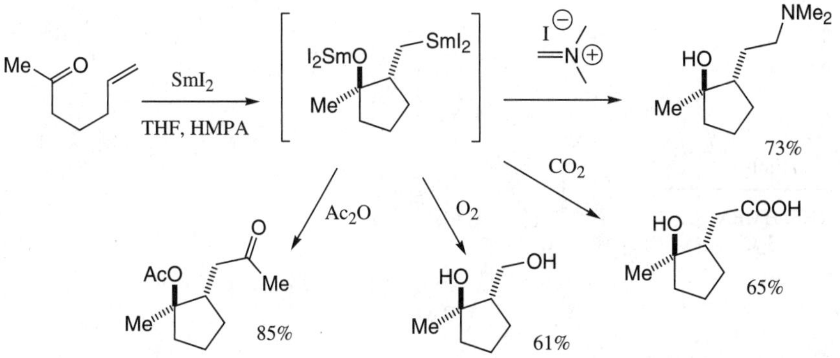

**Scheme 8.30** Further transformations of organosamarium product

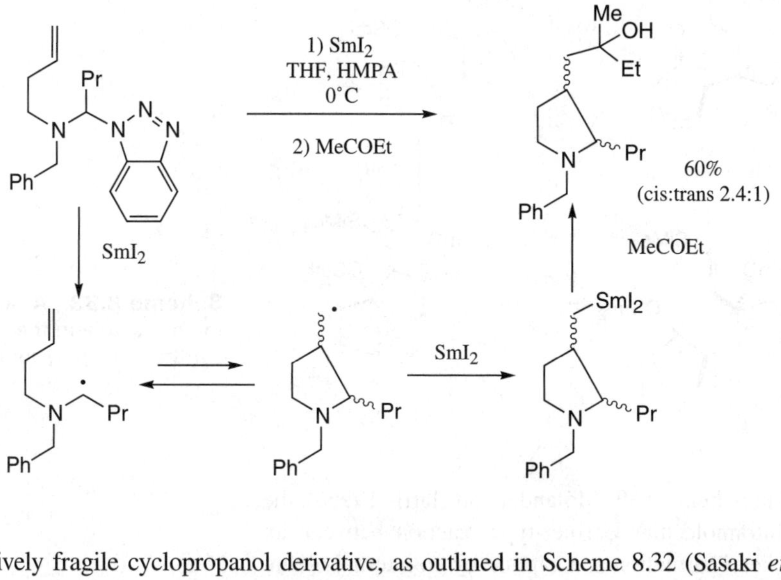

**Scheme 8.31** Reduction of aminoalkyl benzotriazoles with Sm(II) iodide

relatively fragile cyclopropanol derivative, as outlined in Scheme 8.32 (Sasaki *et al.* 1988). The electrophile in this early example is the aromatic ketone, produced upon 6-*exo* cyclization of the acyl radical. The reaction is complete within a minute; aliphatic chlorides, in contrast, are reduced relatively slowly with $SmI_2$.

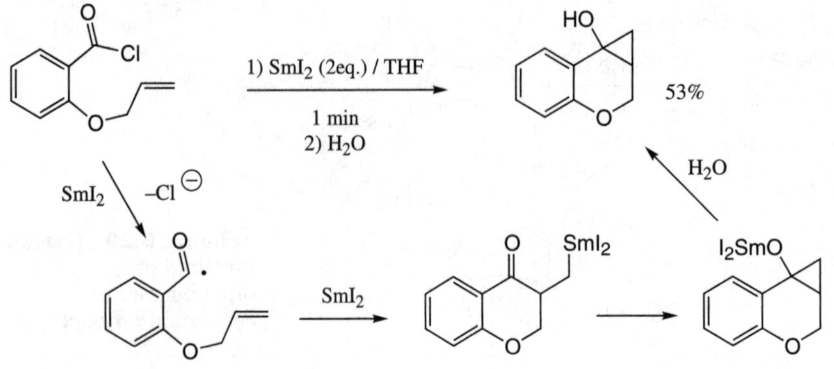

**Scheme 8.32**
Cyclopropanol formation by reductive cyclization of an acid chloride

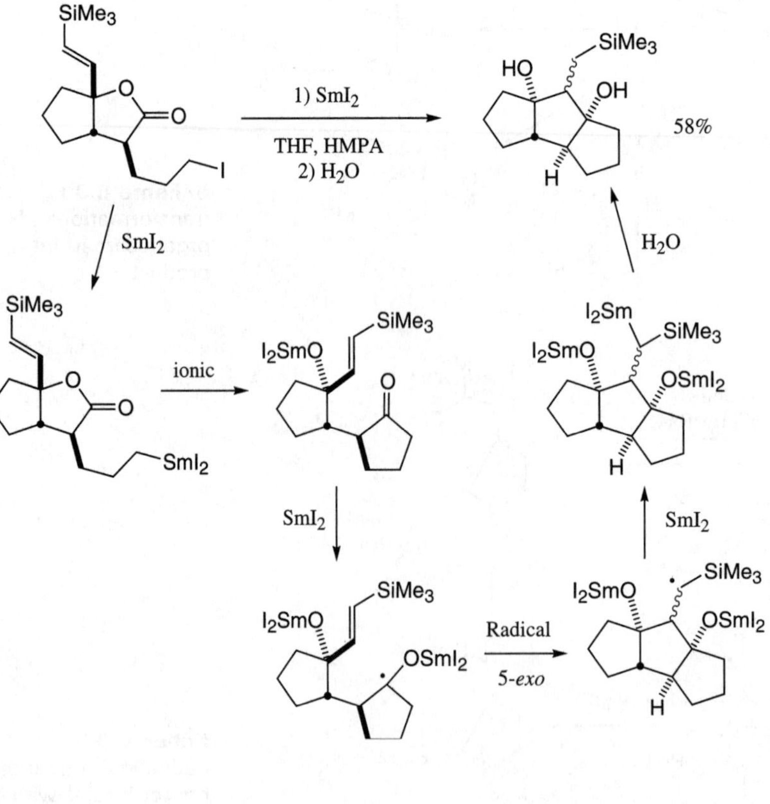

**Scheme 8.33**  A radical ionic cascade in the synthesis of polyquinanes

In another approach, displayed in Scheme 8.33 (Molander and Harris 1996*b*), the ketone is first generated by an intramolecular Barbier-type reaction between an iodide and a lactone, followed by ketyl formation and capture by an internal olefin. In this sequence, which leads to a highly functionalized triquinane containing a *trans-* junction between two of the rings, the ionic process precedes the radical cyclization.

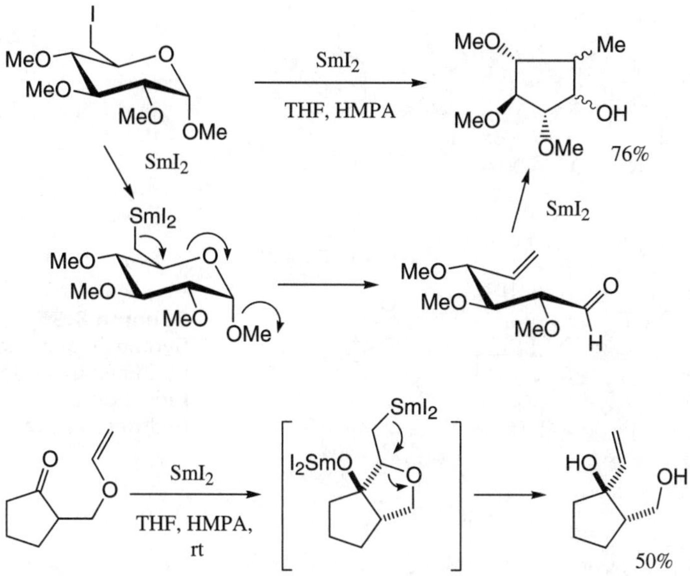

**Scheme 8.34** Sm(II)-induced fragmentations

A β-elimination step may be elegantly exploited, either to unmask a reducible carbonyl group such as the carbohydrate aldehyde in the first example in Scheme 8.34 (Chiara *et al.* 1996), or to allow an intramolecular and highly stereoselective (50:1 ratio of epimers) transfer of a vinyl unit, as in the second transformation (Molander and Harris 1998*b*). In both cases, the substrate is slowly added to the blue SmI$_2$ solution. These and the preceding examples only give a glimpse of the exquisitely rich and variegated chemistry of SmI$_2$, a reagent that was practically unknown two decades ago.

## 8.8 Reductions with iron, copper, and ruthenium salts or complexes

Ferrous and cuprous salts are mild one-electron reducing agents. They are usually incapable of reducing ketones and simple halides, but can reduce peroxides, hydroperoxides, oxaziridines, diazonium salts, and some polyhalo derivatives. One advantage of using iron and copper salts is the possible coexistence in the medium of Fe(II)/Fe(III) and Cu(I)/Cu(II) couples. Ferric and cupric ions are mild one-electron oxidizing agents; thus, both a weak reducing and oxidizing species are present allowing complementary processes to take place, depending on the nature of the radical intermediates. Sometimes, especially in the case of the copper system, a chain reaction may be established whereby only a catalytic amount of the copper salt or complex is needed. The use of F$_e$SO$_4$ to cleave a hydroperoxide, generated by the reversible addition of hydrogen peroxide to cyclohexanone methyl ketal, is illustrated by the transformation in Scheme 8.35 (Minisci *et al.* 1968). The primary radical arising from the ring opening is trapped with pyridinium sulfate. Finally, ferric

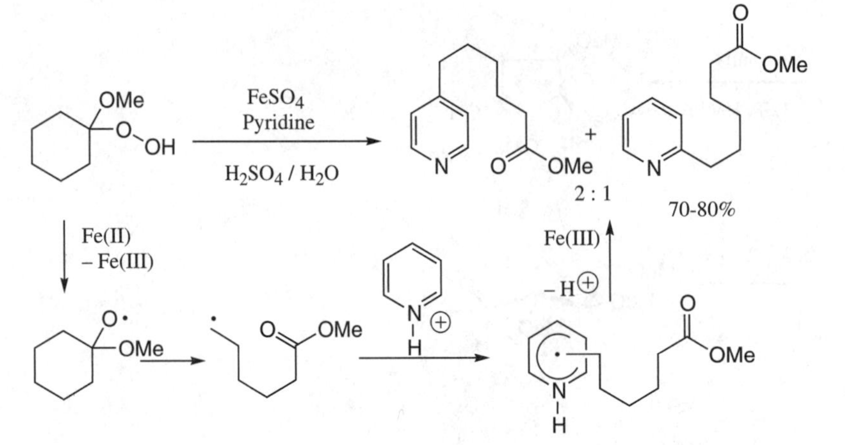

**Scheme 8.35**
Synthesis of substituted pyridines by FeSo$_4$ reduction of hydroperoxides

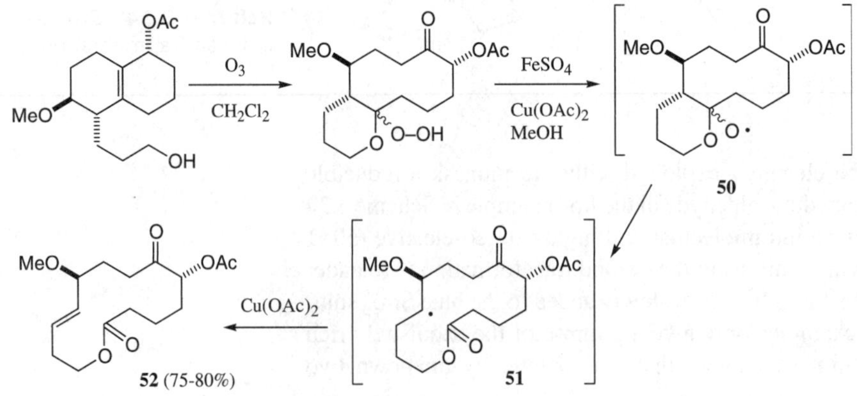

**Scheme 8.36**
FeSo$_4$/Cu(OAC)$_2$-synthesis of a macrolactone

ions, produced in the first step, mediate the oxidative aromatization of the adduct radical. Although in principle the sequence is catalytic in FeSO$_4$, in practice stoichiometric amounts are employed to ensure convenient rates.

Ferrous and cupric ions are mutually compatible, and advantage can be taken of the ability of the latter to rapidly and cleanly oxidise a carbon radical into an olefin (cf. Mn(III)/Cu(II) oxidations). In the ingenious macrolide synthesis displayed in Scheme 8.36, the requisite hydroperoxide is formed by intercepting the ozonide by the side chain hydroxy group (Schreiber and Liew 1985). Treatment with a mixture of FeSO$_4$ and Cu(OAc)$_2$ in methanol leads initially to the alkoxy radical **50**, which undergoes ring cleavage to the more stable secondary radical **51**. Oxidation of this radical by cupric acetate gives the least substituted olefin **52** as the more stable *trans* isomer.

The reduction of diazonium salts with copper powder or copper salts to give aromatic radicals is the basis of several well-known reactions, some of which are used in industry on a large scale (Galli 1988). Thus, the Sandmeyer and related processes allow the conversion of diazonium salts into chlorides, bromides, nitriles, etc. by exposure to an adequate copper salt. The Pschorr reaction is extensively used to

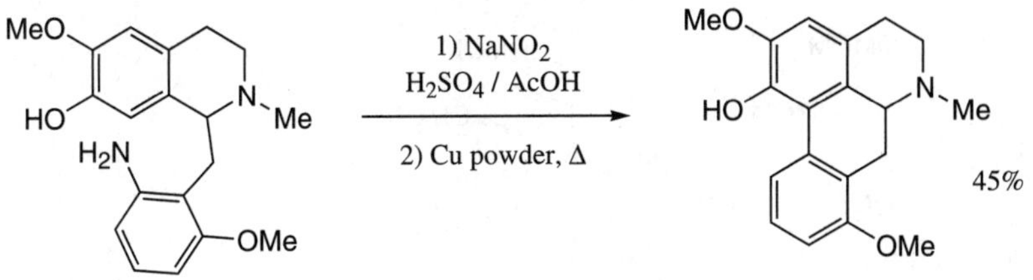

**Scheme 8.37** Biaryl formation by copper powder reduction of a diazonium salt

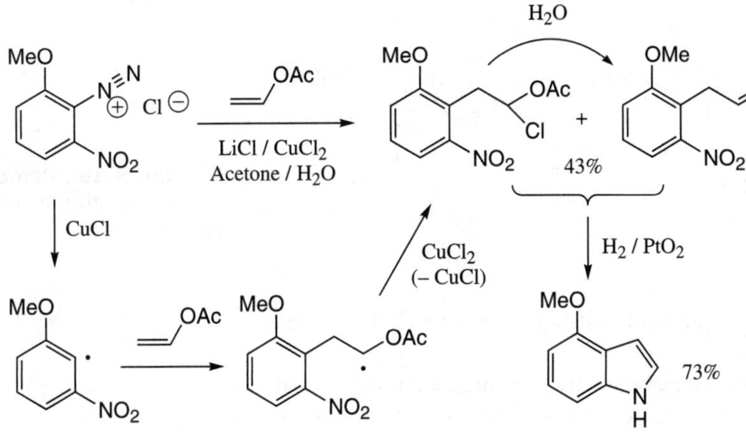

**Scheme 8.38** Synthesis of an indole from a diazonium salt by the Meerwein reaction

couple aromatics (De Tar 1957). One example is the model synthesis of a thalicarpine precursor shown in Scheme 8.37 (Kupchan *et al.* 1973). Copper powder is often used but sometimes mere heating is sufficient; in some rare instances, a 1,5-hydrogen atom shift between the aromatic rings leading to isomeric products has been observed (Karady *et al.* 1995).

Another important copper-mediated process is the Meerwein arylation reaction (Rondestvedt 1960, 1976), an example of which was discussed above involving $TiCl_3$ (Scheme 8.20). The Meerwein reaction is usually performed with cupric chloride in acetone: cupric chloride reacts slowly with acetone to give chloroacetone and cuprous chloride, the actual reducing agent for the diazonium salt. In contrast to the Ti(III) modification, where the last step is a reduction of the adduct radical, the presence of cupric chloride results in a chlorine atom transfer and thus an overall oxidation. This is illustrated by a approach to indoles outlined in Scheme 8.38 (Raucher and Koolpe 1983). The pathway followed corresponds to **M–P** in Scheme 8.17.

The high-energy aromatic radical is capable of abstracting a hydrogen atom from various compounds including ethanol, isopropanol, or hypophosphorous acid. This property is the basis of a number of methods for the reductive deamination of anilines via their diazonium salts. However, if the aromatic radical is generated using the $CuCl/CuCl_2$ system and if a suitable hydrogen is located within the molecule, then a radical translocation occurs, followed by a chloride transfer from cupric chloride (Cohen *et al.* 1971). This sequence has been cleverly exploited to oxidize amides, as shown by the transformation in Scheme 8.39 (Han *et al.* 1994). Under the reaction

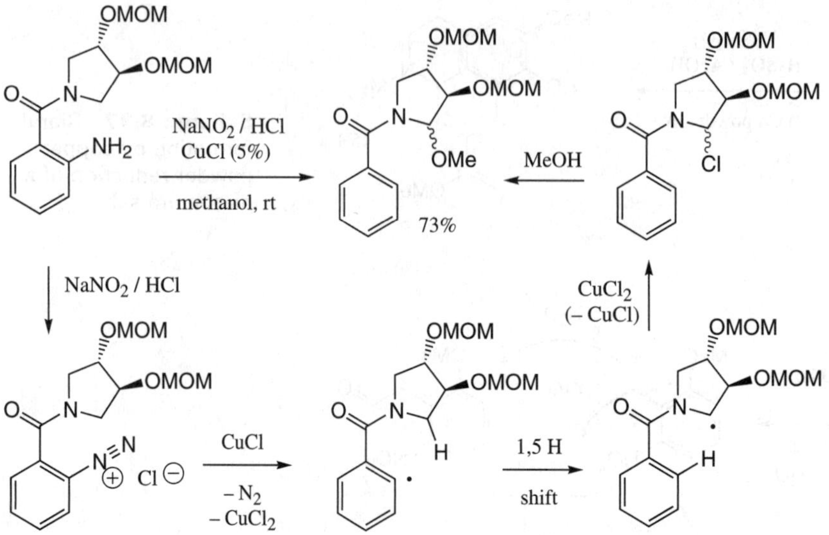

**Scheme 8.39** Remote functionalization using a diazonium salt

conditions, the $\alpha$-chloroamide undergoes an ionic substitution with methanol to give the observed product.

Cuprous salts and complexes are also capable of transferring electrons to easily reducible organic halides and polyhalides such as $\alpha$-halo-ketones, -aldehydes, esters, -nitriles, and -amides (Iqbal *et al.* 1994). Chloroamines and chloroamides can also be reduced with cuprous chloride, in the same manner as with Ti(III) salts. Radical chain processes, similar to some of the above diazonium based sequences, can therefore be implemented and applied to a variety of synthetic problems. This chemistry stems from the early work by Kharasch on the radical chain halogen atom transfer discussed in detail in Chapter 6: the chain is now initiated by electron transfer from Cu(I) and the last propagation step is a ligand transfer from the cupric halide (once again corresponding to steps **M–P** in Scheme 8.17). This is generally more efficient than the direct halide transfer from the starting material, as in the original peroxide-initiated process, at least in the case of chlorides. Cuprous and, but to a lesser extent, Fe(II)- and Ru(II)-mediated radical additions starting with organic halides have since been expanded considerably, largely in an industrial context (Bellus 1985). Indeed, the power of this technology is not widely appreciated because much of the work is buried in the patent literature. The example in Scheme 8.40 illustrates an application to the synthesis of pyridines (Martin *et al.* 1985). CuCl-catalysed addition of trichloroacetaldehyde to methacrylonitrile gives an intermediate adduct which can be converted into 2,5-dichloro-3-methylpyridine by heating with gaseous HCl. If the addition–chlorine-atom transfer is performed at a higher temperature (150 °C), then the pyridine is obtained directly in 65% yield.

The reactivity of the CuCl may be considerably increased by placing suitable nitrogen ligands around the copper. This allows sometimes the use of smaller amounts of copper and lower reaction temperatures. Most of the examples in the literature are typical 5-*exo* cyclizations starting with the appropriate unsaturated trichloroacetamides. An interesting variation, recently reported by two groups

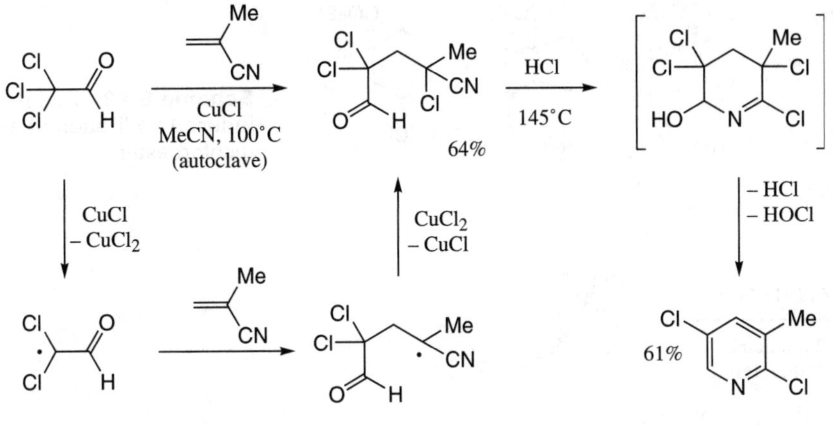

**Scheme 8.40** CuCl mediated addition of chloral to methacrylonitrile

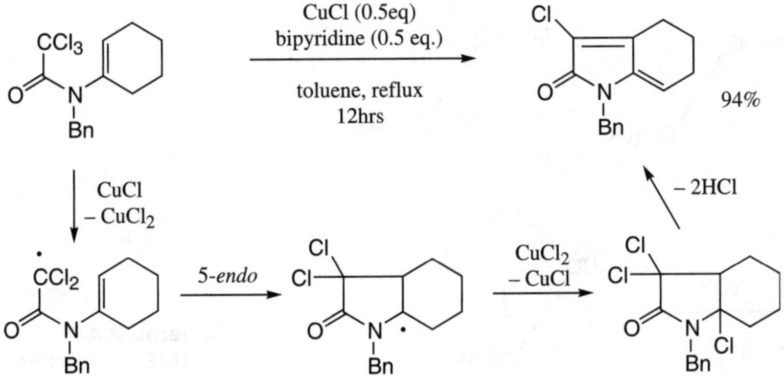

**Scheme 8.41** CuCl-mediated 5-endo ring closure

simultaneously (Clark *et al.* 1999; Davies *et al.* 1999), is outlined in Scheme 8.41. It involves an unusual 5-*endo* cyclization initially discovered using standard stannane chemistry (Ishibashi *et al.* 2002). Replacing the bipyridine in the example shown with a tetra-amine ligand allows the reaction to be performed at room temperature, albeit in somewhat lower yield (Clark *et al.* 1999). Copper(I) complexes are currently undergoing intense scrutiny in the context of controlled radical polymerizations, especially of styrene and acrylate monomers (Patten and Matyjaszewski 1999).

Ruthenium(II) and iron(II) complexes have been less extensively studied as mediators of halogen atom transfers (Iqbal *et al.* 1994), even though they sometimes exhibit comparable, and in some cases even superior, reactivity to Cu(I). The example depicted in Scheme 8.42 illustrates the cyclization of a dichloroester (Hayes *et al.* 1988). In addition to the expected two epimeric 5-*exo* products, a small amount of the 6-*endo* derivative is also formed. Similar results were obtained with the cheap and readily available $FeCl_2[(EtO)_3P]_3$ complex instead of the ruthenium catalyst, and the cumbersome use of benzene in a sealed tube may be replaced by refluxing *t*-butylbenzene. An analogous transformation can be accomplished starting from trichloroketones (Lee and Weinreb 1990).

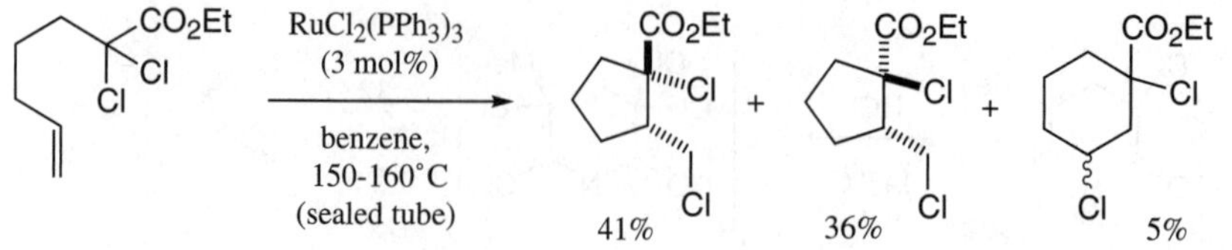

**Scheme 8.42** Ru(II)-induced cyclization of a dichloroester

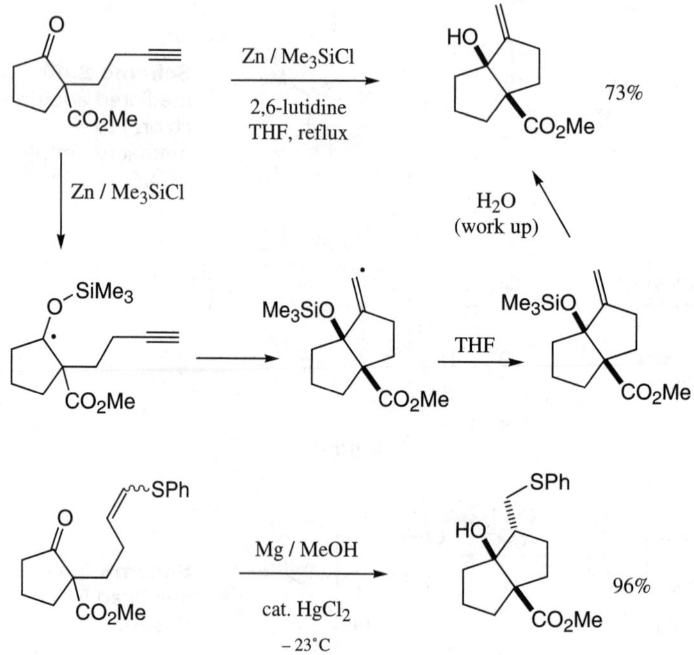

**Scheme 8.43** Zn/TMSCl reductive ring closure of ketones

## 8.9   Dissolving metal reductions

Electron transfer from dissolving metals also allow the generation and capture of radical intermediates. The reducing ability of the metal and the nature of the medium determine the type of functional groups that can be reduced. Ketyl radical formation by electron transfer to a ketone from magnesium metal in an alcoholic solvent has been used to perform pinacolic couplings (Robertson 1991). Finely divided and highly reactive metallic titanium can induce either pinacolic coupling or olefin formation from ketones or aldehydes (the McMurry reaction; McMury 1989). A combination of zinc powder and a chlorosilane has also been devised for similar transformations (Motherwell 1973, 2001). As with SmI$_2$, capture of the ketyl radical with an olefin results in ring formation. One such application using Zn/Me$_3$SiCl is portrayed in Scheme 8.43 (Corey and Pyne 1983). The final hydrogen-atom transfer to the reactive vinylic radical presumably takes place from the solvent. Lutidine is needed to prevent premature cleavage of the trimethylsilyl group. The zinc metal may be replaced with magnesium in some cases (Ikeda *et al.* 1985). A typical example

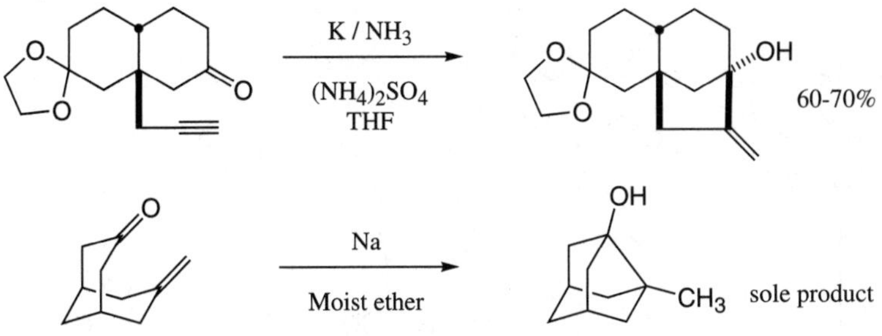

**Scheme 8.44**
Dissolving alkali metal reductive ring closure of ketones

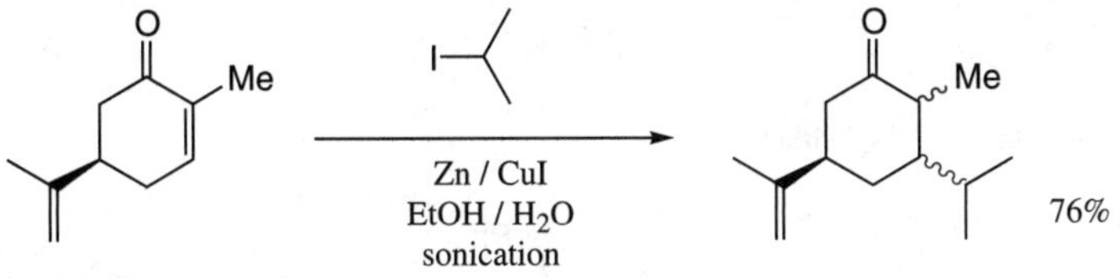

**Scheme 8.45** Zinc-mediated conjugate addition of iodides

appears in the bottom of Scheme 8.43, where magnesium amalgam has been used allowing the reductive cyclization to be performed at low temperature in excellent yield (Lee *et al.* 1999).

The earliest observations related to the interception of a ketyl radical by an internal olefin were simultaneously reported by three research groups and involved alkali metals as the reducing element. The groups of Stork (Stork *et al.* 1965) and Sutherland (Greenwood *et al.* 1965) used lithium or potassium in ammonia. The first example in Scheme 8.44, a model in an approach to gibberellic acid, is taken from a later study by Stork and his group (Stork *et al.* 1979). A related cyclization (the second example in Scheme 8.44) was accidentally observed when a Bouveault–Blanc type reduction was attempted on the [3.3.1]bicyclic enone using sodium in moist ether (Eakin *et al.* 1965). In both cases, the solvent (THF or diethyl ether) is presumably the hydrogen atom donor in the last step of the radical sequence.

Zinc metal, activated by sonication, is capable of producing radicals from aliphatic iodides, as shown by the example in Scheme 8.45 (Luche and Allavena 1988). Even though the reaction may also proceed by way of an organozinc species, evidence has been provided that at least part of the product arises from a radical pathway (Luche *et al.* 1988). It has also been reported that various additives, and especially iron salts, can facilitate the process, but the mechanistic rationale underlying this effect remains to be fully clarified (Blanchard *et al.* 1992, 1993).

A combination of plain nickel powder and a carboxylic acid such as acetic acid is capable of generating radicals from easily reducible halides. This very mild reducing system is compatible with a variety of radical traps, as illustrated by the series of transformations assembled in Scheme 8.46 (Boivin *et al.* 1994a). The bicyclic radical **54**, in contrast to the initial radical **53**, cannot be further reduced by metallic nickel since the corresponding anion is unstabilized. Hydrogen abstraction from a thiol or simply from the solvent, isopropanol (but the yields tend to be lower), gives the reduced bicyclic product. The corresponding bromide, the TEMPO adduct, the selenide, or the unsaturated derivative can also be obtained by using the appropriate radical trap.

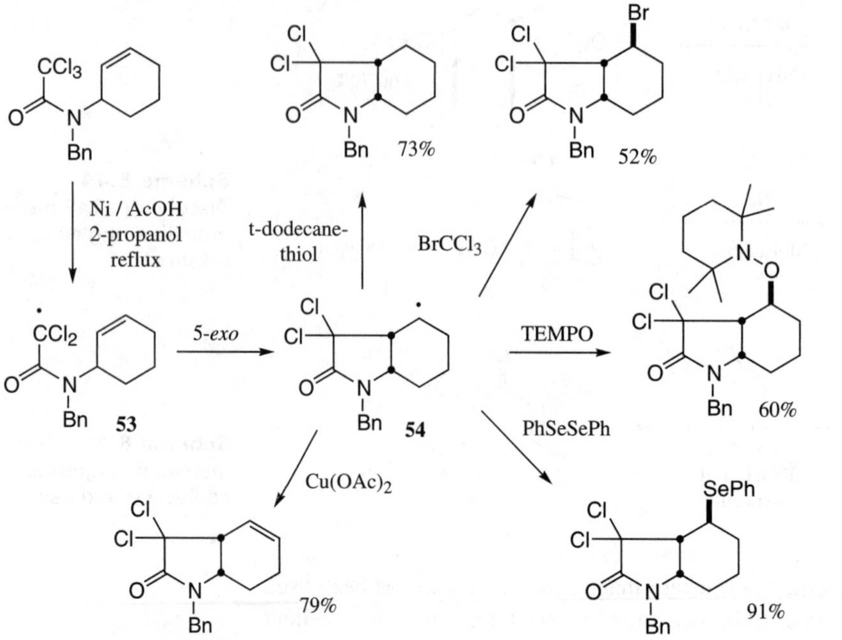

**Scheme 8.46** Synthetic transformations of allylic trichloroacetamides mediated by nickel powder

These transformations illustrate several of the general pathways outlined in Scheme 8.17. It is interesting that the cupric acetate oxidant is tolerated by the reducing metallic nickel/acetic acid reagent (Cassayre *et al.* 2000). It is important to use a co-solvent (e.g. isopropanol) in these reactions; in pure acetic acid the medium becomes too reducing and premature reduction of radical **53** into the corresponding anion equivalent occurs.

The mildness of the Ni/acetic acid combination is demonstrated by its ability to distinguish between the starting trichloroacetamide and the dichlorolactam product. The latter is slightly more difficult to reduce and much longer reaction times are required in order to remove one of the two remaining chlorines. More difficult 4-*exo* ring closures leading to β-lactams (Cassayre *et al.* 1998) or cyclizations to an aromatic ring attest to the relatively long lifetime of the intermediate radicals in the medium. A simple oxindole synthesis, starting from the readily available trichloroacetanilides, is presented in Scheme 8.47 (Boivin *et al.* 1994*b*). In this transformation, the remaining chlorine atoms become benzylic upon cyclization and are therefore easier to reduce than those of the starting material. The process thus leads to chlorine-free products and is mild enough to be compatible with the presence of an aromatic iodide or a free phenol. A small amount of prematurely reduced, uncyclized dichloroacetanilide is observed as a side product.

The exact nature of the oxidant in the aromatization step is not clear. Although disproportionation and intervention of adventitious oxygen or some nickel species can be invoked, it seems more likely that the starting trichloroacetanilide is acting as the oxidant in an otherwise reducing medium (cf. path **V** in Scheme 8.17). Electron transfer from easily oxidised radicals (such cyclohexadienyl or ketyl) to organic halides are known to occur (Wayner *et al.* 1988; Fontana *et al.* 1994; Amoli *et al.* 1995).

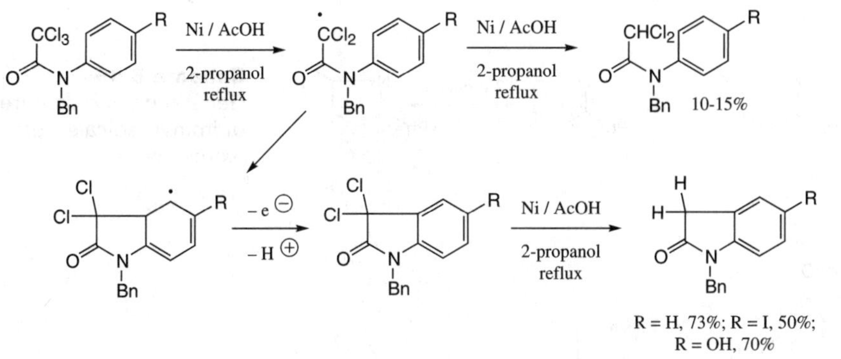

**Scheme 8.47**
Synthesis of oxindoles from trichloroacetanilides

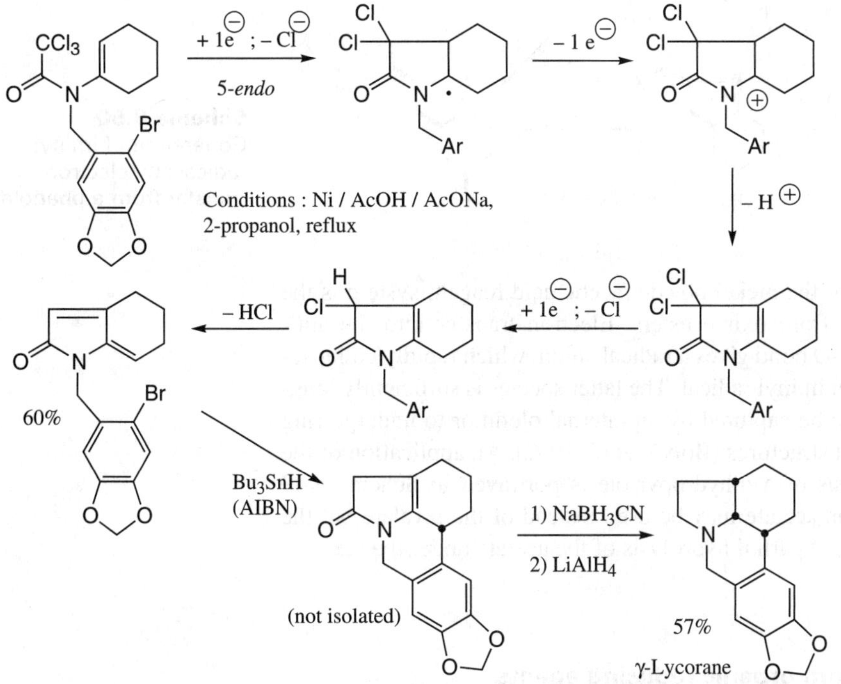

R = H, 73%; R = I, 50%; R = OH, 70%

**Scheme 8.48** Two different radical reactions in the synthesis of γ-lycorane

A similar oxidation step was found to follow a 5-*endo* cyclization of a trichloroaceteanamide, and this property was exploited in a short synthesis of γ-lycorane displayed in Scheme 8.48 (Cassayre and Zard 1999*b*). The α-amidyl radical arising from the ring closure (similar to that in Scheme 8.41 above) is also easily oxidized to give ultimately an olefin, which induces the reductive removal of a second—now allylic—chlorine atom. The last chlorine atom is slightly more difficult to reduce and is simply eliminated as HCl. The overall result is a diene which can be made to undergo a stannane-mediated 6-*endo* cyclization to give the desired lycorane skeleton. The properties of two completely different radical processes were thus combined in this approach.

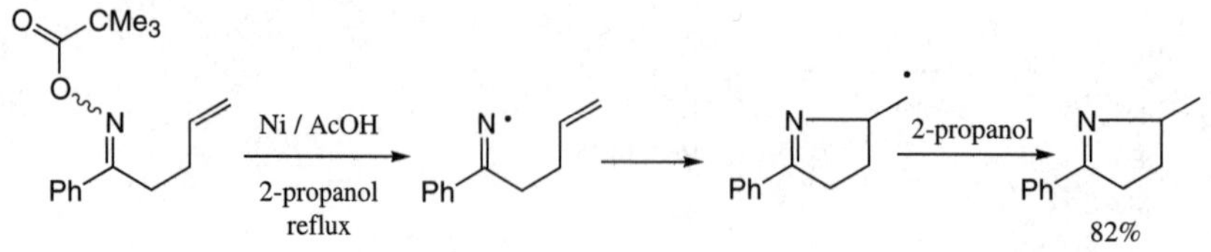

**Scheme 8.49**
Generation and capture of iminyl radicals from oxime esters

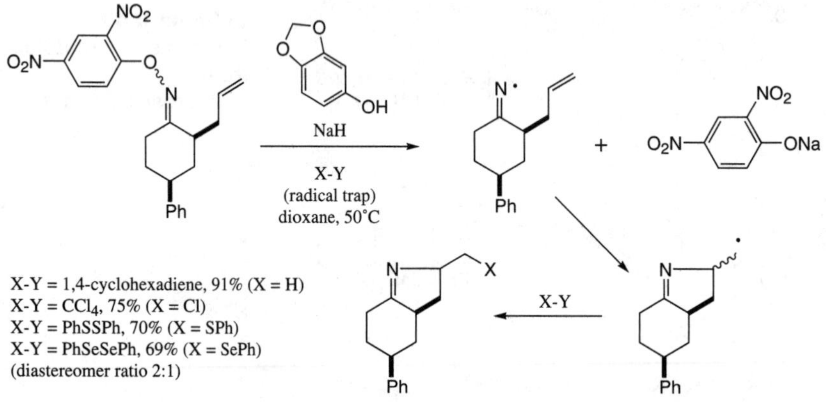

**Scheme 8.50**
Generation of iminyl radicals by electron transfer from a phenolate

One interesting extension of the nickel powder/acetic acid reagent system is the generation of iminyl radicals from oxime esters. Electron transfer into the anti-bonding orbital of the weak N–O bond gives a radical anion, which rapidly collapses into a carboxylate anion and an iminyl radical. The latter species is sufficiently long-lived under these conditions to be captured by an internal olefin or to undergo ring opening in the case of strained structures (Boivin *et al.* 1992). An application of the former process to the synthesis of a dihydropyrrole is portrayed in Scheme 8.49 (Boivin *et al.* 1999). An oxime acetate may be used instead of the pivalate but the yield is slightly lower because of partial hydrolysis of the acetate under the reaction conditions.

## 8.10 Electron transfer from organic reducing agents

In a manner similar to low-valent and dissolving metals, electron-rich organic species can cede an electron to an appropriate substrate to give the corresponding radical anion, which then fragments in the usual fashion. Phenolates, for instance, can reduce *O*-2,4-dinitrophenyloximes to produce ultimately iminyl radicals (and *O*-2,4-dinitrophenoxide as the anion component), much in the same way as the nickel powder/acetic acid-mediated reduction of oxime esters discussed above. The transformations outlined in Scheme 8.50 (Mikami and Narasaka 1999) give an idea of the scope of this versatile process. The phenolate is generated *in situ* from the appropriate phenol (sesamol in this case) and the carbon radical arising from ring closure of the iminyl can be captured with various radical traps placed in the medium.

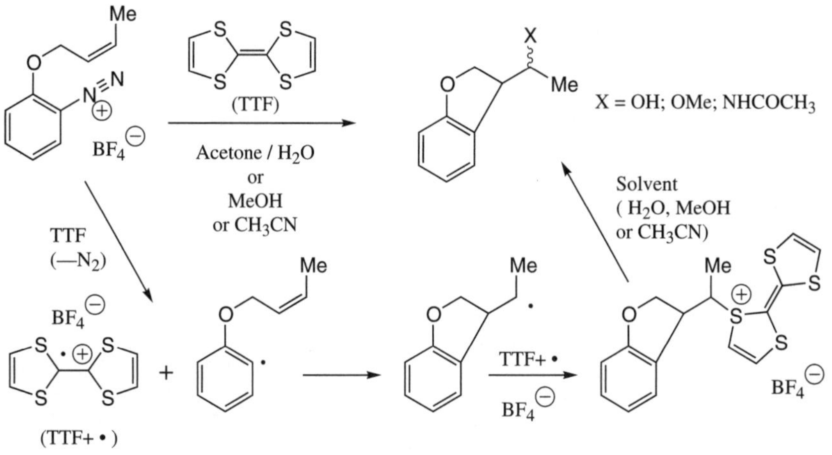

**Scheme 8.51**
Reduction of diazonium salts with tetrathiafulvalene

The electron donor may be a neutral molecule or even a free radical. An important representative of the former case is tetrathiafulvalene (TTF). TTF and its derivatives have often been used as components of organic metals because of their good electron donor ability and because they readily undergo reversible electron transfer. These features were elegantly exploited for the generation of aromatic radicals from diazonium salts (Bashir *et al*. 1999). The transformations in Scheme 8.51 leading to various dihydrobenzofuran derivatives provide a demonstration of the mechanistic versatility of this method, which allows a convenient and efficient crossover from the radical level to that of a *cation* equivalent. Evolution of nitrogen is observed at room temperature, immediately upon addition of TTF to the diazonium salt. Depending on whether the reaction is conducted in wet acetone, methanol, or anhydrous acetonitrile, the TTF salt intermediate is solvolysed to give an alcohol, a methyl ether, or an acetamide. The - formation of the last derivative results from the quenching of the cation with acetonitrile (the Ritter reaction) and gives an idea on the potency of the TTF moiety as a leaving group. The last, ionic substitution step regenerates the TTF and the process is in principle catalytic; in practice, >20 mol% of TTF must be used for complete conversion as some of it is destroyed in as yet undefined side reactions. The reduction of the diazonium salt with TTF is in contradistinction to Ti(III)-mediated reactions, where an organotitanium anion equivalent is often the ultimate product in the sequence.

An interesting application of this chemistry to the synthesis of aspidosperma alkaloids is depicted in Scheme 8.52 (Callaghan *et al*. 1999) and relies on solvolysis of the intermediate TTF salt to give the secondary alcohol with overall retention of configuration. The trifluoroacetamide side chain could be participating in this process.

Finally, electron transfer from easily oxidized radicals, invoked to explain the observed oxidation products in the nickel powder/acetic-acid-mediated reactions, may be harnessed into synthetically useful chain reactions (Fontana *et al*. 1994; Amoli *et al*. 1995). One such example, exploiting the propensity of ketyl radical **55** derived from isopropanol to interact with activated bromides, is outlined in Scheme 8.53 (Kolt *et al*. 1990). The process is triggered by irradiation of benzophenone. Both ketyl radicals **55** and **56** can in fact behave as electron transfer agents allowing, in the latter case, the recycling of benzophenone. Hydrogen abstraction from

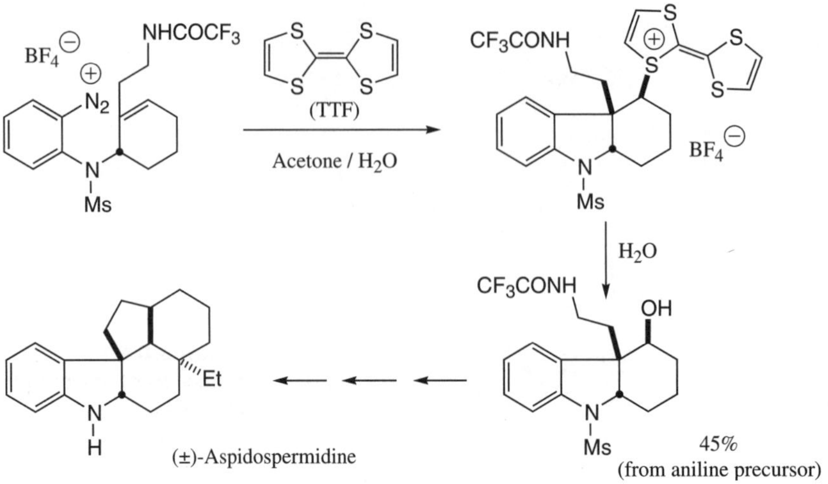

**Scheme 8.52**
Reduction of diazonium salts with tetrathiafulvalene en route to aspidospermidine

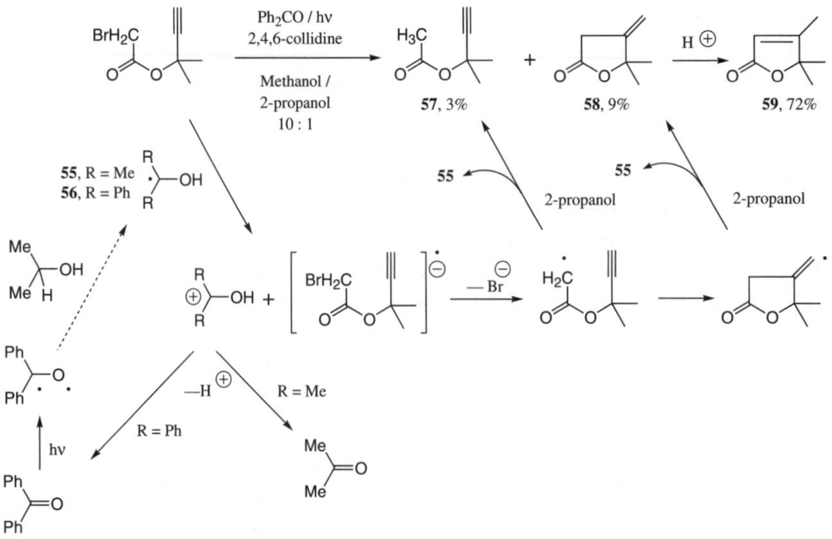

**Scheme 8.53**
Reductive cyclization of bromoesters by electron transfer from ketyl radicals.

2-propanol by the energetic vinyl radical regenerates ketyl radical **55** and propagates the chain. Collidine is required to neutralize the hydrobromic acid co-produced in the sequence. The primary product is the *exo*-isomer **58**, but it is gradually converted into the more stable conjugated *endo*-isomer **59** by a proton shift in the mildly acidic medium. A small amount of the prematurely reduced derivative **57** is also observed. The yield of **57** increases dramatically, to nearly 40%, if the reaction is conducted in pure 2-propanol.

## 8.11  Electrochemical and photochemical redox processes

Electron transfer to and from an electrode is another way to generate radical species (Schäfer 1981). Anodic oxidation of a carboxylate salt, for instance, is the basis of the Kolbe decarboxylation reaction (Schäfer 1991; see Scheme 5.1). In principle, the electrode potential may be fine-tuned to achieve maximum selectivity; however, this favourable element of flexibility is counterbalanced by transport phenomena and overreaction at the electrode surface as well as the usual need to have a supporting electrolyte (which has to be separated from the product). The radicals created by electrochemical electron transfer may thus undergo further reduction (or oxidation) before they have time to diffuse away from the electrode layer; alternatively, the radical production may be too rapid with respect to diffusion, causing radical–radical reactions to occur near the electrode. One widely used contrivance for circumventing some of these complications is to incorporate a mediator molecule, which is electrochemically easier to reduce (or oxidize) than the actual substrate. The radical anion (or radical cation) derived from the mediator then interacts with the substrate in the bulk of the solution. A suitable electrochemical mediator may be selected from a number of substances that readily undergo reversible electrochemical redox processes, and whose reduction potential has been accurately measured, by cyclic voltammetry for example. A case of direct electroreduction of a ketone and intramolecular capture of the ensuing ketyl radical anion is shown in Scheme 8.54 (Shono *et al.* 1976). Its similarity to the Zn/Me₃SiCl and other dissolving metal-mediated transformations discussed earlier highlights the close mechanistic analogy between all of these methods. In this case, too, the solvent is presumably the source of the hydrogen atom in the final product.

The direct electrochemical oxidation involving radicals is illustrated by the second example in Scheme 8.54 (Yoshida *et al.* 1988). This reaction is almost identical to the first transformation in Scheme 8.12.

The beneficial influence of a redox mediator on the course of an electrochemical process is illustrated by the radical addition in Scheme 8.55 (Medebielle 1995). The

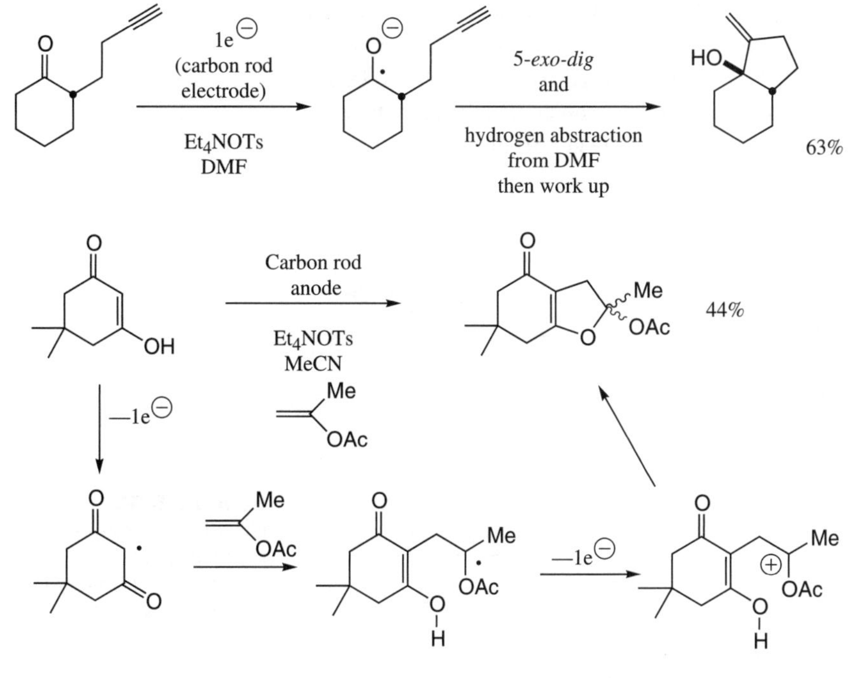

**Scheme 8.54** Ketyl radical formation by electroreduction of ketones

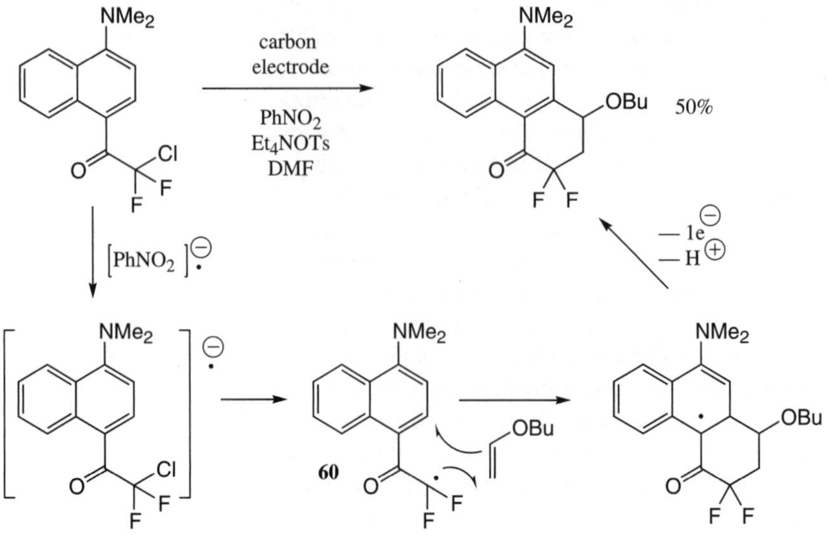

**Scheme 8.55**
Electroreduction of haloketones

radical anion derived from nitrobenzene has the correct redox potential to transfer an electron to the chlorodifluoroketone but not to further reduce the ensuing radical **60** and a clean sequence of addition to butylvinyl ether and ring closure to the naphthalene ring takes place. In the absence of nitrobenzene, extensive premature reduction to the difluoroketone is observed.

A variety of electrochemical redox mediators have been used, including transition metal complexes. In the latter case, single electron transfer at the electrode represents simply a means to generate the low-valent transition metal species, whose reduction potential will of course also depend on the nature of the ligands surrounding it. One electrochemical transformation involving a nickel complex is shown in Scheme 8.56 (Ozaki *et al.* 1993). The Ni(II)/Ni(I) redox couple for this complex was measured by cyclic voltammetry and found to be $-0.70\,\text{V}$ versus a Standard Calomel Electrode. The Ni(I) complexed species, generated by reduction at the cathode, is able in turn to transfer an electron to the bromide. The usual, very rapid collapse of the ensuing transient

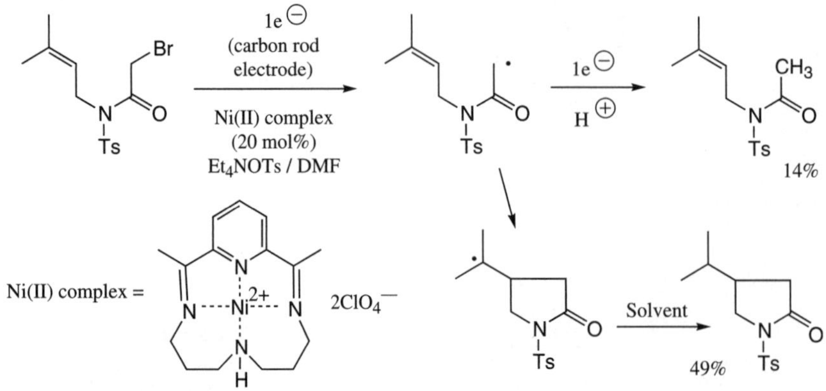

**Scheme 8.56**
Nickel-complex-mediated electroreductive cyclization of an allylic bromoacetamide

radical anion produces a free carbon-centred radical, which may be further reduced before cyclization. Cyclization, on the other hand, followed presumably by hydrogen abstraction from the solvent, furnishes the observed γ-lactam. The present, electrochemically generated Ni(I) reducing species is much more powerful, but also much less selective, than the nickel powder/acetic acid combination discussed above. The latter system, which probably does not involve a discrete Ni(I) intermediate, is not capable of reducing a primary bromoacetamide at convenient rates; it exhibits, however, a more useful selectivity with easily reducible substrates such as trichloroacetamides.

Ideally, the metal complex should only be used in truly catalytic amounts and recycled at the electrodes. In fact, most of the redox reactions we have so far discussed could in principle be made catalytic, consuming only electricity; but despite much effort, few practical procedures have really emerged.

Photochemically induced electron transfers are also well known and, in the same way, often make use of mediators (Julliard and Chanon 1983; Yoon and Mariano 1992). The mediator, in its excited state, acts as the radical acceptor (or donor). An application of a photoinduced SET in the construction of an indolizidine skeleton starting from a silylamine is pictured in Scheme 8.57 (Hoegy and Mariano 1994). The mediator is 9,10-dicyanoanthracene (DCA) which, in its excited state, is capable of extracting an electron from the amine to give a nitrogen-centred radical cation in the vicinity of the silyl group. Delocalization of an electron from the C–Si bond to the positively charged nitrogen atom leads to another resonance form, from which it is

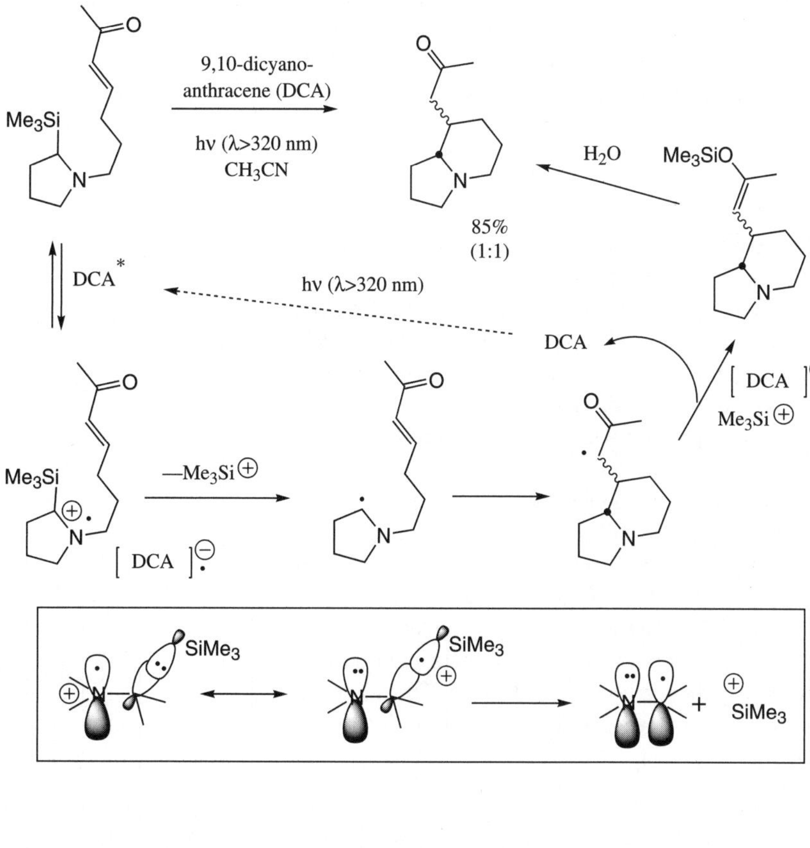

**Scheme 8.57**
Photoinduced oxidation of an α-silylamine

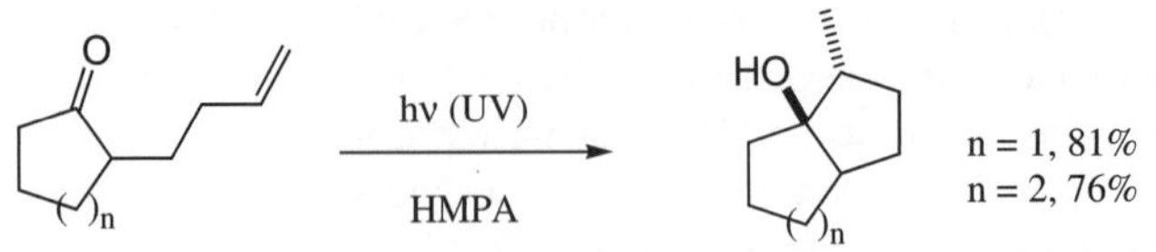

**Scheme 8.58**
Photoreductive cyclization of an $\gamma$–$\delta$-unsaturated ketone

easier to see a fragmentation leading to a trimethylsilyl cation and a carbon-centred free radical. This key step is detailed in the bottom of Scheme 8.57. The photo induced electron transfer is normally reversible and the quantum yield thus depends on the rate of collapse of the radical cation as compared with back electron donation from the DCA radical anion, which returns the system to its initial state (for a recent theoretical study, see: Robert and Savéant 2000). The $\alpha$-carbonyl radical resulting from the ring closure is finally reduced to the corresponding enolate (presumably as the silylated derivative) by electron transfer from DCA radical anion. That an enolate equivalent is involved in the final stages of the process was demonstrated by deuteration experiments. This photochemical route to $\alpha$-amino radicals is an alternative to the reductive process starting with a benzotriazole outlined earlier in Scheme 8.30.

Photoreductive cyclization of $\delta$,$\varepsilon$-unsaturated ketones has been accomplished by irradiation in the presence of HMPA or, in some cases, triethylamine. Two simple examples are given in Scheme 8.58 (Belotti *et al.* 1985). It is assumed that excited HMPA is acting as the electron-transfer agent and presumably as the source of hydrogen atom.

Photochemical reactions very often involve radical species of one type or another. This immense field lies beyond the scope of this book where, in most cases, we have used light merely as an initiator to trigger the various radical processes.

## 8.12   A word on S$_{RN}$1 reactions

As we have seen throughout, the fragmentation of a radical anion into a free radical and an anionic species is normally a favoured and fast process. It is however reversible, in principle. Thus, in special cases, it is possible to generate a radical anion **62** by the combination of a radical R$^{\bullet}$ with an anionic fragment Nu$^{-}$ (Scheme 8.59). This is an electron-rich species capable of acting as a reducing agent if a suitable electron acceptor, such as an organohalide (R–X), is present in medium. Electron transfer leads to another radical anion **61**, which in turn collapses into the initial radical and an anion.

The overall result is a nucleophilic substitution sequence proceeding by what is termed an S$_{RN}$1 chain reaction (or the Russell–Kornblum mechanism), in contradistinction to the more familiar S$_{N}$1 and S$_{N}$2 ionic reactions (for reviews, see: Bunnett 1978; Bowman 1988; Rossi *et al.* 1990, 1999, 2003). Two radical anions are involved: **61**, which has a fleeting existence because X$^{-}$ is a very good leaving group (HX is normally a very strong acid), and **62**, which is much more persistent because Nu$^{-}$ is a poor leaving group (HNu corresponds to a very weak acid). The latter radical anion has therefore enough lifetime to participate in an electron-transfer process to the alkylating agent R–X.

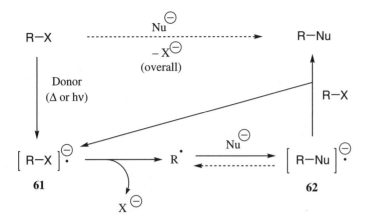

**Scheme 8.59** General mechanism of $S_{RN}1$ processes

An $S_{RN}1$ chain usually need to be started through an initiation step (the 'donor' could be Nu⁻). They are accelerated by irradiation—indeed, photostimulation is the most common way of promoting $S_{RN}1$-based transformations—and inhibited by typical radical inhibitors. 1,4-Dinitrobenzene is often used a diagnostic inhibitor for an $S_{RN}1$ process: it is generally a more powerful electron acceptor than the substrate, but its corresponding radical anion does not propagate the chain.

An $S_{RN}1$ pathway elegantly explains substitution reactions on aromatic halides, for example, which cannot be easily rationalized by any of the more traditional mechanisms. Two such transformations are outlined in Scheme 8.60 (Wong *et al.* 1997). The ready substitution of the bromine, despite the presence of the methyl groups, demonstrates the usefulness of this approach for the obtention of sterically congested products (incidentally, this example also eliminates completely the possibility of a benzyne intermediate). Not unexpectedly, the iodide, with its lower-lying LUMO, is more reactive than the bromide and some double arylation is observed.

Strong evidence for the intermediacy of an aryl radical is provided by the major formation of a cyclic product if a suitably located olefin is present in the starting material, as in the example shown in Scheme 8.61 (Beckwith and Palacios 1991). The major product arises from capture of the cyclized radical by diphenyl disulfide, which is produced in the medium by oxidation of the phenylthiolate through electron transfer to the aryl iodide (i.e. in the first step of the $S_{RN}1$ process). The study of such systems has allowed the obtention of approximate rate constants for the reaction of radicals with various anions.

Hopefully, the examples discussed in the present chapter will give an idea of the tremendous synthetic potential of redox processes for the generation and capture of various types of radicals. Numerous parameters can be modified to accomplish the desired transformations, not least as concerns the nature of the metal or combination of metal salts or complexes. Understanding the subtle interplay of the various factors in this vast field of investigation is a never ending source of excitement and fun, for the unexpected is often the outcome of a supposedly well-thought experiment.

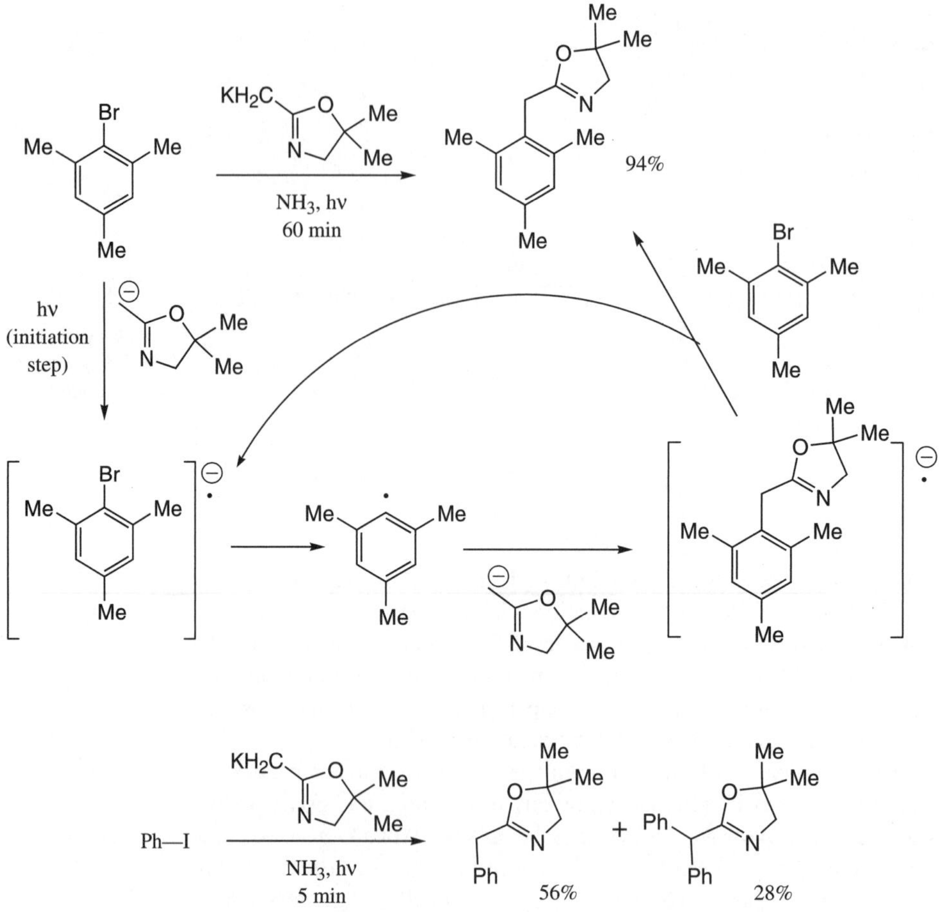

**Scheme 8.60**   Examples of $S_{RN}1$ transformations.

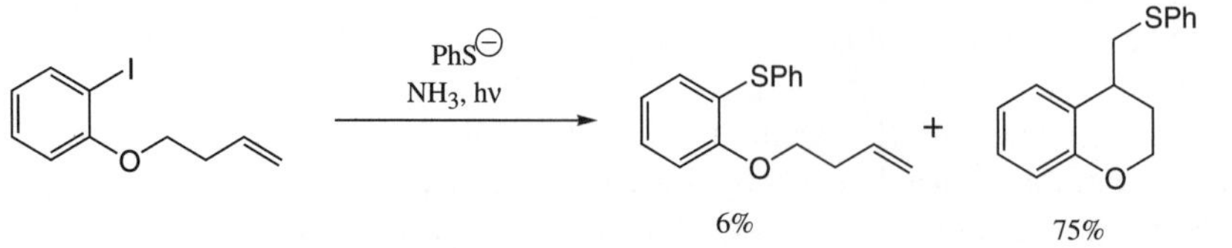

**Scheme 8.61**   Capture of the radical intermediate in an $S_{RN}1$ transformation

# References

Amoli, M., Workentin, M. S., and Wayner, D. D. M. (1995). Organic reducing agents. Reduction of electron deficient bromides by 1,2,2,6,6-pentamethylpiperidine (PMP)/mercaptoethanol. *Tetrahedron Letters*, **36**, 3997–4000.

Andrus, M. B. and Lashley, J. C. (2002). Copper-catalyzed allylic oxydation with peresters. nol. *Tetrahedron*, **58**, 845–866.

Barton, D. H. R., Basu, N. K., Hesse, R. H., Morehouse, F. S., and Pechet, M. M. (1966). Radical mechanisms in chromous ion reductions. An improved synthesis of 11β-hydroxy steroids. *Journal of the American Chemical Society*, **88**, 3016–3021.

Barton, D. H. R., Wozniak, J., and Zard, S. Z. (1989). A short and efficient degradation of the bile acid side chain. Some novel reactions of sulfines and α-ketoesters. *Tetrahedron*, **45**, 3741–3754.

Bashir, N., Patro, B., and Murphy, J. A. (1999). Reactions of arenediazonium salts with tetrathiafulvalene and related electron donors: A study of 'radical-polar crossover' reactions. In *Advances in free radical chemistry*, Vol. 2 (ed. S. Z. Zard), pp. 123–150. JAI Press Inc. Stamford.

Beckwith, A. L. J. and Palacios, S. M. (1991). $S_{RN}1$ reactions of nucleophiles with radical clocks: rate constants for some radical nucleophile reactions. *Journal of Physical Organic Chemistry*, **4**, 404–412.

Bellus, D. (1985). Copper-catalyzed additions of organic polyhalides to olefins: a versatile synthetic tool. *Pure and Applied Chemistry*, **57**, 1827–1838.

Belotti, D., Cossy, J., Pete, J.-P., and Portella, C. (1985). Photoreductive cyclization of δ, ε-unsaturated ketones. *Tetrahedron Letters*, **26**, 4591–4594.

Blanchard, P., Da Silva, A. D., Fourrey, J.-L., Machado, A. S., and Robert-Géro, M. (1992). Zinc–copper couple promoted C-branching in the carbohydrate series. *Tetrahedron Letters*, **33**, 8069–8072.

Blanchard, P., El Kortbi, M. S., Fourrey, J.-L., and Robert-Géro, M. (1993). Zinc–iron couple induced addition of alkyl halide derived radicals to activated olefins. *Journal of Organic Chemistry*, **58**, 6517–6519.

Boivin, J., Schiano, A.-M., and Zard, S. Z. (1992). A novel and practical synthesis of 13-epi-17-ketosteroids. *Tetrahedron Letters*, **33**, 7849–7852.

Boivin, J., Yousfi, M., and Zard, S. Z. (1994*a*). A versatile radical based synthesis of γ-lactams using nickel powder and acetic acid. *Tetrahedron Letters*, **35**, 9553–9556.

Boivin, J., Yousfi, M., and Zard, S. Z. (1994*b*). A new and practical synthesis of indolones. *Tetrahedron Letters*, **35**, 9553–9556.

Boivin, J., Schiano, A.-M., Zard, S. Z., and Zhang, H. (1999). A new method for the generation and capture of iminyl radicals. *Tetrahedron Letters*, **40**, 4531–4534.

Booker-Milburn, K. I. (1992). Tandem ring expansion cyclisation reactions: a novel method for the rapid construction of the bicyclo[5.3.0]decane ring system. *Synlett*, 809–810.

Bowman, W. R. (1988). Reactivity of substituted aliphatic nitro-compounds with nucleophiles. *Chemical Society Reviews*, **17**, 283–316.

Bunnett, J. F. (1978). Aromatic substitution by the $S_{RN}1$ mechanism. *Accounts of Chemical Research*, **11**, 413–420.

Callaghan, O., Lampard, C., Kennedy, A. R., and Murphy, J. A. (1999). A novel synthesis of (±) aspidospermidine. *Journal of the Chemical Society, Pertkin Transactions 1*, 995–1001.

Cassayre, J. and Zard, S. Z. (1999*b*). A short synthesis of (±)-γ-lycorane using Ni/AcOH-mediated radical cyclisation. *Synlett*, 501–503.

Cassayre, J., Dauge, D., and Zard, S. Z. (2000). Influence of copper(II) acetate on Ni/AcOH-promoted 5-*endo* and 5-*exo* radical cyclisations of trichloroacetenamides. *Synlett*, 471–474.

Cassayre, J., Quiclet-Sire, B., Saunier, J.-B., and Zard, S. Z. (1998). β- And γ-lactams by a nickel powder mediated 4-exo and 5-endo radical cyclisations. A concise construction of the mesembrine skeleton. *Tetrahedron*, **54**, 1029–1040.

Chiara, J. L., Cabri, W., and Hanessian, S. (1991). The stereocontrolled formation of cyclic vicinal cis-diols via a samarium diiodide pinacol coupling of dialdehydes. *Tetrahedron Letters*, **32**, 1125–1128.

Chiara, J. L., Martinez, S., and Barnabé, M. (1996). Cascade reactions of 6-deoxy-6-iodo-hexopyranosides promoted by samarium diiodide: a new ring contraction of carbohydrate derivatives. *Journal of Organic Chemistry*, **61**, 6488–6489.

Citterio, A. (1984). Reductive arylation of electron deficient olefins. *Organic Syntheses*, **62**, 67–73.

Citterio, A. and Vismara, E. (1980). 2-Arylalkyl ketones and 3-arylalkanals from arenediazonium salts and $\alpha,\beta$-unsaturated carbonyl compounds. *Synthesis*, **62**, 291–292.

Citterio, A., Cerati, A., Sebastiano, R., and Finzi, C. (1989). Oxidative deprotonation of carbonyl compounds by Fe(III) salts. *Tetrahedron Letters*, **30**, 1289–1292.

Clark, A. J., Dell, C. P., Ellard, J. M., Hunt, N. A., and McDonagh, J. P. (1999). Efficient room temperature copper (I) mediated 5-*endo* radical cyclisations. *Tetrahedron Letters*, **40**, 8619–8623.

Cohen, T., Smith, K. W., and Swerdloff, M. D. (1971). Isotope effects after the rate determining step. The role of rotational isomerism in a hydrogen transfer. *Journal of the American Chemical Society*, **93**, 4303–4304.

Corey, E. J. and Pyne, S. G. (1983). Conversion of ketnoes having $\delta,\varepsilon$-$\pi$-functions to cyclopentanols by zinc-trimethylchlorosilane. *Tetrahedron Letters*, **24**, 4591–4594.

Costentin, C., Robert, M., and Savéant, J.-M. (2003). Activation barriers in the homolytic cleavage of radicals and ion radicals. *Journal of the American Chemical Society*, **125**, 105–112.

Curran, D. P. and Totleben, M. J. (1992). The samarium Grignard reaction. *In situ* formation and reactions of primary and secondary alkylsamarium(III) reagents. *Journal of the American Chemical Society*, **114**, 6050–6058.

D'Annibale, A., Pesce, A., Resta, S., and Trogolo, C. (1997). Ceric ammonium nitrate promoted free radical reactions leading to $\beta$-lactams. *Tetrahedron Letters*, **38**, 1829–1832.

Dalko, P. I. (1995). Redox induced radical and radical ionic carbon–carbon bond forming reactions. *Tetrahedron*, **51**, 7579–7653.

Davies, D. T., Kapur, N., and Parsons, A. F. (1999). Copper (I) reactions in *N*-heterocycle synthesis: efficient preparation of substituted pyrrolidinones. *Tetrahedron Letters*, **40**, 8615–8618.

De Tar, D. F. (1957). The Pschorr synthesis and related diazonium ring closure reactions. *Organic Reactions*, **9**, 409–462.

Doll, M. K.-H. (1999). A short synthesis of the 8-azaergoline ring system by intramolecular tandem decarboxylation–cyclization of the Minisci-type reaction. *Journal of Organic Chemistry*, **64**, 1372–1374.

Eakin, M., Martin, J., and Parker, W. (1965). Transannular reactions in the bicyclo[3.3.1] nonane system. *Journal of the Chemical Society, Chemical Communications*, 206.

Eberson, L. (1987). *Electron transfer reactions in organic chemistry*. Springer Verlag, Berlin.

Enemaerke, R. J., Daasbjerg, K., and Skrydstrup, T. (1999). Is samarium diiodide an inner or outer sphere electron-donating agent? *Journal of the Chemical Society, Chemical Communications*, 343–344.

Enemaerke, R. J., Hertz, T., Skrydstrup, T., and Daasbjerg, K. (2000). Evidence for ionic samarium(II) species in THF/HMPA solution and investigatioin of their electron donating properties *Chemistry A European Journal*, **6**, 3747–3754.

Fernández-Mateos, A., Martin de la Nava, E., Pascual Coca, G., Ramos Silvo, A., and Rubio González, R. (1999). Radicals from epoxides. Intramolecular addition to aldehyde and ketone carbonyls. *Organic Letters*, **1**, 607–609.

Fontana, F., Kolt, R. J., Huang, Y., and Wayner, D. D. M. (1994). Organic reducing agents. Some radical chain reactions of ketyl and 1,3-dioxolanyl radicals with activated bromides. *Journal of Organic Chemistry*, **59**, 4671–4676.

Fristad, W. E., Peterson, J. R., and Ernst, A. B. (1985). Manganese(III) γ-lactone annulation with substituted acids. *Journal of Organic Chemistry*, **50**, 3143–3148.

Galli, C. (1988). Radical reactions of arenediazonium ions: an easy entry into the chemistry of aryl radicals. *Chemical Reviews*, **88**, 765–792.

Gansäuer, A. and Bluhm, H. (2000). Reagent-controlled transition-metal-catalyzed radical reactions. *Chemical Reviews*, **100**, 2771–2788.

Gansäuer, A. and Pierobon, M. (2000). Titanocene-catalyzed 5-exo cyclizations of epoxides. *Synlett*, 1357–1359.

Gansäuer, A., Bluhm, H., and Pierobon, M. (1998). Emergence of a novel catalytic radical reaction: titanocene-catalyzed reductive opening of epoxides. *Journal of the American Chemical Society*, **120**, 12849–12859.

Girard, P., Namy, J.-L., and Kagan, H. B. (1980). Divalent lanthanide derivatives in organic synthesis.1. Mild preparation of $SmI_2$ and $YbI_2$ and their use as reducing agents. *Journal of the American Chemical Society*, **102**, 2693–2698.

Goldman, J., Jacobsen, N., and Torsell, K. (1974). Syntheses in the camphor series. Alkylation of quinones with cycloalkyl radicals. Attempted syntheses of lagopodin A and desoxyhelicobasidin. *Acta Chemica Scandinavica B*, **B28**, 492–500.

Greenwood, J. M., Qureshi, I. H., and Sutherland, J. K. (1965). Transannular reactions in the caryophyllene series. *Journal of the Chemical Society*, 3154–3159.

Hackmann, C. and Schäfer, H.-J. (1993). New methods for reductive free-radical cyclizations of α-bromoacetals to 2-alkoxytetrahydrofurans with activated chromium(II) acetate. *Tetrahedron*, **49**, 4559–4574.

Hagesawa, E. and Curran, D. P. (1993). Rate constants for the reaction of primary alkyl radicals with $SmI_2$ in THF/HMPA. *Tetrahedron Letters*, **34**, 1717–1720.

Han, G., McIntosh, M. C., and Weinreb, S. M. (1994). A convenient method for amide oxidation. *Tetrahedron Letters*, **35**, 5813–5816.

Hayes, T. K., Villani, R., and Weinreb, S. M. (1988). Exploratory studies of the transition-metal-catalyzed intramolecular cyclization of unsaturated α,α-dichloro esters, acids, and nitriles. *Journal of the American Chemical Society*, **110**, 5533–5543.

Heiba, E. I. and Dessau, R. M. (1971). Oxidation by metal salts. VII. Synthesis based on the selective oxidation of organic radicals. *Journal of the American Chemical Society*, **93**, 524–527.

Heiba, E. I., Dessau, R. M., and Koehl, W. J. Jr. (1968). Oxidation by metal salts. IV. A new method for the preparation of γ-lactones by the reaction of manganic acetate with olefins. *Journal of the American Chemical Society*, **90**, 5905–5906.

Heiba, E. I., Dessau, R. M., and Rodewald, P. G. (1974). Oxidation by metal salts. X. One-step synthesis of γ-lactones from olefins. *Journal of the American Chemical Society*, **96**, 7977–7981.

Hoegy, S. E. and Mariano, P. S. (1994). Indolizidine and quinolizidine ring formation in the SET-photochemistry of silylamines. *Tetrahedron Letters*, **35**, 8319–8322.

Hou, Z. and Wakatsuki, Y. (1994). Isolation and X-ray structure of the hexamethylphosphoramide (hmpa)-coordinated lanthanide(II) diiodide complexes $[SmI_2(hmpa)_4]$ and $[Yb(hmpa)_4(thf)_2]I_2$. *Journal of the Chemical Society, Chemical Communications*, 1205–1206.

Ikeda, I., Yue, S., and Hutchinson, C. R. (1985). Reductive, radical-induced cyclizations of 5-hexenals as a biomimetic model of the chemistry of secologanin formation. *Journal of Organic Chemistry*, **50**, 5193–5199.

Inanaga, J., Ishikawa, M., and Yamaguchi, H. (1987). A mild and convenient method for the reduction of organic halides by using a $SmI_2$–THF solution in the presence of hexamethylphosphoric triamide (HMPA). *Chemistry Letters*, 1485–1486.

Inanaga, J., Ujikawa, O., and Yamaguchi, H. (1991*a*). SmI$_2$-promoted aryl radical cyclisation. A new synthetic entry into heterocycles. *Tetrahedron Letters*, **32**, 1737–1740.

Inanaga, J., Katsuki, J., Ujikawa, O., and Yamaguchi, H. (1991*b*). Carbon–carbon bond formation by intermolecular radical reactions. SmI$_2$-promoted carbonyl-alkyne reductive coupling. *Tetrahedron Letters*, **32**, 4921–4924.

Iqbal, J., Bhatia, B., and Nayyar, N. K. (1994). Transition metal-promoted free-radical reactions in organic synthesis: the formation of carbon–carbon bonds. *Chemical Reviews*, **94**, 519–564.

Ishibashi, H., Sato, T., and Ikeda, M. (2002). 5-*Endo-trig* radical cyclisations. *Synthesis*, 695–713.

Iwasawa, N., Funahashi, M., Hayakawa, S., and Narasaka, K. (1993). Synthesis of medium-sized bicyclic compounds by intramolecular cyclization of cyclic $\beta$-keto radicals generated from cyclopropanols using manganese(III) tris(2-pyridinecarboxylate). *Chemistry Letters*, 545–548.

Julliard, M. and Chanon, M. (1983). Photoelectron transfer catalysis: its connections with thermal and electrochemical analogues. *Chemical Reviews*, **83**, 425–506.

Kagan, H. B. and Namy, J.-L. (1986). Lanthanides in organic synthesis. *Tetrahedron*, **46**, 6573–6614.

Kan, T., Hosokawa, S., Nara, S., Oikawa, M., Ito, S., Matsuda, F., and Shirahama, H. (1994). Total synthesis of Grayanotoxin III. *Journal of Organic Chemistry*, **59**, 5532–5534.

Karady, S., Abramson, N. L., Dolling, U.-H., Douglas, A. W., and McManemin, G. J. (1995). Intramolecular aromatic 1,5-hydrogen transfer in free radical reactions.1. Unprecedented rearrangements in Pschorr cyclization, Sandmeyer, and hydro, hydroxy and iododediazoniation reactions. *Journal of the American Chemical Society*, **117**, 5425–5426.

Karim, M. R. and Sampson, P. (1988). Transannular aldol condensation within macrocyclic lactones: a novel approach to 8-membered carbocyclic rings. *Tetrahedron Letters*, **29**, 6897–6900.

Kates, S. A., Dombroski, M. A., and Snider, B. B. (1990). Mn(III)-based oxidative free-radical cyclizations of $\beta$-ketoesters, 1,3-diketones, and malonate diesters. *Journal of Organic Chemistry*, **55**, 2427–2436.

Katritzky, A. R., Feng, D., Qi, M., Aurrecoechea, J. M., Suero, R., and Aurrekoetxea, N. (1999). Novel tandem cyclization/reaction with electrophiles of $\alpha$-aminoalkyl radicals. *Journal of Organic Chemistry*, **64**, 3335–3338.

Kawatsura, M., Matsuda, F., and Shirahama, H. (1994). Samarium(II) iodide promoted intermolecular ketone–olefin couplings chelation controlled by $\alpha$-hydroxy groups. *Journal of Organic Chemistry*, **59**, 6900–6901.

Kochi, J. K. (1973). Oxidation–reduction reactions of free radicals and metal complexes. In *Free radicals*, Vol. 1 (ed. J. K. Kochi), pp. 591–683. Wiley Interscience, New York.

Kolt, R. J., Griller, D., and Wayner, D. D. M. (1990). Unsaturated spiro-$\gamma$-lactone formation by the dissociative reduction of bromoacetates. *Tetrahedron Letters*, **31**, 7539–7540.

Kuhlman, M. L. and Flowers, II, R. A. (2000). Aggregation state and reducing power of the samarium diiodide–DMPU complex in acetonitrile. *Tetrahedron Letters*, **41**, 8049–8052.

Kupchan, S. M., Kameswaran, V., and Findlay, J. W. A. (1973). Aporphine synthesis by Pschorr cyclization of aminophenols. An improved synthesis of a thalicarpine precursor. *Journal of Organic Chemistry*, **38**, 406–407.

Lee, G. M. and Weinreb, S. M. (1990). Transition metal catalyzed intramolecular cyclizations of (trichloromethyl)alkenes. *Journal of Organic Chemistry*, **55**, 1281–1285.

Lee, G. H., Ha, S. J., Yoon, I. K., and Pak, C. S. (1999). Intramolecular ketyl-olefin cyclization mediated by magnesium method. *Tetrahedron Letters*, **49**, 2581–2584.

Lübbers, T. and Schäfer, H.-J. (1990). Reductive cyclizations of ethyl 3-allyloxy-2-bromo-propionates with chromium(II) acetate to tetrahydrofurans. *Synlett*, 44–46.

Luche J.-L. and Allavena, C. (1988). Ultrasound in organic synthesis.16. Optimisation of the conjugate additions to $\alpha,\beta$-unsaturated carbonyl compounds in aqueous media. *Tetrahedron Letters*, **29**, 5369–5372.

Luche, J.-L., Allavena, C., Petrier, C., and Dupuy, C. (1988). Ultrasound in organic synthesis.17. Mechanistic aspects of the conjugate additions to $\alpha$-enones in aqueous media. *Tetrahedron Letters*, **29**, 5373–5374.

Machrouhi, F., Hamann, B., Namy, J. L., and Kagan, H. B. (1996). Improved reactivity of diiodosamarium by catalysis with transition metal salts. *Synlett*, 633–634.

Martin, P., Steiner, E., Streith, J., Winkler, T., and Bellus, D. (1985). Convenient approaches to heterocycles via copper catalysed additions of organic polyhalides to activated olefins. *Tetrahedron*, **41**, 4057–4078.

McMurry, J. E. (1989). Carbonyl-coupling reactions using low valent titanium. *Chemical Reviews*, 89, 1513–1524.

Medebielle, M. (1995). Electrochemical addition of chlorodifluoroacetyl aromatic compounds to electron-rich olefinic substrates. A convenient synthesis of *gem*-difluoro heterocyclic compounds. *Tetrahedron Letters*, **36**, 2071–2074.

Melikyan, G. G. (1993). Manganese(III)-mediated reactions of unsaturated systems. *Synthesis*, 833–850.

Merlic, C. A. and Walsh, J. C. (1998). Completely diasterioselective radical reactions using arenechromium tricarbonyl complexes. *Tetrahedron Letters*, **39**, 12083–12086.

Mikami, T. and Narasaka, K. (1999). Generation of radical species by single electron transfer reactions and their application to the development of synthetic reactions. In *Advances in free radical chemistry*, Vol. 2 (ed. S. Z. Zard), pp. 45–88. JAI Press Inc. Stamford.

Miller, R. S., Sealy, J. M., Shabangi, M., Kuhlman, M. L., Fuchs, J. R., and Flowers, III, R. A. (2000). Reactions of SmI$_2$ with alkyl halides and ketones: inner-sphere vs outer-sphere electron transfer in reactions with Sm(II) reductants. *Journal of the American Chemical Society*, **122**, 7718–7722.

Minisci, F. (1975). Free-radical additions to olefins in the presence of redox systems. *Accounts of Chemical Research*, **8**, 165–171.

Minisci, F., Citterio, A., and Giordano, C. (1983). Electron-transfer processes: peroxydisulfate, a useful and versatile reagent in organic chemistry. *Accounts of Chemical Research*, **16**, 27–32.

Minisci, F., Galli, R., Cecere, M., Malatesta, V., and Caronna, T. (1968). Nucleophilic character of alkyl radicals: new syntheses by alkyl radicals generated in redox processes. *Tetrahedron Letters*, **9**, 5609–5612.

Molander, G. A. (1992). Application of lanthanide reagents in organic synthesis. *Chemical Reviews*, **92**, 29–68.

Molander, G. A. (1994). Reductions with samarium(II) iodide. *Organic Reactions*, **46**, 211–367.

Molander, G. A. and Harris, C. R. (1996*a*). Sequencing reactions with samarium(II) iodide. *Chemical Reviews*, **96**, 307–338.

Molander, G. A. and Harris, C. R. (1996*b*). Sequenced reactions with samarium(II) iodide. Tandem nucleophilic acyl substitution/ketyl olefin coupling reaction. *Journal of the American Chemical Society*, **118**, 4059–4071.

Molander, G. A. and Harris, C. R. (1998*a*). Sequenced reactions with samarium(II) iodide. *Tetrahedron*, **54**, 3321–3354.

Molander, G. A. and Harris, C. R. (1998*b*). Sequential ketyl-olefin coupling/$\beta$-elimination reactions mediated by samarium(II) iodide. *Journal of Organic Chemistry*, **63**, 812–816.

Molander, G. A. and Kenny, C. (1987). Stereocontrolled intramolecular ketone–olefin reductive coupling reactions promoted by samarium diiodide. *Tetrahedron Letters*, **28**, 4367–4370.

Molander, G. A. and Kenny, C. (1989). Intramolecular coupling reactions promoted by samarium diiodide. *Journal of the American Chemical Society*, **111**, 8236–8246.

Molander, G. A. and McKie, J. A. (1992). Samarium(II) iodide induced reductive cyclization of unactivated olefinic ketones. Sequential radical cyclization/intermolecular nucleophilic addition and substitution reactions. *Journal of Organic Chemistry*, **57**, 3132–3139.

Molander, G. A. and McKie, J. A. (1995). Stereochemical investigations of samarium(II) iodide promoted 5-exo and 6-exo ketyl-olefin radical cyclization reactions. *Journal of Organic Chemistry*, **60**, 872–882.

Monovich, L. G., Le Huérou, Y., Rönn, M., and Molander, G. A. (2000). Total synthesis of (−)-steganone utilizing a samarium (II) iodide promoted 8-endo ketyl-olefin cyclization. *Journal of the American Chemical Society*, **122**, 52–57.

Motherwell, W. B. (1973). Direct deoxygenation of alicyclic ketones. A new olefin synthesis. *Journal of the Chemical Society, Chemical Communications*, 935.

Motherwell, W. B. (2001). On the evolution of organozinc carbenoid chemistry from carbonyl compounds—A personal account. *Journal of Organometallic Chemistry*, **624**, 41–46.

Nugent, W. A. and Rajanbabu, T. V. (1988). Transition-metal centered radicals in organic synthesis. Titanium(III) induced cyclisation of epoxy olefins. *Journal of the American Chemical Society*, **110**, 8561–8562.

Ozaki, S., Matsushita, H., and Ohmori, H. (1993). Indirect electroreductive cyclisation of *N*-allylic and *N*-propargylbromo amides and *o*-bromoacroylanilides using nickel(II) complexes as electron-transfer catalysts. *Journal of the Chemical Society, Pertkin Transactions 1*, 2339–2344.

Patten, T. E. and Matyjaszewski, K. (1999). Copper(I)-catalyzed atom transfer radical polymerization. *Accounts of Chemical Research*, **32**, 895–903.

Péralez, E., Négrel, J.-C., and Chanon, M. (1994). New perspectives in the formation of Grignard reagents. *Tetrahedron Letters*, **35**, 5857–5860.

Raucher, S. and Koolpe, G. A. (1983). Synthesis of substituted indoles via Meerwein arylation. *Journal of Organic Chemistry*, **48**, 2066–2069.

Rajanbabu, T. V. and Nugent, W. A. (1994). Selective generation of free radicals from epoxides using a transition metal radical. A powerful tool for organic synthesis. *Journal of the American Chemical Society*, **116**, 986–997.

Robert, M. and Savéant, J.-M. (2000). Photoinduced dissociative electron transfer: is the quantum yield theoretically predicted to equal unity? *Journal of the American Chemical Society*, **122**, 514–517.

Robertson, G. M. (1991). Pinacol coupling reactions. In *Comprehensive organic synthesis*, Vol. 3 (ed. B. M. Trost and I. Fleming), pp. 563–611. Pergamon Press, Oxford.

Rondestvedt, C. S., Jr. (1960). Arylation of unsaturated compounds by diazonium salts (the Meerwein arylation reaction). *Organic Reactions*, **11**, 189–260.

Rondestvedt, C. S., Jr. (1976). Arylation of unsaturated compounds by diazonium salts (the Meerwein arylation reaction). *Organic Reactions*, **24**, 225–259.

Rossi, R. A., Pierini, A. B., and Palacios, S. M. (1990). Nucleophilic substitution by the $S_{RN}1$ mechanism on alkyl halides. (1990). In *Advances in free radical chemistry*, Vol. 1 (ed. D. D. Tanner), pp. 193–252. JAI Press Inc. Stamford.

Rossi, R. A., Pierini, A. B., and Penénory, A. B. (2003). Nucleophilic substitution reactions by electron transfer. *Chemical Reviews*, **103**, 71–167.

Rossi, R. A., Pierini, A. B., and Santiago, A. N. (1999). Aromatic substitution by the $S_{RN}1$ reaction. *Organic Reactions*, **54**, 1–271.

Sasaki, M., Collin, J., and Kagan, H. B. (1988). Double cyclization of allyloxybenzoic acid chlorides mediated by samarium diiodide giving cyclopropanols. *Tetrahedron Letters*, **29**, 6105–6106.

Savéant, J.-M. (1993). Electron transfer, bond breaking, and bond formation. *Accounts of Chemical Research*, **26**, 455–461.

Schäfer, H. J. (1981). Anodic and cathodic CC-bond formation. *Angewandte Chemie International Edition in English*, **20**, 911–934.

Schäfer, H. J. (1991). Kolbe reactions. In *Comprehensive organic synthesis*, Vol. 3 (ed. B. M Trost and I. Fleming), pp. 633–658. Pergamon Press, Oxford.

Schreiber, S. L. and Liew, W.-F. (1985). Iron / copper promoted fragmentation reactions of $\alpha$-hydroxy peroxides. The conversion of octalins into14-me1mbered ring macrolides. *Journal of the American Chemical Society*, **107**, 2980–2982.

Shabangi, M., Kuhlman, M. L., and Flowers, II, R. A. (1999). Mechanism of the reduction of primary radicals by $SmI_2$–HMPA. *Organic Letters*, **1**, 2133–2135.

Shabangi, M., Sealy, J. M., Fuchs, J. R., and Flowers, II, R. A. (1998). The effect of cosolvent on the reducing power of $SmI_2$ in tetrahydrofuran. *Tetrahedron Letters*, **39**, 4429–4432.

Sheldon, R. A. and Kochi, J. K. (1972). Oxidative decarboxylation of acids by lead tetraacetate. *Organic Reactions*, 19279–19421.

Shono, T., Nishigushi, I., and Omizu, H. (1976). Reductive cyclization of nonconjugated ketones to 2-methylenecyclopentanols. *Chemistry Letters*, 1233–1236.

Skene, W. G., Scaiano, J. C., and Cozens, F. L. (1996). Fluorescence from samarium(II) iodide and its electron transfer quenching: dynamic of the reaction of benzyl radicals with Sm(II). *Journal of Organic Chemistry*, **61**, 7918–7921.

Skrydstrup, T., Mazéas, D., Elmouchir, M., Doisneau, G., Riche, C., Chiaroni, A., and Beau, J.-M. (1997). 1,2-*cis*-C-glycoside synthesis by samarium diiodide-promoted radical cyclization. *Chemistry A European Journal*, **3**, 1342–1356.

Snider, B. B., Mohan, R. N., and Kates, S. A. (1985). Manganese(III)-based oxidative free-radical cyclizations. Synthesis of ($\pm$)-podocarpic acid. *Journal of Organic Chemistry*, **50**, 3659–3661.

Snider, B. B. (1996). Manganese(III)-based oxidative free-radical cyclizations. *Chemical Reviews*, **96**, 339–363.

Snider, B. B. and Cole, B. M. (1995). Mn(III)-based oxidative free-radical cyclizations of unsaturated ketones. *Journal of Organic Chemistry*, **60**, 5376–5377.

Snider, B. B. and McCarthy, B. A. (1993*a*). Ligand, solvent, and deuterium isotope effects in Mn(III)-based oxidative free-radical cyclizations. *Journal of Organic Chemistry*, **58**, 6217–6223.

Snider, B. B. and McCarthy, B. A. (1993*b*). Oxidative free-radical cyclizations of allylic $\alpha$-chloromalonates. *Tetrahedron*, **49**, 9447–9452.

Snider, B. B., Kieselgof, J. Y., and Foxman, B. M. (1998). Total synthesis of ($\pm$)-isosteviol and ($\pm$)-beyer-15-ene-3b,19-diol by a manganese(III)-based oxidative quadruple free-radical cyclization. *Journal of Organic Chemistry*, **63**, 7945–7952.

Snider, B. B., Patricia, J. J., and Kates, S. A. (1988). Mechanism of manganese(III)-based oxidations of $\beta$-ketoesters. Synthesis of ($\pm$)-podocarpic acid. *Journal of Organic Chemistry*, **53**, 2137–2143.

Snider, B. B., Vo, N. H., and Foxman, B. M. (1993). Mn(III)-based oxidative fragmentation–cyclization reactions of unsaturated cyclobutanols. *Journal of Organic Chemistry*, **58**, 7228–7237.

Stella, L. (1983). Homolytic cyclisation of *N*-chloroalkenylamines. *Angewandte Chemie International Edition in English*, **22**, 337–350.

Stella, L., Raynier, B., and Surzur, J.-M. (1977). Synthesis of 2-methyl-6,7-benzomorphane via radical cyclisation. *Tetrahedron Letters*, 2721–2724.

Stork, G., Malhorta, S., Thompson, H., and Uchibayashi, M. (1965). A new cyclization. 2-Methylenecyclopentanols by the chemical reduction of γ-ethynyl ketones. *Journal of the American Chemical Society*, **87**, 1148–1149.

Stork, G., Boeckman, R. K., Jr., Taber, D. F., Still, W. C., and Singh, J. (1979). A Reductive cyclization of ethynyl ketones in the construction of a significant tricyclic intermediate for the synthesis of gibberellic acid. *Journal of the American Chemical Society*, **101**, 7107–7109.

Takai, K., Nitta, K., Fujimura, O., and Utimoto, K. (1989). Preparation of alkylchromium reagents by reduction of alkyl halides with chromium(II) chloride under Co catalysis. *Journal of Organic Chemistry*, **54**, 4732–4764.

van Rheenen, V. (1969*a*). Copper-catalyzed oxygenations of branched aldehydes. An efficient ketone synthesis. *Tetrahedron Letters*, **10**, 985–988.

van Rheenen, V. (1969*b*). Copper-catalyzed oxygenations of enamines. *Journal of the Chemical Society, Chemical Communications*, 314–315.

Vinogradov, M. G., Kondorsky, A. E., and Nikishin, G. I. (1988). Oxidative addition of 1, 3-dicarbonyl compounds to conjugated olefins. *Synthesis*, 60–62.

Vo, N. H. and Snider, B. B. (1994). Total synthesis of (−)-silphiperfol-6-ene and (−)-methyl cantabradienate. *Journal of Organic Chemistry*, **59**, 2427–2436.

Volger, H. C. and Brackman, W. (1965). The copper-catalysed oxidation of unsaturated carbonyl compounds. Part I. Catalysed oxidation of $\Delta^5$-cholestenone into $\Delta^4$-cholesten-3, 6-dione. *Recueil des Travaux Chimiques des Pays-Bas*, **84**, 579–580.

Wayner, D. D. M., McPhee, D. J., and Griller, D. (1988). Oxidation and reduction potentials of transient free radicals. *Journal of the American Chemical Society*, **110**, 132–137.

Wong, J.-W., Natalie, K. J., Jr., Nwokogu, G. C., Pisipati, J. S., Flaherty, P. T., Greenwood, T. D., and Wolfe, J. F. (1997). Compatibility of various carbanion nucleophiles with heteroaromatic nucleophilic substitution by the $S_{RN}1$ mechanism. *Journal of Organic Chemistry*, **62**, 6152–6159.

Yoon, U. C. and Mariano, P. S. (1992). Mechanistic and synthetic aspects of amine–enone single electron transfer photochemistry. *Accounts of Chemical Research*, **25**, 233–240.

Yoshida, J., Sakaguchi, K., and Isoe, S. (1988). Oxidative [3 + 2] cycloaddition of 1,3-diketones and olefin using electroorganic chemistry. *Journal of Organic Chemistry*, **53**, 2525–2533.

# 9 Some concluding remarks

## 9.1 A brief overview and some practical considerations

> Reactions are ever so quizzical
> Some gentle, some wild, some hysterical,
> But the ones I've seen
> So seldom are clean,
> And the clean ones so seldom are radical

This modification of an anonymous limerick (Parrott 1983) sums up humourously a prevailing view of radical reactions that will fade away eventually when radical chemistry takes up its legitimate position in the curriculum of undergraduate and graduate students. Indeed, the foregoing chapters give only a brief glimpse of the synthetic opportunities accruing from the implementation of radical processes. It is hoped that this survey of the main reactions and methods will encourage synthetic chemists to use this powerful arsenal for solving their synthetic problems. Stannane-based methods are the most familiar and are employed extensively in academic laboratories. Applications in the medicinal and agrochemical fields are hampered to a certain extent by cost but mostly by the potential toxicity of organotin derivatives and the formidable challenge of eliminating tin residues from the product. It is not surprising, therefore, that much effort has been devoted in recent times to finding alternative, tin-free procedures (Baguley and Walton 1998; Studer and Amrein 2002). No single reagent has so far demonstrated the breadth of reactivity exhibited by tributylstannane. Some of the organosilane derivatives discussed in Chapter 4 have a promising reactivity profile but are still too costly for industrial applications. If the demand for such reagents increases in the future, then industrial availability may not be a problem any more. In some particular cases, radical reactions typical of stannanes can be accomplished by cheaper and ecologically more acceptable reagents. For example, bromides, iodides, and xanthates (and related dithiocarbonyl congeners such as thiocarbonyl imidazolides) may be reduced with hypophosphorus and some of its derivatives (Chapter 6). Isopropanol, in combination with a suitable peroxide, can sometimes effectively accomplish the Barton–McCombie deoxygenation.

Methods involving organomercury intermediates and redox process based on transition metals may also present environmental problems in large-scale work; nevertheless, their potential for organic synthesis is tremendous. The development of catalytic cycles whereby the reacting metallic species is regenerated electrochemically or through the agency of an ecologically and economically acceptable stoichiometric oxidant or reducing agent could provide a viable solution to an industrial problem. In this respect, non-metallic reducing agents such as tetrathiafulvalene (Scheme 8.52) offer exciting possibilities. The Barton decarboxylation using thiohydroxamate esters discussed in Chapter 5 is also an extremely powerful reaction which certainly deserves a much greater attention on the part of synthetic organic chemists, especially that the key reagent, the sodium salt of *N*-hydroxy-2-thiopyridone, is an industrially available chemical.

In terms of ease of scale-up, it is perhaps the Kharasch-type processes that show the greatest promise. The starting materials are in most cases cheap and readily available, and the reactions often proceed better under concentrated conditions. Indeed, several atom-transfer processes are used industrially to produce intermediates for agrochemicals, some involving catalytic amounts of copper or iron salts to improve the halogen atom exchange in the case of chlorides or bromides. This is not necessary for xanthate transfer. Both the halogen atom and dithiocarbonyl group transfer techniques, as well as the persistent radical effect, are being intensely studied in the context of controlled radical polymerizations. The near future will certainly see custom-designed polymers and resins, made by such processes, appear on the market.

'Never say, "I tried it once and it did not work"' is a favourite quotation of Lord Rutherford. If an attempted radical transformation fails the first time, it is often worthwhile not to be discouraged but to take the trouble of analysing the products and the reaction parameters carefully. Sometimes a radical chain fails because impurities in the substrate and/or reagents are acting as inhibitors (inhibitors can sometimes leach from the septum through the action of the hot solvent). Another, more common and not surprising problem, is that the intermediate radical is pursuing a different, undesired pathway. This is revealed by determining the structure of the main side products. Several types of modifications may then be contemplated to remedy the situation. The following are some points for consideration:

1 The presence of dissolved oxygen may be the complicating factor, especially when working on a small scale, *below the boiling point of the solvent*. A more efficient degassing of the system is then necessary.
2 If the desired transformation is an intramolecular process, then an increase in the dilution will give it an advantage with respect to bimolecular side reactions. If the key step is a fragmentation, implying a large positive entropy term, then an increase in the temperature will have a beneficial effect in speeding the fragmentation with respect to other inter- or even intramolecular pathways, which usually have a negative transition state entropy. Sometimes a change in the order of addition of the reactants may be beneficial.
3 One frequent observation is that the radical gets reduced instead of undergoing the expected reaction. The source of the hydrogen atom may of course be the reagent (stannane, silane, organomercury hydride, thiol, etc.) or the solvent. Altering the mode of addition, the dilution of the medium, and changing the solvent are simple remedies. A more insidious cause might be the translocation of the radical centre by an unexpected 1,5-hydrogen shift. This may not be obvious upon examination of the side products when working with stannanes for example, since the final hydrogen abstraction from the reagent will give what appears to be simply the prematurely reduced material. A classical example is the 6-*exo*-ring closure, which is often complicated by an allylic hydrogen abstraction (see Section 2.8). If a hydrogen shift is suspected, it may be unmasked by a deuteration experiment. It may then be necessary to block the unwanted hydrogen shift by modifying the substituents/ protecting groups in order to alter the folding of the molecule and perhaps slow down the internal migration of the hydrogen atom. A more drastic solution is to replace the hydrogen in question with some temporary substituent.

4 Temporary groups may be used to accelerate or disfavour a given reaction. The whole gamut of polar and steric properties of substituents (including the Thorpe–Ingold effect) may be exploited to speed up or slow down one transformation with respect to another. The radical may thus be made more or less electrophilic in character by attaching appropriate substituents to be removed at a later stage; the olefinic trap may be activated at one end, hindered at another etc. The use of a temporary tether has been used to great effect in many situations to control the regio- and stereochemistry of radical additions.

5 If the problem appears to be due to unwanted radical–radical interactions, then it is necessary to slow down the initiation process. This may be done by matching the temperature of the reaction with the decomposition rate of the initiator or by modifying the rate of its introduction into the system. The power of the lamp may be decreased in the case of a photochemically induced transformation or, for redox processes, one of the reagents may be introduced slowly.

6 Unruly radical–radical interactions may also become important if one step in the desired sequence is too slow, leading to a build-up in the concentration of radicals. The step causing the bottle-neck should be identified and modified to increase its efficiency, for example by improving the reactivity of the radical or by activating the trap through the use of appropriate temporary substituents.

7 The choice of the initiator or initiating system is important. For instance, AIBN leads to stabilized tertiary isobutyronitrile radicals. These are reactive enough to start a stannane radical chain but are often incapable of triggering atom or group transfer reactions. For the latter processes, lauroyl peroxide, which decomposes at a similar temperature to give the much more reactive primary undecyl radicals, is usually a better choice.

As in other activities in life, practising (and making mistakes!) is still the best way to learn. It is hoped that this book will help organic chemists gain a feel for radical reactions, allowing them perhaps to select the best tool for a given synthetic job or, even better, to design their own new radical process to complement existing methods.

## 9.2 Nature's radical chemistry

This book is only concerned with synthetic aspects of radical chemistry. Yet, the way nature exploits and controls radical processes is a never ending source of wonder. Although many of the reactions discussed so far have no parallel in nature (triorganotin hydrides have not yet emerged as natural products!), a large number of important biosynthetic pathways appear to involve free radicals (Frey 1990; Stubbe and van der Donk 1998). The impression of most people is that radicals are deleterious to living organisms and this is certainly true in some instances. Hydroxyl radicals have thus been implicated in ageing and cancerogenesis. Irradiation and over-exposure to sunlight can cause skin cancer through, among other things, the initial photochemical generation of free radicals which can cause damage to DNA. To counteract the undesired behaviour of such reactive species, nature has developed defence systems based for example on vitamins C and E, which act as anti-oxidants and radical scavengers. Both lead to relatively harmless, stabilized radicals.

Superoxide dismutase, peroxidases and catalases are enzymatic systems that play a crucial role in defending against the detrimental effect of oxygen-derived species. One example is glutathione and the enzyme glutathione peroxidase, which relies on the radical chemistry of selenium, present in the active site as selenocystein.

But radicals are also involved in many biochemical pathways essential to life. A key step in the biosynthesis of DNA is the deoxygenation of ribonucleotides by various classes of ribonucleotide reductases (Stubbe and van der Donk 1998). The iron-containing cytochrome P450 family of enzymes is capable of replacing unactivated C–H with C–O bonds. This process can involve radicals and is used both to selectively functionalize metabolites and to degrade them through extensive oxidation. It also serves to detoxify xenobiotics (e.g. foreign substances, like drugs or toxins), again by oxidation. One of organic chemistry's holy grails is indeed to be capable, like nature, of selectively functionalizing unactivated C–H bonds. Probing the mechanism of action of cytochromes P450 is, not surprisingly, an area of intense research (Newcomb and Toy 2000).

Oxidation of phenols and coupling of the resulting phenoxyl radicals can lead to highly complex structures. This process constitutes a major biosynthetic pathway for many alkaloids and can be mimicked in the laboratory, albeit inefficiently in most cases. The sequence in Scheme 9.1 represents a portion of the biosynthesis of morphine, where the di-phenolic radical derived from (−)-reticuline undergoes an *o,p*-coupling to give salutaridine, after tautomerization of the initial coupling product to the phenol.

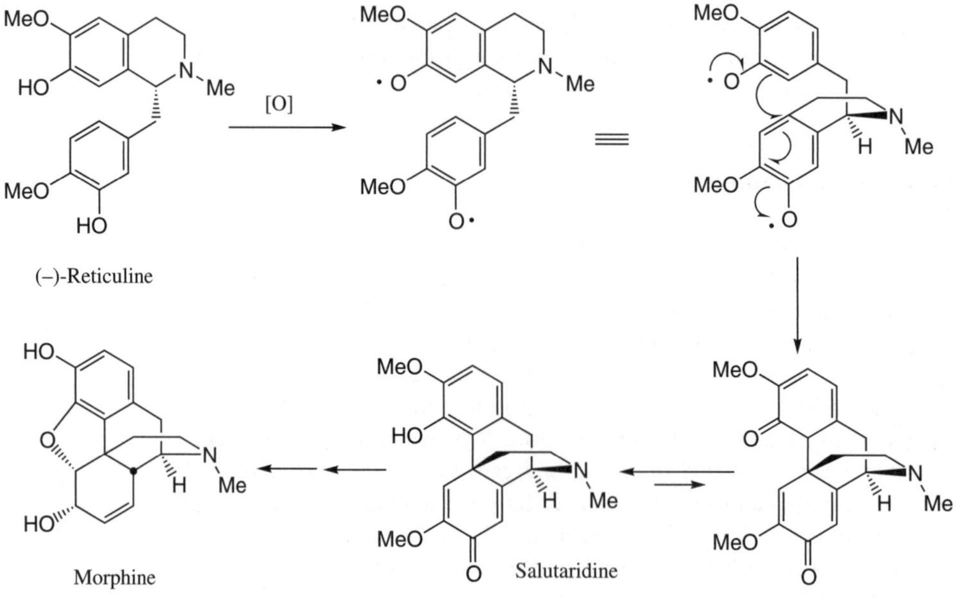

**Scheme 9.1**　Oxidative phenolic coupling in the biosynthesis of morphine

Radical biohalogenations, needed in the biosynthesis of halogenated, mostly marine natural products, rely on haloperoxidases, some of which also require a transition metal (Hartung 1999). Nature has also learnt to take advantage of the persistent radical effect. Nitric oxide (NO) is a persistent free radical that serves as a messenger in various biochemical processes. The exquisite radical chemistry mediated by vitamin B12 has fascinated chemists and biochemists for decades, both in terms of mechanistic intricacy and synthetic applications, some of which have been detailed in Chapter 7. Understanding all these radical processes that nature has elaborated over the aeons is a tremendous endeavour. What has been revealed so far is a marvellous combination of simplicity and extraordinary sophistication, and in any case a wonderful source of inspiration for the organic chemist.

## References

Baguley, P. A. and Walton, J. C. (1998). Flight from the tyranny of tin: the quest for practical radical sources free from metal encumbrances. *Angewandte Chemie International Edition in English*, **37**, 3072–3082.

Frey, P. A. (1990). Importance of organic radicals in enzymatic cleavage of unactivated C–H bonds. *Chemical Reviews*, **90**, 1343–1357.

Hartung, J. (1999). The biosynthesis of barbamides—a radical pathway for 'biohalogenation'? *Angewandte Chemie International Edition*, **38**, 1209–1211.

Newcomb, M. and Toy, P. H. (2000). Hypersensitive radical probes and the mechanisms of cyctochrome P450-catalyzed hydroxylation reactions. *Accounts of Chemical Research*, **33**, 449–455.

Parrott, E. O. (1983). *Limericks*, p. 21. Penguin Books Ltd, London.

Stubbe, J. and van der Donk, W. A. (1998). Protein radicals in enzyme catalysis. *Chemical Reviews*, **98**, 705–762.

Studer, A. and Amrein, S. (2002). Tin hydride substitutes in reductive radical chain reactions. *Synthesis*, 835–849.

## Further useful general references on radical chemistry

Alfassi, Z. B. (ed.) (1999). *General aspects of the chemistry of free radicals*. Wiley, Chichester.

Curran, D. P., Porter, N. A., and Giese, B. (1996). *Stereochemistry of radical reactions*. VCH, Weinheim.

Fossey, J., Lefort, D., and Sorba, J. (1995). *Free radicals in organic chemistry*. Wiley, New York.

Giese, B. (1986). *Radicals in organic synthesis: formation of carbon–carbon bonds*. Pergamon Press, Oxford.

Giese, B., Kopping, B., Göbel, T., Dickhaut, J., Thoma, G., Kulicke, K. J., and Trach, F. (1996). Radical cyclisation reactions. *Organic Reactions*, **48**, 301–856.

Kochi, J. K. (ed.) (1973). *Free radicals*, Vols 1 & 2. Wiley Interscience, New York.

Leffler, J. E. (1993). *An introduction to free radicals*. Wiley, New York.

Motherwell, W. B. and Crich, D. (1992). *Free radical chain reactions in organic synthesis*. Academic Press, New York.

Nonhebel, D. C., Tedder, J. M., and Walton, J. C. (1979). *Radicals*. Cambridge University Press, Cambridge.

Perkins M. J. (2000). *Radical chemistry: the fundamentals*. Oxford University Press, Oxford.

Regitz, M. and Giese, B. (eds) (1989). *C-Radikale*. Houben-Weyl Methoden der Organische Chemie Band E19a, Vols 1 & 2. Georg Thieme Verlag, Stuttgart.

Renaud P. and Sibi, M. (eds) (2001). *Radicals in organic synthesis*, Vols 1 & 2. Wiley-VCH, Weinheim.

# Index